T0177051

TOPOLOGICAL PHASES OF MATTER

Topological phases of matter are an exceptionally dynamic field of research: several of the most exciting recent experimental discoveries and conceptual advances in modern physics have their origins in this field. These have generated new – topological – notions of order, interactions, and excitations. This text provides an accessible, unified, and comprehensive introduction to the phenomena surrounding topological matter, with detailed expositions of the underlying theoretical tools and conceptual framework, alongside accounts of the central experimental breakthroughs. Among the systems covered are topological insulators, magnets, semimetals, and superconductors. The emergence of new particles with remarkable properties such as fractional charge and statistics is discussed alongside possible applications such as fault-tolerant topological quantum computing. Suitable as a textbook for graduate or advanced undergraduate students, or as a reference for more experienced researchers, the book assumes little prior background, providing self-contained introductions to topics as varied as phase transitions, superconductivity, and localization.

RODERICH MOESSNER is director at the Max Planck Institute for the Physics of Complex Systems in Dresden. His theoretical discoveries include classical and quantum spin liquids, emergent magnetic monopoles, and nonequilibrium spatiotemporal ordering phenomena. He is recipient of the Leibniz Prize and the Europhysics Prize, and an Honorary Fellow of Hertford College, Oxford.

JOEL E. MOORE is Chern-Simons Professor of Physics at the University of California, Berkeley, and Senior Faculty Scientist at Lawrence Berkeley National Laboratory. His research interests include topological insulators, semimetals, and semiconductors, along with the application of quantum information concepts to many-body physics. He is a Simons Investigator and a Fellow and former elected Member-at-Large of the American Physical Society.

TOPOLOGICAL PHASES OF MATTER

RODERICH MOESSNER

Max-Planck-Institut für Physik komplexer Systeme, Dresden

JOEL E. MOORE

University of California, Berkeley, and Lawrence Berkeley National Laboratory

CAMBRIDGE
UNIVERSITY PRESS

CAMBRIDGE
UNIVERSITY PRESS

University Printing House, Cambridge CB2 8BS, United Kingdom

One Liberty Plaza, 20th Floor, New York, NY 10006, USA

477 Williamstown Road, Port Melbourne, VIC 3207, Australia

4843/24, 2nd Floor, Ansari Road, Daryaganj, Delhi – 110002, India

79 Anson Road, #06–04/06, Singapore 079906

Cambridge University Press is part of the University of Cambridge.

It furthers the University's mission by disseminating knowledge in the pursuit of education, learning, and research at the highest international levels of excellence.

www.cambridge.org
Information on this title: www.cambridge.org/9781107105539
DOI: 10.1017/9781316226308

© Roderich Moessner and Joel E. Moore 2021

First published 2021

Printed in the United Kingdom by TJ Books Limited, Padstow Cornwall

A catalogue record for this publication is available from the British Library.

ISBN 978-1-107-10553-9 Hardback

Contents

List of Tables	*page*	viii
List of Boxes		ix
Preface		xi
Acknowledgments		xiii
1	Introduction	1
2	Basic Concepts of Topology and Condensed Matter	11
2.1	Berry Phases in Quantum Mechanics	12
2.2	One Electron in a Magnetic Field: Landau Levels	20
2.3	One Electron in a Crystal: Bloch's Theorem	24
2.4	The Simplest Tight-Binding Model	26
2.5	Dirac Band Structure of the Honeycomb Lattice	29
2.6	Landau Theory of Symmetry-Breaking Phases	32
2.7	Two Mathematical Approaches to Topology	40
2.8	Topological Defects in Symmetry-Breaking Phases	50
3	Integer Topological Phases: The Integer Quantum Hall Effect and Topological Insulators	58
3.1	IQHE: Basic Phenomena and Theory	60
3.2	Two Lattice Models of the IQHE, and Chern Number	69
3.3	Time-Reversal Symmetry in Classical and Quantum Physics	74
3.4	Topological Insulators in 2D: Basic Phenomena and Theory	76
3.5	A Lattice Model of the 2D Topological Insulator	82
3.6	3D Topological Insulators: Basic Phenomena	84
3.7	Skyrmions in the Quantum Hall Effect	88

4   Geometry and Topology of Wavefunctions in Crystals   95
     4.1   Inversion Symmetry, Electrical Polarization, and Thouless
            Pumping   97
     4.2   The Integer Quantum Hall Effect and Topological Invariants
            of Energy Bands   104
     4.3   Many-Particle Interpretation of Topological Invariants   107
     4.4   Time-Reversal Invariance and $\mathbb{Z}_2$ Invariants   109
     4.5   Axion Electrodynamics, Non-Abelian Berry Phase, and
            Magnetoelectric Polarizability   118

5   Hydrogen Atoms for Fractionalization   124
     5.1   The Fractional Quantum Hall Effect   126
     5.2   Fractionalization, Order, and Topology in $d = 1$ 145
     5.3   The Resonating Valence Bond Liquid   156
     5.4   Spin Ice   167

6   Gauge and Topological Field Theories   179
     6.1   Pure Ising Gauge Theory and Absence of Local Order   181
     6.2   Ising Gauge Theory with Matter   187
     6.3   Kitaev's Toric Code   192
     6.4   Maxwell Electromagnetism   194
     6.5   Tensor Gauge Theories and Fractons   198
     6.6   Long-Wavelength and Topological Field Theories   201
     6.7   Mutual Statistics and the Quantum Hall Hierarchy   214
     6.8   *BF* Theory   215

7   Topology in Gapless Matter   218
     7.1   Geometric Quantities in the Semiclassical Theory of
            Metals   220
     7.2   Dirac and Weyl Semimetals   225
     7.3   Electromagnetic Response of Topological Semimetals   229
     7.4   Kitaev Honeycomb Model   233

8   Disorder and Defects in Topological Phases   239
     8.1   Introduction to Disorder and Localization   241
     8.2   A Semiclassical Model of Quantum Hall Transitions   247
     8.3   Adding Quantum Mechanics: Network Models   251
     8.4   Basic Ideas of Random Matrix Theory and the Tenfold
            Way   253
     8.5   Vortices in Conventional Superconductors   260

	8.6	Flux and Crystalline Defects in Integer Topological Phases	266
	8.7	Vortices in Quantum Hall States and Composite Fermions	268
	8.8	Spin Liquids and Disorder	271
9		Topological Quantum Computation via Non-Abelian Statistics	285
	9.1	Quantum Computation: Universality and Complexity	286
	9.2	Error Correction versus Fault-Tolerance	289
	9.3	Nonlocal Operations for Quantum Computing	292
	9.4	Majoranas in One Dimension: The Kitaev Chain	297
	9.5	Majoranas in Two Dimensions	301
	9.6	Universal Computation and the Read–Rezayi States	310
	9.7	Experimental Implementations of Majorana Modes	311
10		Topology out of Equilibrium	316
	10.1	Time-Dependent and Time-Periodic (Floquet) Hamiltonians	317
	10.2	Floquet Basics	318
	10.3	Floquet Topological Insulators	324
	10.4	Anomalous Floquet–Anderson Insulator	326
	10.5	Driven Kitaev Chain and π-Majorana Fermions	329
	10.6	Many-Body Floquet Discrete Time Crystal	334
11		Symmetry, Topology, and Information	340
	11.1	Symmetry-Protected Topological Phases	341
	11.2	Entanglement Entropy in Topological States	349
	11.3	The Universe of Topological Materials; Closing Remarks	352
		Appendix: Useful Sources, Quantities, and Equations	355
		References	358
		Index	371

Tables

4.1   Comparison of Berry phase theories of polarization and magnetoelectric
      polarizability   *page* 121
8.1   Ten symmetry classes of free-fermion Hamiltonians in dimensions 1–4
      and their topological possibilities   259
A.1   Some useful quantities and equations   357

viii

Boxes

2.1	The Berry Phase of the Adiabatic Dynamics of a Spin	*page* 17
2.2	Topology from Geometry: The Gauss–Bonnet Theorem	40
2.3	The Berezinskii–Kosterlitz–Thouless Transition	53
3.1	One Particle on a Ring Pierced by Magnetic Flux	62
3.2	Modulation Doping	67
4.1	Tight-Binding Chain with Two Orbitals per Unit Cell	102
4.2	The Wess–Zumino–Witten Model	110
5.1	Single-Mode Approximation	134
5.2	Fractional Statistics of Particles in Two Dimensions	140
5.3	Fractional Quantum Numbers	146
5.4	Klein Models	151
5.5	Classical Dimer Models and Their Correlations	162
6.1	Quantum IGT in d Dimensions and Classical IGT in $d+1$	182
6.2	Bound States of the Dirac Equation: Jackiw–Rebbi Model	202
7.1	Semiclassical Equilibrium	224
8.1	One-Parameter Scaling Approach to Anderson Localization	243
8.2	Bogoliubov–de Gennes Formalism of Superconductivity	254
8.3	The Josephson Effect and Gauge Invariance	263
9.1	The No-Cloning Theorem	290
9.2	What Is a Majorana Fermion or Zero Mode?	294
9.3	The Jordan–Wigner Transformation and Statistics in 1D	299
9.4	Solution and Phase Diagram of the Kitaev Honeycomb Model	306
10.1	Phase Structure in and out of Equilibrium	320

Preface

Topological condensed matter physics presents an embarrassment of riches, both in the bewildering variety of phenomena that fall under this heading and in the sheer volume of publications in the field. In addition, the breadth of the field is reflected in a diversity of backgrounds of its practitioners: it is a place where immigrants from high-energy physics collaborate with materials chemists. This poses the twin challenges of selection and organization of material for a book.

Against this backdrop, the content of this book reflects our vision of the field, in particular what we feel might in the long run form part of the canon of many-body physics. We have tried to emphasize conceptual and historical milestones. We focus on topological phases in solids, rather than on similar phases of neutral atoms in either helium or ultra-low-density atomic gases, which couple very differently to electromagnetic fields, even though these have certainly contributed to central developments of the field.

We have limited ourselves to a relatively small number of references. Given that the material of the book corresponds to such a vast body of work, we would otherwise have ended up with an unpalatably long but necessarily still woefully incomplete list of references. We have tried to include scholarly review articles where hundreds more references can be found in a more structured fashion. We apologize to those objecting to this or any other aspect of the presentation for our shortcomings; they are warmly encouraged to contribute their own versions, as condensed matter physics does not have the literature it deserves.

Few are the book projects that are swiftly concluded, and ours is not one of them. Without the collapse of our travel schedules due to the present pandemic, perhaps this manuscript would still not be finished. Of course, had we completed the book more swiftly, we would have missed a number of exciting developments, such as the discoveries of Floquet time crystals and Weyl and Dirac semimetals. At any rate, we do not expect a letup of the sequence of discoveries any time soon.

Indeed, overall condensed matter physics continues to have a refreshingly unmodern flavor to it. Most of its discoveries are made by small collaborations of creative individuals, often entirely serendipitously – just think of the integer quantum Hall effect, the cuprate superconductors, the isolation of graphene, or the prediction of topological insulators. This is a far cry from purportedly goal-oriented large-scale research programs, the organization of which both of us are admittedly also guilty of. At the same time, it is undeniably true that the use of large-scale facilities like modern neutron and light sources has advanced the field tremendously.

The passage of time has also asserted itself in several other ways. Topological condensed matter physics has in the meantime been recognized by the 2016 Nobel Prize for Duncan Haldane, Michael Kosterlitz, and David Thouless. Neither was sad news in short supply, with Thouless passing away only a few months after the untimely death of Shoucheng Zhang. The passing of Phil Anderson in March 2020 concluded the extraordinary career of arguably the most influential scientist of the second half of the twentieth century.

Finally, it might be useful to explain our approach to pedagogy, since it is intended that this book will be useful also for courses of self-study. Our feeling is that an encyclopedic list of results without derivations is unlikely to help readers understand the material for themselves, while too detailed a presentation tends to obscure the underlying conceptual structure of the material. In compromise, we have tried to explain a moderate number of central results, with a key example where appropriate; most of these do not require a great deal of technical background or impose greatly on the reader's patience. We would like to think, of course, that the book can thus also be useful as an initial reference that can be consulted for the basics of a subject and as a source for more comprehensive reading.

For the nonexpert reader, although much interesting material has had to be omitted, what has been included may still be hard to navigate initially. To lower the entry bar, besides providing background in Chapter 2 to make the book reasonably self-contained, we have collated the most fundamental material in two chapters, which we recommend as an entry point. These are Chapter 3, on integer topological phases, and Chapter 5, on fractionalization.

Ideally, the book will be useful for active practitioners in the field as well as for newcomers, to whom we would like to extend a heartfelt welcome.

Acknowledgments

The embarrassment of riches of topological condensed matter physics mentioned in the preface is matched by the number and generosity of the people to whom we are indebted on our scientific journey to date. The first mention should be of those scientists who guided our steps during our scientific childhood and adolescence. Added to this are the collaborators with whom we began and sustained a research effort in the field.

In this spirit, J.E.M. thanks Duncan Haldane and Xiao-Gang Wen for explaining the beauty of topological order; Leon Balents and Cenke Xu for collaborations in the early days of topological materials; and Joseph Orenstein and David Vanderbilt for adding a dose of realism, in their distinct ways, later on. Similarly, R.M. is grateful to John Chalker and Shivaji Sondhi, especially for the early work on frustrated magnetism that was finding its place in the fabric of topological condensed matter physics as it was being woven. Since then, he has benefited greatly from the time and enthusiasm of many collaborators. Feeling distinctly uncomfortable singling out any individually, he would like to hide behind the Europhysics Prize Committee to signal his gratitude for satisfying theory–experiment collaborations like the one involving the discovery of emergent magnetic monopoles with Claudio Castelnovo and the groups of Alan Tennant and Santiago Grigera.

Both of us have had the privilege of watching junior scientists from our research groups develop into independent contributors to the field of topological physics. We particularly thank our students and postdocs, too many to list, not just for their collaboration on research but for making the job of professor a worthwhile one. Early versions of this material were inflicted upon students, and refined based on their questions, not just at our institutions but at a number of advanced schools. J.E.M. acknowledges Oxford University Press for permission to reuse material in Section 2.7, Chapter 4 and Section 6.6.1 from his Les Houches lecture notes (Chamon et al., 2017).

J.E.M. acknowledges primary research support from the US government, chiefly the Division of Materials Research of the National Science Foundation, currently via grant DMR-1918065, and the Office of Science of the Department of Energy. He is also grateful for support from a Simons Investigator award and the Emergent Phenomena in Quantum Systems program of the Gordon and Betty Moore Foundation, and for the support of his colleagues at the University of California, Berkeley, and Lawrence Berkeley National Laboratory. R.M. thanks the Max Planck Society for providing steadfast support and, to his mind enviable, academic freedom, which underpin the stimulating scientific atmosphere of the Max Planck Institute for the Physics of Complex Systems in Dresden.

For the immediate realization of this manuscript, we are grateful for comments, feedback, encouragement, and help with figures, from Noelle Blose, John Chalker, Tom Fennell, Peter Fulde, Andrew Green, Matteo Ippoliti, Vedika Khemani, Kai Klocke, Yu-Hsuan Lin, Netanel Lindner, Jonathan Nilsson-Hallen, Benedikt Placke, Regine Schuppe, Vijay Shenoy, Kirill Shtengel, Jurgen Smet, Veronika Sunko, and Ruben Verresen, and Kohtaro Yamakawa. Portions of this work were completed at the Kavli Institute for Theoretical Physics, the Aspen Center for Physics, and the Galileo Galilei Institute. We are grateful to the staff of Cambridge University Press, and to Nick Gibbons in particular, for their steady encouragement and patience. Finally, we would like to thank our families for their sacrifices, before and during the writing of this book, toward enabling us to enjoy the life of the mind.

R.M. dedicates this book to the memory of Philip W. Anderson. As the founding pioneer of much of condensed matter physics, his intellectual influence pervades this field, and this book, on many levels. We mourn the loss of an outstanding person who was a role model for many of us, not just as a scientist, and feel privileged to have known him.

J.E.M. dedicates this book to another native of Illinois, William Moore, for his contributions in other spheres, including as a father.

1

Introduction

An abiding feature of physics is the search for universality. The goal of research is not primarily to accumulate laundry lists of observed phenomena – although this is how most science starts – but to deduce the underlying organizing principles, that is, to develop concepts in terms of which to formulate a deeper understanding.[1] "Condensed matter physics" is a catchall term for systems of many constituent particles in which the fundamental objects are electrons, nuclei, and photons, described by electromagnetism and quantum mechanics. The existence of subatomic particles such as quarks is remarkable but irrelevant to condensed matter physics, except insofar as it determines which nuclei appear in the periodic table, while concepts of collective behavior from condensed matter have proven useful in modeling animate matter and even sociological structures. Condensed matter physics deals with the question of how phenomena such as crystalline order, or waves, or superconductivity, emerge from a collection of particles for which individually these concepts do not even have a meaning: no water molecule will ever form a wave, nor even a ripple, on its own.

This book introduces the reader to the young and rapidly growing field of topological phases in condensed matter. The terms "phase" and "state" are used more or less interchangeably to denote one of the basic notions of condensed matter: a class of systems whose physical properties are, at least in principle, adiabatically connected. Before 1980, there was hardly a hint of the existence of topological phases, not unlike the situation with quantum physics in the late 1800s. There were important precursors to the revolutionary developments that were to follow, but it was only with the discovery of the quantum Hall effects that the importance and universality of the topological phenomena we discuss were beginning to be appreciated. Their importance was immediately apparent from the striking novelty of the observed phenomena: a simply measurable quantity such as electrical resistance

---

[1] In Ernest Rutherford's pithy summary, "all science is either physics or stamp collecting."

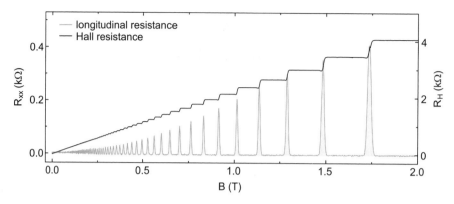

Fig. 1.1 Longitudinal and Hall resistances of a two-dimensional electron gas as a function of magnetic field. The plateaux around integer filling factor ν are exquisitely quantized in units of h/e^2, a macroscopic quantum effect involving only the fundamental constants e, the electronic charge, and h, Planck's constant. This information can be used to infer the value of the fine-structure constant α of quantum electrodynamics. From Friess (2016). Reprinted with permission by Springer International Publishing.

could be quantized with unprecedented accuracy, independent of "details" such as purity, or even size, of the material under investigation.

Since then, the list of observed topological phenomena in quantum matter has grown considerably, as has our grasp of the underlying organizing principles. This book aims to make the reader familiar with the items on the former and therefore receptive of the understanding provided by the latter. Figures 1.1, 1.2, and 1.3 display three instances of such new types of behavior: the quantized Hall resistance, a topologically protected conducting state at the surface of an insulator, and a neutron scattering plot reflecting an emergent gauge field in a magnetic compound known as spin ice.

The first of these, the quantum Hall effect, is the observation that when an electron gas is confined to two dimensions and subjected to a large perpendicular magnetic field, there are finite ranges of field, known as plateaux, over which the transverse conductance (and also conductivity) is simply

$$G_{xy} \equiv \frac{I_x}{V_y} = ne^2/h = \frac{n}{25812.807\Omega} \tag{1.1}$$

for integers $n = 1, 2, 3, \ldots$. The startling aspect of this observation is its precision: the quantization is observed to better than one part per billion. Note that the conductance values on plateaux are "universal" in the sense of being independent of the material used, since e and h are fundamental constants. This precise

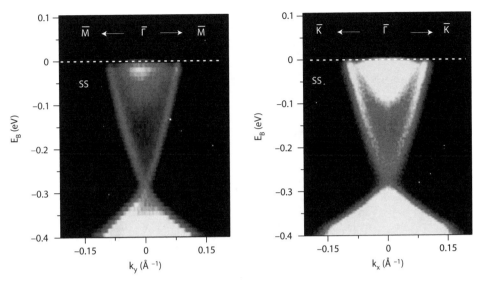

Fig. 1.2 The energy-momentum dispersion relation of occupied states at the surface of Bi_2Se_3, a three-dimensional topological insulator, as imaged by angle-resolved photoemission, along two different cuts in the Brillouin zone. The characteristic signature of the topologically protected surface state is the single Dirac cone. Image from Xia et al. (2009). Reprinted with permission by Nature Publishing Group.

quantization is observed in samples that are macroscopic in two directions and are subject to both defects and thermal fluctuations.

The conductance quantization in the QHE is related to the existence of conducting one-dimensional (1D) edge channels at the boundary of the 2D electron droplet. More generally, many topological states show a "bulk-edge correspondence" by which edge or surface phenomena are essentially determined by what the particles are doing in the bulk of the material, rather than by edge or surface details. An exciting recent development along these lines is that there are some 3D bulk insulators that, in zero magnetic field, have special surface states. An experimental example is shown in Figure 1.2: the bulk insulator Bi_2Se_3 has, on the surface being probed, a single branch of massless electronic excitations; in the language of the field, this material is a 3D topological insulator with a single linearly dispersing cone on its surface. Actually, all surfaces are conducting and have an odd number of Dirac cones. If the material is cut, metallic layers form at the new surface.

There are now several materials in this category, and they show topological behavior at energy scales comparable to room temperature. The source of the topological behavior is the intrinsic coupling between spin and orbital angular momentum of electrons in solids. It is quite subtle to explain under what conditions this surface state leads to a quantized physical response analogous to the

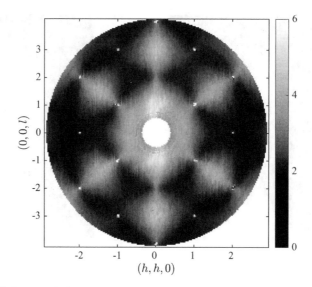

Fig. 1.3 Spin correlations measured by neutrons in the spin ice compound $Dy_2Ti_2O_7$ (Fennell et al., 2009). The characteristic pinch-points (Figure 5.14) at [002] and [111] in reciprocal space are signatures of an emergent gauge field. This distinguishes this state from conventional ordered phases, which would go along with sharp, divergent Bragg peaks. Simple paramagnets, by contrast, have completely smooth neutron scattering profiles. Figure courtesy of Tom Fennell.

Hall conductance in the previous example. It is now understood theoretically, but not yet confirmed experimentally, that under some conditions, the surface state shown in Figure 1.2 can develop an energy gap and lead to a quantized magneto-electric polarizability in the material; that is, an applied magnetic field produces a proportionate electrical polarization, with a quantized proportionality constant.

To appreciate the novelty of the quantized conductance in the QHE and the quantized magnetoelectric effect in topological insulators, it may be warranted to compare them to a different, by now familiar, kind of universality, that was conclusively formulated in the 1970s. This appears in crystals, magnets, superfluids, and many other kinds of ordered phases that form the bread and butter of condensed matter physics. The organizing principle underpinning this universality is the concept of symmetry breaking: the low-temperature phase of the system is less symmetric than the Hamiltonian that describes the interactions between the individual constituents. For example, the direction in which a permanent magnet points could be reversed to form a new state with the same energy.

Universality appears in quantities that are only sensitive to the form of the symmetry breaking and not to microscopic details. An example is shown in Figure 1.4. The liquid–gas transition of a simple fluid such as water disappears at a critical

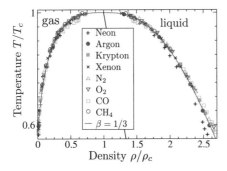 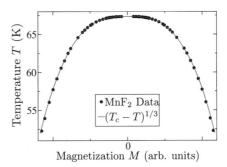

Fig. 1.4 Universality of critical behavior in two distinct systems. (left) Liquid–gas transition (Guggenheim, 1945). (right) Magnetization at the ordering transition in an Ising magnet (Heller and Benedek, 1962). Figure adapted from Sethna (2006). Reprinted with permission by American Institute of Physics and Oxford University Press.

point in the pressure–temperature plane, where the liquid and gas phases cease to be distinct. The difference in density between the liquid, ρ_L, and gas, ρ_G, at the boiling transition vanishes as the temperature T approaches the critical value T_c according to a power law:

$$\rho_L - \rho_G \propto (T_c - T)^{\beta}, \quad \beta \approx 0.33. \tag{1.2}$$

Not only is this value of critical exponent β common to all fluids reasonably similar to water but it appears in what prima facie seems like a totally different context. A ferromagnet with a preferred axis (in technical terms, an Ising magnet) eventually loses its spontaneous magnetization when heated above a Curie temperature T_c. Just below the Curie temperature, the magnetization difference vanishes as

$$M_\uparrow - M_\downarrow \propto (T_c - T)^{\beta'}, \quad \beta' \approx 0.33. \tag{1.3}$$

Within experimental accuracy, all the distinctions between the liquid–gas transition and the ferromagnetic transition are "irrelevant" for quantities such as critical exponents.

Work in the second half of the last century explained the ubiquity of such universality at continuous phase transitions and in several other contexts; this is formalized in the framework known as Landau–Ginzburg–Wilson theory, for which we later summarize some crucial facts relevant to understanding topological phenomena. While some organizing principles are specific to one scale of phenomena – we usually describe animate matter rather differently than we do subatomic particles – the ideas of Landau–Ginzburg–Wilson theory are so fundamental that they are used everywhere from particle physics to astrophysics.

However, the type of universality reflected in conductance quantization and other properties of topological phases is equally remarkable but comes about for totally different reasons: in general, it is not useful to think of topological phases as resulting from a broken symmetry, and the search is ongoing for the proper abstract description (the topological analogue of Landau–Ginzburg–Wilson theory) of some topological phases.

The last example involves magnetic behavior beyond the Landau–Ginzburg–Wilson description. Figure 1.3 shows neutron scattering data on a magnetic compound known as spin ice. Neutron scattering measures the Fourier transform of the spin correlations: the sharper the neutron scattering signal is, the longer ranged are the spin correlations. Symmetry breaking thus appears in the form of a narrow, intense Bragg peak, signaling long-range order, while a paramagnet goes along with a smooth pattern. Spin ice corresponds to neither: one finds a pinch-point, signaling more structure than the latter but less than the former. This, we will see, reflects topological features emerging at low temperatures.

Shortly after the discovery of the integer quantum Hall effect, observation of a fractional version added an entirely new layer of complexity. Most remarkably, the fractional quantum Hall state was found to support quasiparticle excitations that carry a fraction of an electronic charge. Counterintuitive though this may be – given that the microscopic degrees of freedom, the electrons, all carry integer charge – the phenomenon of fractionalization was already known from one dimension. What was entirely new was the observation that even the statistics of the electrons fractionalized, yielding entities obeying neither Fermi nor Bose but rather "anyonic" statistics. The complex of new physics then grew to include a new type of order – topological order. This unifies several remarkable features of topological phases: for example, fractional quantum Hall liquids are (in rough terms) insensitive to the size of the surface they inhabit but sensitive to its topology, for example, whether it is a sphere or torus, and this topological dependence is intimately connected to the existence of fractional quasiparticles.

As our understanding of this new set of phenomena grew, it turned out that a range of other phenomena were intimately related on a conceptual level. These include a class of unconventional magnetic states, called spin liquids, where fractionalized excitations appear in the form of holons and spinons. These independently of each other carry the electric charge and the magnetic moment, respectively, of the electron. The low-energy description of such spin liquids differs fundamentally from conventional magnets, where spin flips, domain walls, or magnons are more familiar. Instead, the appropriate emergent degree of freedom turns out to be a gauge field, reflected in the neutron scattering pattern in Figure 1.3.

It is the aim of this book to give the reader a solid, essentially self-contained introduction to this set of phenomena and the concepts underpinning our understanding of them. There exist very good specialized texts for some of the older topics covered here. Our two primary motivations in writing this book are, first, to point out just how much interesting physics is accessible once a few basic unifying concepts are learned and, second, to collect some exciting theoretical and experimental developments from the past few years.

We have attempted to organize this book so that it can be read sequentially as a textbook or intermittently as a reference for specific topics. The following chapter contains several mathematical and physical preliminaries that are intended to make the book reasonably self-contained. Bloch's theorem, Landau levels, and the basics of Landau–Ginzburg–Wilson theory are covered in standard condensed matter physics courses but might be useful to voyeurs from high-energy physics or materials science. Berry phases are a fundamental aspect of quantum mechanics and central to several chapters of this book. The basics of two flavors of topology, homotopy ("winding numbers") and de Rham cohomology ("ambiguous integrals"), are presented as they are frequently used in the more theoretical sections. Rigor is an early casualty in the mathematical discussions as we focus on examples and intuition.

The heart of the book begins in Chapter 3, which covers the basic phenomena of the integer quantum Hall effect (IQHE) and topological insulators. This chapter describes the experimental facts and an intuitive picture of what distinguishes these topological phases of noninteracting particles from more familiar phases. More precisely, these "integer topological phases" are those that, although they frequently persist to strong interactions, can be realized even with independent electrons. The two-dimensional and three-dimensional topological insulators have generated considerable excitement in the past few years. These are similar to the IQHE in that they do not require strong interactions but differ in that spin-orbit coupling rather than magnetic field is the source of the topological phase. For the three-dimensional topological insulator, two alternate pictures are presented. These materials are time-reversal-symmetric bulk insulators with an odd number of Dirac cones on their surfaces; if a weak magnetic perturbation is added to break time-reversal symmetry, they exhibit a quantized magnetoelectric effect.

There are now several ways to understand integer topological phases theoretically, and Chapter 4 presents a few that have proven to be useful either for classifying possible phenomena or for practical calculations. It starts from the picture of the integer quantum Hall effect developed by Thouless and collaborators in terms of the Berry phase of Bloch electrons. This formulation is fruitful not just for the quantum Hall effect but for several other quantities in solids, such as the electrical polarization, and its non-Abelian generalization appears later in the description of topological insulators and the orbital magnetoelectric effect. An interesting question for future theory is whether this non-Abelian gauge structure on the Brillouin zone has other physical consequences. Another way to understand the topological invariants underlying integer topological phases is via the analysis of simple model Hamiltonians, which also proves useful in analyzing the transitions between topological phases.

Chapter 5 covers the basic examples ("hydrogen atoms") of fractionalized topological phases, in $d = 1, 2$, and 3, in which interactions play the central role. Together with Chapter 3, it forms the core explanation of the most important concepts and experimental phenomena in a nontechnical manner, which we hope is appropriate not only for an introductory graduate-level course but also for the casual reader interested in gaining familiarity with the phenomenology in order to read research literature. Specifically, it contains an introduction to the fractional quantum Hall effect, one-dimensional fractionalization, resonating valence bond liquids, and spin ice. It explains the central concepts of quantum number fractionalization in its incarnations as fractional charge, spin–charge separation, and magnetic monopoles; connections of fractionalization to topological order and topological degeneracy are sketched here.

These topics are covered more formally in Chapter 6, which takes a more theory-oriented approach and develops a unifying field-theoretical framework of the material of the previous chapters. The goal is to find descriptions of topological phases comparable to the description of symmetry-breaking phases by Landau–Ginzburg–Wilson theories, which capture the essentials (and only the very essentials). Starting from a gauge-theoretic formulation of the spin liquid physics in the quantum dimer model, we identify its emergent low-energy degrees of freedom and their relative statistics. We then start the study of continuum theories with an explanation of how spin ice gives rise to Maxwell gauge theory, while fracton models yield higher-rank generalizations. Following this, Chern–Simons theory, the topological field theory for both integer and fractional quantum Hall effects, is discussed, enabling a more abstract and universal discussion of fractionalization, fractional statistics, ground-state degeneracy, and edge states. The chapter closes with a brief description BF theory as a topological field theory description, in the same spirit as Chern–Simons theory, for the spin liquid of the quantum dimer model and other states.

Chapter 7 starts from the realization that similar concepts of topology also govern the behavior of several kinds of materials without an energy gap, or even a mobility gap, a gap to delocalized states. The number of experimental candidates in this category has greatly increased even in the few years since this book was begun. Indeed, in hindsight, we can say that possibly the first consequence of the Berry curvature of Bloch wavefunctions was not in the integer Hall effect but in the analysis of semiclassical contributions to an electron's velocity in metallic crystals of low symmetry. Without an energy gap, concepts such as adiabatic continuity and quasiparticle braiding are absent, or at least strongly modified, but other aspects such as edge states and topological invariants can remain relevant in materials such as the recently discovered topological semimetals. We close by covering the physics of gapless spin liquids, which will reappear in the discussion of non-Abelian phases in Chapter 9.

The influence of disorder on topological phases is covered in Chapter 8, with disorder being not only an inevitable fact of life but also a source of interesting new physics and a potentially useful probe. In particular, we start with the role of disorder in generating the conductance plateaux in the quantum Hall effects to introduce the deep connection between localization physics and integer topological phases. Defects in topological phases have interesting properties and combine two different senses of topology: the nontrivial homotopy of topological defects such as vortices, which was the original (essentially classical) appearance of topology in condensed matter, leads to trapped quasiparticles and protected conductors when the topological defects exist in a topological phase.

In Chapter 9, topological quantum computing provides a motivation to discuss various examples of non-Abelian topological states. The goal of building a particularly stable type of quantum computer using topological phases is probably the most ambitious of several potential technological applications of the newly discovered topological phases. Topological quantum computing is of interest to condensed matter physicists in large part because of the elegant way in which several different ingredients come together: the novel properties of fractionalized particles, their robustness to disorder and decoherence, and the demanding conditions for their realization.

The relatively young, but rapidly growing, field of topological phenomena in nonequilibrium settings is covered in Chapter 10. This focuses on periodically driven systems, also known as Floquet systems. It covers how periodic driving can generate so-called Floquet topological insulators and emphasizes new phenomena that cannot be realized in equilibrium: the anomalous Floquet–Anderson insulator, and the Floquet discrete time crystal, which represents a fundamentally new type of spatiotemporal order.

Finally, the material covered in this book is part of a living, rapidly developing research field. While some of its aspects are mature and well understood, others are clearly important but have not yet found their place in the conceptual fabric of the field. The final chapter groups together some emerging areas. The book concludes with an appendix that includes a survey of literature that we believe might be useful for the reader interested in a greater level of detail. Interspersed through the text are self-contained boxes that provide additional information on various topics.

Also included are discussions of a few classic materials that enabled the first identifications of topological behavior. While a reasonable survey of the materials aspects would be a book in itself and would require much more knowledge of solid state physics and chemistry than we have assumed, the combination of hard work and inspiration that identified various topological materials certainly deserves our appreciation. As topological phenomena move closer to engineering applications, the materials aspects are likely to become even more important.

We should emphasize that we have focused on topological phases in solids, rather than on similar phases of neutral atoms in either helium or modern ultra-low-density atomic gases. Studies of helium contributed significantly to the understanding of several aspects discussed here, such as topological superfluidity or Weyl points, but there are important differences. Chief among these are that electrons couple differently to electromagnetic fields than do neutral atoms, and solids originate from a lattice rather than being naturally described in the continuum. The reader wishing to know more about topological aspects of helium and other neutral atom systems is encouraged to look at the book by Volovik (2012).

We close the introduction with some comments on how different categories of readers might use this book. One of us has taught a semester-long course on topological phases by following Chapters 2–7 in reasonable detail. An experimentally inclined reader might wish to start with Chapters 3 and 5, with occasional reference to Chapter 2 as needed. Readers specifically interested in strong correlation physics could try jumping directly to Chapters 5 and 6. We would like to think that the whole book could be profitably read by theorists planning to work in this field, but Chapters 8 and 9 on topological quantum computing and topology out of equilibrium could perhaps be omitted on a first reading. Ideally, readers at all levels of sophistication will emerge appreciating something of the beauty that is a hallmark of this field.

2

Basic Concepts of Topology and Condensed Matter

In these days the angel of topology and the devil of abstract algebra fight for
the soul of each individual mathematical domain.

Hermann Weyl (1939)

The goal of this chapter is to introduce some of the key concepts and tools that
are used repeatedly in the analysis of topological phases of matter. We start with
two basic results from nonrelativistic quantum mechanics. The Berry phase is a
geometric phase in quantum mechanics that can be introduced simply as resulting
from adiabatic evolution in time. An example of its physical meaning is that, when
the Hamiltonian of a system is brought adiabatically around a closed path in parameter space, a system initially in a nondegenerate energy eigenstate accumulates a
physically meaningful phase relative to its initial wavefunctions, even when there
are no transitions to other states. The same mathematical structure appears whenever wavefunctions evolve smoothly as a function of some other parameter, such
as momentum or position. Clearly this structure is related to gauge fields and other
physical examples of holonomy. As an example of how the Berry phase appears
outside its original context of adiabatic evolution, we introduce the coherent-state
representation of a quantum spin.

The second basic quantum result introduced is Bloch's theorem, which is the
statement that wavefunctions of a single particle in a periodic potential take a particularly simple form: they are products of a plane wave and a part with the same
periodicity as the potential. Before considering a periodic potential, we warm up
with the specific problem of single-particle energy levels (Landau levels) of a particle in the continuum in a constant magnetic field, as these states wind up being
widely used in quantum Hall physics. Bloch's theorem yields two important differences between electrons in a solid and in vacuum. In a solid, the wavefunction has
additional material-dependent structure beyond that of a plane wave, and momentum is also modified as only crystal momentum (momentum modulo a reciprocal

lattice vector) is defined. In more mathematical language, Bloch's theorem naturally gives to wavefunctions of a particle in a periodic potential (e.g., a crystal) a fiber bundle structure similar to that of a gauge theory: a band structure associates every value of the crystal momentum with a wavefunction with the periodicity of the crystal. We discuss how this works in the tight-binding models beloved by physicists working on topological phases.

We then review the basic picture of "conventional" order in solids, via the Landau theory of symmetry-breaking phases. The concept of an order parameter remains essential for many of the contexts where topology arises in condensed matter physics, particularly for the discussion of topological defects that follows. The next section uses simple examples to introduce two flavors of topology: the cohomology of differential forms and the basics of homotopy. This section is primarily for theoretically inclined readers who would like to understand the mathematical description of topological phases. As there are excellent textbooks specifically devoted to explaining these mathematical methods for physicists (Nakahara, 1998; Stone and Goldbart, 2009), our emphasis is on conveying the minimum necessary to profit from discussions later in the book; we have often preferred the specific example to the general classification and have sought accessibility at some cost in rigor.

Topological defects in ordered phases breaking continuous symmetries were the first nontrivial application of topology to condensed matter, and we discuss vortices in a superfluid or XY model in some detail. Even for readers who skip the introduction to topological methods, the idea of homotopy groups is quite useful for complicated topological defects and also for defect-free topological configurations like skyrmions. The Berezinskii–Kosterlitz–Thouless transition from vortex unbinding in superfluid films is discussed at a nontechnical level (i.e., without renormalization group ideas) in a sidebar, as an example of the emergent physics of many vortices.

2.1 Berry Phases in Quantum Mechanics

We start with a beautiful geometric property of quantum mechanics whose full significance was understood only in the early 1980s: the geometric or Berry phase. In the course of this book we will frequently be interested in how the eigenstates of a Hamiltonian vary as a function of some parameters. The Berry phase is a subtle and physically important consequence of such variation. One way to introduce the geometric phase is by considering adiabatic changes in time of the Hamiltonian's parameters. We will primarily be interested in this book in cases where the same quantity appears even though the change in Hamiltonian parameters is not a function of time. The first examples of this type of geometric phase in physics were found more than fifty years ago in optics (Pancharatnam, 1956) and chemical dynamics (Longuet-Higgins et al., 1958), and the classic paper of Berry (1984) was the first to identify the concept in its full generality.

An important result from undergraduate quantum mechanics is the adiabatic approximation. Suppose that a system is prepared in a nondegenerate eigenstate of a time-dependent Hamiltonian H. For later reference, we will write H as a function of some parameters λ_i that depend on time: $H(t) = H(\lambda_1(t), \lambda_2(t), \ldots) = H(\boldsymbol{\lambda})$. If the eigenstate remains nondegenerate, then the adiabatic theorem states that if the Hamiltonian changes slowly in time (how slowly depends primarily on the energy gap between adjacent eigenstates), then there are no transitions between eigenstates.

This approximation, when correct, actually only gives part of the story: it describes the probability to remain in the eigenstate that evolved from the initial eigenstate, but there is actually nontrivial information in the *phase* of the final state as well. This result may seem quite surprising because the overall phase in quantum mechanics is in general independent of observable quantities. However, the Berry phase from an adiabatic evolution is observable: more precisely, phase differences are observable, such as the phase difference between one system taken around a closed path in parameter space and another system initially identical to the first whose parameters remain unchanged. Conceptually and mathematically, the Berry phase will turn out to be closely related to the notion of how the electromagnetic vector potential in quantum mechanics gives a relative phase between wavefunctions at different points, with physical consequences such as the Aharonov-Bohm effect.

Berry's result for a closed path is deceptively simple to state. In moving a system adiabatically around a closed path in parameter space, the final wavefunction is in the same eigenstate as the initial one (again, under the assumptions of the adiabatic approximation as stated above), but its phase has changed:

$$|\psi(t_f)\rangle = e^{-(i/\hbar) \int_{t_i}^{t_f} E(t')\, dt'} e^{i\gamma} |\psi(t_i)\rangle. \tag{2.1}$$

Here $E(t')$ means the corresponding eigenvalue of the Hamiltonian at that time, and γ is the Berry phase, expressed as an integral over a path in *parameter* space with no time dependence:

$$\gamma = i \oint \langle \psi(\boldsymbol{\lambda}) | \nabla_\lambda | \psi(\boldsymbol{\lambda}) \rangle \cdot d\boldsymbol{\lambda}. \tag{2.2}$$

Note that there are two different arguments of ψ in the above formulas. When ψ has a time argument, it means the wavefunction of the system at that time. When ψ has a parameter argument, it means the reference wavefunction we have chosen for that point in the parameter space of the Hamiltonian. A key assumption of the derivation is that there is some smooth choice of the $\psi(\lambda)$ throughout a surface in parameter space with the loop as boundary.

For an open path, we need to describe the phase of the wavefunction relative to this reference set, so the expression becomes more complicated (for the closed path, we could simply compare the initial and final wavefunctions, without needing the reference set at these points). We will show that, assuming $\psi(t_i) = \psi(\lambda(t_i))$ so that the initial wavefunction is equal to the reference state at the corresponding value of parameters,

$$\langle \psi(\lambda(t)) | \psi(t) \rangle = e^{-(i/\hbar) \int_0^t E(t')\,dt'} e^{i\gamma}, \tag{2.3}$$

that is, the Berry phase appears when comparing the actual time-dependent evolved state $\psi(t)$ to the reference state at the point in parameter space $\lambda(t)$. We can take the time derivative of both sides and use the time-dependent Schrödinger equation

$$i\hbar \frac{\partial \psi}{\partial t} = H(t)\psi. \tag{2.4}$$

The two sides agree initially if we choose the appropriate boundary condition on the Berry phase. The time derivative of the left side of (2.3) is

$$\langle \psi(\lambda(t)) | \frac{-iE(t)}{\hbar} | \psi(t) \rangle + \frac{d\lambda_j}{dt} \langle \partial_{\lambda_j} \psi(\lambda(t)) | \psi(t) \rangle, \tag{2.5}$$

so writing $e^{i\theta(t)} = \langle \psi(\lambda(t)) | \psi(t) \rangle$, we have computed

$$\frac{d}{dt} e^{i\theta(t)} = \left(\frac{-iE(t)}{\hbar} + \frac{d\lambda_j}{dt} \langle \partial_{\lambda_j} \psi(\lambda(t)) | \psi(\lambda(t)) \rangle \right) e^{i\theta(t)}. \tag{2.6}$$

Note that in the second term on the right side, it is $\psi(\lambda(t))$ that appears in the ket, and we have used $|\psi(\lambda(t))\rangle e^{i\theta(t)} = |\psi(\lambda(t))\rangle \langle \psi(\lambda(t)) | \psi(t) \rangle = |\psi(t)\rangle$, because the projection operator onto $|\psi(\lambda(t))\rangle$ must act trivially on a state already in the same one-dimensional subspace (i.e., differing only by a phase), which we remain in because of the adiabatic approximation. This time evolution of the phase is satisfied if (note that for E we do not need to distinguish between time and λ dependence)

$$\dot{\theta}(t) = -\frac{E(t)}{\hbar} - i \frac{d\lambda_j}{dt} \langle \partial_{\lambda_j} \psi(\lambda(t)) | \psi(\lambda(t)) \rangle, \tag{2.7}$$

which is our desired conclusion after noting that (dropping the explicit t dependence)

$$\frac{d\gamma}{dt} = i \frac{d\lambda_j}{dt} \langle \psi(\lambda) | \partial_{\lambda_j} \psi(\lambda) \rangle = -i \frac{d\lambda_j}{dt} \langle \partial_{\lambda_j} \psi(\lambda) | \psi(\lambda) \rangle. \tag{2.8}$$

The negative sign in the last equality, which shows that the Berry connection or Berry vector potential

$$\mathcal{A}_j = i \langle \psi(\lambda) | \partial_{\lambda_j} \psi(\lambda) \rangle \tag{2.9}$$

is real, follows from noting that $\partial_{\lambda_j}\langle\psi|\psi\rangle = 0$ by constancy of normalization. It is crucial for a nonzero Berry phase that the eigenstates of H change somewhere on the path beyond just a phase factor (i.e., they are physically inequivalent vectors in Hilbert space). If this were not the case, as must happen for a Hilbert space of dimension 1, then the Berry phase is just the phase difference between the reference choice at the initial and final points, which is zero for a closed path. The spin example below shows that a Hilbert space of dimension 2 is sufficient for the Berry phase to be significant. So although the rate of change in H dropped out, as long as the system remains adiabatic, and only the path taken by H enters the Berry phase, the eigenstates of H must have a genuine evolution, not just a phase change, for the Berry phase to be nontrivial.

Now one can ask whether the Berry phase is independent of how we chose the reference wavefunctions (in this case, the U(1) degree of freedom in the wavefunction at each λ, where U(1) means the group of one-by-one unitary matrices or phase factors $e^{i\phi}$). While for an open path it clearly is not gauge-independent (here meaning that we change the phase of the wavefunction at different points in parameter space), the Berry phase is gauge-independent for a closed path, for exactly the same reasons as a closed line integral of $\mathcal{A}$ is gauge-independent in electrodynamics: we can integrate the Berry flux or Berry curvature $\mathcal{F}_{ij} = \partial_i\mathcal{A}_j - \partial_j\mathcal{A}_i$ (which the reader can check is gauge-independent, just like the field strength tensor $F_{\mu\nu}$ in electrodynamics, by recalling that gauge transformation adds a gradient to A) on the surface bounded by the path. The Berry curvature is a field strength in parameter space, which can have any number of dimensions, but when working in three dimensions we often convert it to a vector, the curl of $\mathcal{A}$.

Note, however, that this gauge invariance assumed that the loop used to compute the Berry phase encircled a surface on which the Berry curvature $\mathcal{F}$ was defined. If we had a closed loop on a manifold that is topologically nontrivial, in a sense that will be clearer by the end of the chapter, then this argument doesn't work. For an example, consider the circle as an abstract object without an interior, which is the right picture of the Brillouin zone (set of inequivalent momenta) of a one-dimensional crystal. Note that a gauge transformation changes $\mathcal{A}$ by the gradient of a scalar phase, from the definition (2.2). Even if the reference wavefunctions are uniquely defined, however, that phase could change by a multiple of 2π on circling the path, so that only $e^{i\gamma}$ would be well defined; this object is the Berry phase version of a Wilson loop in gauge theories. The possibility of having a Berry vector potential that, when integrated along paths, gives quantities that are almost but not completely gauge-invariant (i.e., gauge-invariant except for discrete jumps) will be explained more mathematically in Section 2.7.

To get some geometric intuition for what the Berry phase means, we explain why the Berry connection $\mathcal{A}$ is sometimes called a connection, and the flux $\mathcal{F}$ is sometimes called a curvature. A connection is a way to compare vector spaces that are attached to different points of a manifold, forming a vector bundle. In our case, there is a one-dimensional complex vector space attached at each point in parameter space, spanned by the local eigenstate. The inner product lets us compare vectors at the same point in parameter space, but the Berry connection appears when we try to compare two vectors from slightly different points.

A useful example of a real vector bundle is the tangent bundle to a Riemannian manifold (say, a sphere), made up of tangent vectors at each point, which have a dot product corresponding to the inner product in quantum mechanics. The connection in this case, which gives rise to parallel transport of tangent vectors, determines the same curvature that appears below in the example of the Gauss–Bonnet theorem. Consider an airplane moving around the surface of the Earth and carrying a gyroscope that is fixed to lie in the tangent plane to the Earth's surface (i.e., free to rotate around the normal axis to the tangent plane). If the airplane follows a great circle, then it will appear to be going straight ahead to a passenger on board, and the gyroscope will not rotate relative to the plane's axis.

However, if the airplane follows a line of latitude other than the equator, or any other path that is not a geodesic (see a differential geometry text for details), it will feel constantly as though it is turning, and the gyroscope will appear to rotate relative to the airplane's direction. After going around a closed path with the airplane, the gyroscope may have rotated compared to a stationary gyroscope (the same physics that underlies Foucault's pendulum). As an exercise, one can work out that the total angle of rotation in circling a line of latitude is $2\pi \sin \phi$, where ϕ is the latitude. At the equator this gives no rotation, while at the North Pole this gives a 2π rotation. This is a geometrical version of the same idea of holonomy (here, failure of a gyroscope to return to its initial direction in going around a closed path) that underlies the Berry phase. Note that a vector potential in a gauge theory and the associated Wilson loop are also examples of the concept of holonomy in a (now complex) vector bundle. Instead of the vector bundle consisting of the tangent vectors through a point, the Berry phase comes from the Hilbert space of wavefunctions at a certain value of some parameters in the Hamiltonian. More precisely, the Berry connection tells us how to compare two states (Hilbert-space vectors) at different places in parameter space, in the same way as parallel transport tells us how to compare two tangent vectors at two points on a curved manifold.

We make two remarks in passing about generalizations. First, the Abelian U(1) Berry phase described above generalizes immediately to a non-Abelian Berry phase factor U(N), that is, an N-by-N unitary matrix, when instead of a single nondegenerate eigenstate, we consider a subspace of dimension N. Here Abelian means that two elements of U(1) commute, while two elements of U(N) for $N > 1$ do not necessarily commute. This has some important applications to topological phases that were discovered only recently. Bloch's theorem, introduced in a moment, will later be used to discuss how Abelian and non-Abelian connections arise naturally in a solid.

Second, while we have focused on continuous paths in the above, it is possible to make a similarly gauge-invariant object from wavefunctions at as few as three points. Consider the complex number $U = \langle u_1|u_2\rangle\langle u_2|u_3\rangle\langle u_3|u_1\rangle$. This is independent under gauge changes at any of the points; indeed it can be written in terms of the manifestly gauge-invariant projection operators $P_i = |u_i\rangle\langle u_i|$ as $U = Tr(P_1 P_2 P_3)$. The magnitude of U is generally less than unity, so it is not purely a phase, but as we increase the number of points and make them fill out a closed path, U becomes essentially the Wilson loop $e^{i\gamma}$ made from the Berry phase along that path. We demonstrate a use of this in the explicit example of a Berry phase in Box 2.1.

Box 2.1 The Berry Phase of the Adiabatic Dynamics of a Spin

Let us consider how the state of a quantum spin evolves if the spin always points in the direction of an applied magnetic field and the magnetic field varies smoothly. This will give some intuitive understanding of the Berry phase, and the mathematics involved will be useful later when we discuss the topological underpinnings of the integer and fractional quantum Hall effects. We first need to define a representation of the spin Hilbert space that is more symmetric than the usual S_z basis (the basis of states with definite values of the projection of spin along some fixed axis).

The coherent-state representation of a quantum spin of magnitude s is useful for many spin problems where having a fixed reference axis is undesirable. The basic idea is to represent spins in a basis of states $|\hat{\Omega}\rangle$ that in the classical limit behave like classical vectors: the states are obtained by rotating the North Pole $|s, s\rangle$ with a rotation operator $R(\theta, \phi, \chi)$ that is a function of three Euler angles. The vector $\langle \mathbf{S}\rangle$ of expectation values of the spin operators in the basis state $|\hat{\Omega}\rangle$ points along the unit vector $\hat{\Omega}$.

While χ is essentially an arbitrary phase convention, keeping careful track of the spin wavefunction shows that there is a physically meaningful Berry phase for any smooth choice of the spin reference wavefunctions. Define, for a unit vector $\hat{\Omega} = (\sin\theta\cos\phi, \sin\theta\sin\phi, \cos\theta)$,

$$|\hat{\Omega}\rangle = R(\chi, \theta, \phi)|s, s\rangle = e^{iS^z\phi}e^{iS^y\theta}e^{iS^z\chi}|s, s\rangle. \tag{2.10}$$

It is intentional that there are two S^z operators appearing in this formula, as the Euler angles are defined using one rotation around the original z-axis and one around the final one. One can use the explicit representation of the coherent states and some algebra to compute the inner product

$$\langle\hat{\Omega}|\hat{\Omega}'\rangle = \left(\frac{1 + \hat{\Omega}\cdot\hat{\Omega}'}{2}\right)^s e^{-is\psi} \tag{2.11}$$

with

$$\psi = 2 \arctan \left[\tan \left(\frac{\phi - \phi'}{2} \right) \frac{\cos[\frac{1}{2}(\theta + \theta')]}{\cos[\frac{1}{2}(\theta - \theta')]} \right] + \chi - \chi'. \tag{2.12}$$

For further details of the coherent-state representation, see the book of Auerbach (1994). Since (2.11) becomes zero as $s \to \infty$ unless $\hat{\Omega} = \hat{\Omega}'$, coherent states also justify the claim that as $s \to \infty$, the spins are like classical unit vectors (the classical limit of a spin).

Note that the coherent states are not orthogonal for finite s. The completeness relation is

$$\frac{2s + 1}{4\pi} \int d\hat{\Omega} |\hat{\Omega}\rangle \langle \hat{\Omega}| = \mathbf{1}. \tag{2.13}$$

Now specialize to $s = 1/2$ and consider dynamics in the two-by-two Hamiltonian given by

$$H = -\hat{\mathbf{n}} \cdot \boldsymbol{\sigma} \tag{2.14}$$

where $\boldsymbol{\sigma}$ is a vector of Pauli matrices. At time $t = 0$, suppose that the spin is initially prepared in the ground state $|\hat{\Omega}\rangle$, with $\hat{\Omega} = \hat{\mathbf{n}}$. Let the unit vector $\mathbf{n}$ be time-dependent, but assume that it changes sufficiently slowly that there are no transitions out of the lower-energy spin state. In this problem the energies are constant, so we can neglect the energetic part of the phase change around a path, assuming the path is traced out adiabatically.

Note that we follow tradition in using a more compact notation than in the general derivation of Berry's phase: the reference wavefunction $|\psi(\hat{\Omega})\rangle$ is now just written as $|\hat{\Omega}\rangle$. We also give a slightly different perspective on the Berry phase

$$\gamma = i \int \langle \psi(\lambda) | \nabla_\lambda | \psi(\lambda) \rangle \cdot d\lambda = \int_{t_i}^{t_f} \langle \hat{\Omega}(t) | i \frac{d}{dt} | \hat{\Omega}(t) \rangle \, dt. \tag{2.15}$$

Consider the overlap of wavefunctions at two slightly different times t and $t + dt$. The magnitude of this overlap is less than 1, but only by an amount of order dt^2. At order dt, the change in the overlap is purely imaginary. We can use this to build up the Berry phase over a segment of path as a product of a large number of overlaps, using

$$\gamma = -\Im \log \Big[\langle \hat{\Omega}(t_i) | \hat{\Omega}(t_i + dt) \rangle \langle \hat{\Omega}(t_i + dt) | \hat{\Omega}(t_i + 2\,dt) \rangle \dots$$

$$\dots \langle \hat{\Omega}(t_f - 2\,dt) | \hat{\Omega}(t_f - dt) \rangle \langle \hat{\Omega}(t_f - dt) | \hat{\Omega}(t_f) \rangle \Big]. \tag{2.16}$$

Recall that the phase factor between original and final states used in our earlier definition of the Berry phase was $\exp(i\gamma)$, which is why the logarithm appears here. Actually, one can start (Vanderbilt, 2018) from this discrete expression as the definition of the Berry phase for any number of intermediate steps, and then obtain the continuum limit we have focused on here by having a large number of steps. Clearly the 2π

ambiguity of the imaginary part of the logarithm is consistent with the interpretation of γ as a phase. To see the equivalence, write

$$- \Im \log\langle \hat{\Omega}(t_i)|\hat{\Omega}(t_i + dt)\rangle \approx -\Im \log(1 + dt \, \langle \hat{\Omega}|\frac{d}{dt}\hat{\Omega}\rangle)$$

$$\approx -dt \, \Im\langle \hat{\Omega}|\frac{d}{dt}\hat{\Omega}\rangle = i \, dt \, \langle \hat{\Omega}|\frac{d}{dt}\hat{\Omega}\rangle, \qquad (2.17)$$

where we have used that the inner product in the last step is purely imaginary.

Using the explicit coherent state representation, one finds for the overlap (Auerbach, 1994)

$$\langle \hat{\Omega}(t + dt)|\hat{\Omega}(t)\rangle = e^{-is\,dt\,\dot{\phi}\cos(\theta(t))-is\dot{\chi}}, \qquad (2.18)$$

where $\dot{\chi}$ results from whatever phase convention was chosen in the coherent state via

$$\dot{\chi} = \frac{d\chi}{d\hat{\Omega}}\frac{d\hat{\Omega}}{dt}. \qquad (2.19)$$

So the change in the Berry phase is

$$\dot{\gamma} = -s\dot{\chi} - s\,dt\,\dot{\phi}\cos(\theta(t)). \qquad (2.20)$$

Around a closed path $\dot{\chi}$ must integrate to zero, since χ only changes through the change in $\hat{\Omega}$, which returns to its initial value. The other part need not be zero and has a simple geometrical interpretation. We see that for an *open* path the phase change is not directly meaningful since it depends on the arbitrary phase convention in χ.

The phase gained around a closed path $\mathcal{P}$ on the unit sphere is

$$\gamma_{\mathcal{P}} = -s \int_{t_i}^{t_f} dt \, \dot{\phi}\cos\theta = -s \oint_{\phi_i}^{\phi_f = \phi_i} d\phi \, \cos(\theta(\phi)). \qquad (2.21)$$

The point of the second rewriting is that the integral depends only on the closed path traced out by the magnetic field direction, not by how it moves along that path. This integral just measures the area traced out on the unit sphere, since

$$-\oint_{\phi_i}^{\phi_i} d\phi \, \cos(\theta(\phi)) = -\oint_{\phi_i}^{\phi_i} d\phi \, [1 - \cos(\theta(\phi))]. \qquad (2.22)$$

Here we assume that the path did not encircle the North Pole so that ϕ remained well defined. Now the integrand is just the integral of $d(\cos\theta)$ as θ runs from the North Pole to the present point.

To summarize, the net effect of taking the spin around a closed path is to induce a phase proportional to s and to the area enclosed. This can be written as the loop integral of a "magnetic monopole" vector potential on the sphere with constant field strength:

$$\gamma = s \int_{t_i}^{t_f} dt \, \mathbf{A}(\hat{\Omega})\dot{\hat{\Omega}}. \qquad (2.23)$$

One gauge choice for this vector potential is

$$\mathcal{A} = -\frac{1 - \cos\theta}{\sin\theta}\hat{\phi}, \tag{2.24}$$

which has a singularity at the South Pole ($\theta = \pi$).

A subtlety is that any gauge choice for a nonzero monopole vector potential has to have a singularity somewhere on the sphere, because of the nonzero integral of the curl of $\mathcal{A}$, which is gauge-independent. For example, consider integrating the vector potential in (2.24) over a small circle around the South Pole. The result will be nearly equal to the area of the sphere (because we set up this gauge choice to capture areas starting from the North Pole, as in our explicit calculation), which explains why the vector potential had to diverge in order to get a finite answer over a tiny circle. At the North Pole, there is a Dirac string containing (Berry) magnetic flux entering the sphere, in order for the flux elsewhere to be uniformly directed outward as if a magnetic monopole were located at the center of the sphere. An alternative gauge choice would define areas starting from the South Pole and be singular at the North Pole. Note that the observable Berry phase factor $e^{i\gamma}$ is unchanged under this difference of 4π, though, because even after multiplying by the spin quantum number s, the ambiguity is a multiple of 2π.

We will explore some further interesting consequences of this spin problem later on. There are several problems where the integral of a Berry curvature over a two-dimensional manifold, such as the sphere in the above example, is a topological invariant and physically observable. The Berry phase studied here turns out to be essential in developing the path integral representation of a quantum spin, and the degeneracy of the lowest-energy eigenstates in the monopole field (the lowest Landau level as explained in the following section) on the sphere is $2s + 1$, consistent with the number of states in the spin multiplet.

2.2 One Electron in a Magnetic Field: Landau Levels

The first topological phases to be discovered were found in devices known as quantum wells or two-dimensional electron gases, in which electrons are constrained to move in a plane by a confining potential in the $\hat{\mathbf{z}}$ direction (Box 3.2). Looking ahead to our discussion of energy bands in the following chapter, it is typically a good approximation to assume that the electron density in such materials is small enough that electrons are near a band extremum and have a quadratic dispersion like free electrons. The key discoveries known as the integer and fractional quantum Hall effects emerged when a strong magnetic field was applied perpendicular to the plane of motion of the electrons, and a crucial ingredient we will use in Chapters 3 and 5 is the quantum mechanics of a free particle in a magnetic field.

To start, the classical equations of motion for a particle of charge q and mass m moving in two dimensions in a magnetic field applied in the perpendicular direction, $\mathbf{B} = B\hat{\mathbf{z}}$, are

$$m\frac{d^2x}{dt} = qB\frac{dy}{dt},$$
$$m\frac{d^2y}{dt} = -qB\frac{dx}{dt}. \tag{2.25}$$

Defining the cyclotron frequency

$$\omega_c = \frac{qB}{m} \tag{2.26}$$

allows us to write down the general solution of this equation in an elegant form by switching to a complex parameterization of the two-dimensional plane, $z \equiv x + iy$ (with $\dot{z} = dz/dt$):

$$\ddot{z}(t) = i\omega_c \dot{z}(t),$$
$$\dot{z}(t) = \dot{z}(0)e^{i\omega_c t},$$
$$z(t) = z(0) + \frac{\dot{z}(0)}{i\omega_c}\left(e^{i\omega_c t} - 1\right). \tag{2.27}$$

This describes the circular motion of the particle. Note that the frequency of this motion, ω_c, is independent of the radius of the circle, $R_L = \frac{|\dot{z}(0)|}{\omega_c}$: the particle's speed is proportional to R_L. Its kinetic energy is time-independent because the Lorentz force, $\mathbf{F} = q\mathbf{v} \times \mathbf{B}$, only acts in a direction perpendicular to the particle's motion. It is given by

$$E_{\text{kin}} = \frac{q^2B^2}{2m}R_L^2 = \frac{1}{2}m\omega_c^2 R_L^2. \tag{2.28}$$

Such periodic motion with frequency independent of amplitude and an energy proportional to its square is reminiscent of a harmonic oscillator. However, there is a *four*-dimensional phase space, as real space is two-dimensional, but it will turn out that in detail things are not very different from a two-dimensional simple harmonic oscillator: the energy levels are indeed given by $E_n = \hbar\omega_c(n + 1/2)$, with n a nonnegative integer, the Landau level index.

The degeneracy of each level is macroscopic, proportional to the system's area A. This accounts for the remaining pair of degrees of freedom. To obtain this degeneracy, which will be verified in a concrete calculation momentarily, let us posit in an *ad hoc* way that the number of states in the system needs to remain invariant when the field is switched on. In zero field, the number of states is given by $N = \int \frac{d^2p\, d^2x}{h^2}$, where x and p are canonically conjugate position and momentum coordinates. With $p^2/2m = E$, one finds for a system of size $\int d^2x = A$, a density of states $\rho(E) = \frac{1}{A}\frac{dN}{dE} = \frac{2\pi m}{h^2}dE$. A Landau level groups together states over an energy interval $\Delta E = \hbar\omega_c$,

$$\int_{E}^{E+\Delta E} \rho(E)dE = \frac{qB}{h} = \frac{1}{2\pi\ell^2}, \tag{2.29}$$

where $\ell^2 = \hbar/qB$ is called the magnetic length. It is the fundamental length scale in the problem, which, for example, encodes the size of the smallest wavepacket of an eigenstate that can be constructed in the presence of a field. The areal density of states of each Landau level is $\frac{1}{2\pi\ell^2}$, or one electron in the area that encloses one single-electron quantum $\phi = \frac{h}{e}$ of flux from the applied magnetic field.

To confirm these results formally, and as we will need explicit wavefunctions, let us next derive these properties quantum-mechanically. The Hamiltonian reads, now allowing for an effective mass m^*,

$$\mathcal{H} = \frac{1}{2m^*}(-i\hbar\nabla - q\mathbf{A})^2 \equiv \frac{\hat{p}^2}{2m^*}. \tag{2.30}$$

We will pick gauges for $\mathbf{A}$ and find associated wavefunctions momentarily, but first we point out some general symmetry aspects.

For $\mathbf{A} = 0$, this is the Hamiltonian of a free particle. Translations, generated by $\hat{p}_x$ and $\hat{p}_y$, commute: $[\hat{p}_x, \hat{p}_y] = 0$. The resulting group structure is thus Abelian, and the concomitant irreducible representations are therefore one-dimensional, labeled by the usual momentum $\mathbf{k} = (k_x, k_y)$. In nonzero field, $\nabla \times \mathbf{A} = B\hat{z}$, this is no longer the case:

$$[\hat{p}_x, \hat{p}_y] = -i\hbar^2/\ell^2 , \tag{2.31}$$

unlike in a circularly symmetric harmonic oscillator, so x- and y- coordinates are no longer independent.

Indeed, a second set of canonical commutation relations is defined for the so-called guiding center coordinates

$$c_x = x - p_y/m^*\omega_c, \quad c_y = y + p_x/m^*\omega_c , \tag{2.32}$$

for which $[c_x, c_y] = i\ell^2$. Here, ℓ^2 plays the role of the volume element of phase space normally supplied by h. As we will see below, the Hamiltonian is independent of the value of the guiding center variables, and to quantize them semiclassically, one cuts up the c_x–c_y phase into segments of area $2\pi\ell^2$. There are many ways to do this, the most convenient being ones which respect the symmetry of the Hamiltonian implied by the choice of gauge. In general this symmetry is lower than of the physical problem: in the presence of magnetic field, a symmetry operation in general maps the Hamiltonian onto one which is equivalent *up to a gauge-transformation.* The resulting symmetry group in a field is known as a magnetic translation group (see Eq. 5.10 for more details), an instance of a projective symmetry group.

Two gauge choices are particularly useful: the Landau gauge, which preserves explicit translational symmetry in one direction while apparently discarding isotropy and the perpendicular translations: $\mathbf{A} = (0, x, 0)$; and the circular or rotational gauge, which chooses a preferred origin but preserves isotropy: $\mathbf{A} = (-\frac{y}{2}, \frac{x}{2}, 0)$. The concomitant semiclassical quantizations are, firstly, areas bordered by adjacent parallel lines at $x_n = \frac{2\pi\ell^2}{L_y}$ for the Landau case; or by adjacent concentric circles of radius $\ell\sqrt{2n}$ for the circular gauge. In either case, there is one state for each flux quantum threading the system. An interesting modification occurs when the kinetic term is not quadratic but rather linear and Dirac-like, as for the electrons in graphene (Section 2.5.2).

The eigenfunctions can be constructed by defining the ladder operators

$$a^\dagger = \frac{\ell}{\sqrt{2}\hbar} \left(p_x + i p_y \right)$$

$$b = \frac{1}{\sqrt{2}\ell} \left(c_x + i c_y \right) \tag{2.33}$$

where $[a, a^\dagger] = [b, b^\dagger] = 1$, with $b, b^\dagger$ commuting with $a, a^\dagger$ as well as the Hamiltonian

$$\mathcal{H} = \hbar\omega_c \left(a^\dagger a + 1/2 \right) . \tag{2.34}$$

As advertised, the spectrum is that of a harmonic oscillator not depending on the guiding center ladder operators b at all. A complete set of states is then given by

$$|n, m\rangle = \frac{a^{\dagger n} b^{\dagger m}}{\sqrt{n! m!}} |0, 0\rangle . \tag{2.35}$$

2.2.1 Symmetric and Landau Gauge Wavefunctions

For a study of the quantum Hall effect, it is easiest to follow Laughlin's choice of symmetric gauge, which preserves the isotropy obeyed by a radial interaction. For this choice, one obtains $|0, 0\rangle = \frac{1}{\sqrt{2\pi\ell^2}} \exp(-|z|^2/4\ell^2)$ and, for the nth Landau level, an expression involving associated Laguerre polynomials

$$|n, m\rangle = \frac{\left(\frac{z}{\sqrt{2}\ell} \right)^m \exp(-|z|^2/4\ell^2) L_n^m \left(\frac{|z|^2}{4\ell^2} \right)}{\sqrt{2\pi\ell^2 (n+m)!/n!}} . \tag{2.36}$$

The $n = 0$ case of this formula will be used heavily in Section 3.1 and Section 5.1 for the integer and fractional quantum Hall effects, respectively. In Landau gauge, the unnormalized wavefunctions read

$$|n, k\rangle_L = e^{iky} H_n \left(x + k\ell^2 \right) \exp\left[-\frac{1}{2\ell^2} \left(x + k\ell^2 \right)^2 \right] \qquad (2.37)$$

with $k = \frac{2\pi}{L_y} \cdot s$, and $s = 0, 1, \ldots, L_y - 1$ fixing the center of the wavefunctions in the x-direction spaced by $\frac{2\pi}{L_y}\ell^2$ as in the semiclassical quantization. The Hermite polynomial H_n is a constant for the lowest Landau level ($n = 0$). This form is straightforward to see directly from the Hamiltonian in Landau gauge, as the momentum in one direction (here $\hat{\mathbf{y}}$) commutes with the Hamiltonian. Finding simultaneous eigenstates of $\hat{\mathbf{y}}$-momentum and energy then gives plane-wave behavior along y and a harmonic oscillator problem for the x coordinate, consistent with the harmonic oscillator spectrum and the wavefunctions in (2.37).

What if the magnetic field were so strong that we could not get away with using continuum limit of the kinetic term of free electrons? As a starting point to answer that question, let us drop the magnetic field and obtain a general result for one-particle wavefunctions in periodic potentials.

2.3 One Electron in a Crystal: Bloch's Theorem

One of the cornerstones of the theory of solids is Bloch's theorem for electrons in a periodic potential such as provided by the ions of crystal. While stable phases should not depend very much on whether the underlying solid is a perfect crystal, many important developments have started from Bloch's theorem, which naturally gives electronic wavefunctions the structure of a bundle over the set of inequivalent momenta (the Brillouin zone). The classic text of Ashcroft and Mermin is a good source for a detailed treatment. Here we confine ourselves to justifying this theorem in the following form: given a potential invariant under a set of lattice vectors $\mathbf{R}$, $V(\mathbf{r} + \mathbf{R}) = V(\mathbf{r})$, closed under addition, the electronic eigenstates can be labeled by a crystal momentum $\mathbf{k}$ and written in the form

$$\psi_{\mathbf{k}}(\mathbf{r}) = e^{i\mathbf{k}\cdot\mathbf{r}} u_{\mathbf{k}}(\mathbf{r}), \qquad (2.38)$$

where the function u has the periodicity of the lattice. Note that the crystal momentum $\mathbf{k}$ is only defined up to addition of reciprocal lattice vectors, that is, vectors whose dot product with every one of the original lattice vectors is a multiple of 2π.

We give a quick proof of Bloch's theorem in one spatial dimension, then consider the Berry phase of the resulting wavefunctions. For the former, we essentially use two theorems. First, that Abelian groups (ones where multiplication is commutative) have one-dimensional representations: in this case, lattice translations form an Abelian group as they have the same net result regardless of the order in which they are effected, so that they can be labeled by a crystal momentum k. Second, in quantum mechanics two operators which commute can be simultaneously diagonalized, so that the label k can also be attached to the eigenfunctions of the Hamiltonians. This latter ceases to apply in the presence of a magnetic field, when instead of Bloch states we encounter Landau levels as a result of a non-Abelian (i.e., noncommuting) magnetic translation algebra (Eq. 5.10).

Starting with the second item, we note that in the problem at hand, we have a non-Hermitian operator (lattice translations by the lattice spacing a: $(T_a \psi)(x) = \psi(x+a)$) that commutes with the Hamiltonian. It turns out that only one of the two operators needs to be Hermitian for simultaneous eigenstates to exist if the other operator is normal (commutes with its adjoint), which is true for the translation operator since $T_a^\dagger = T_{-a}$, and two translations commute. Therefore we can find wavefunctions that are energy eigenstates and satisfy

$$(T_a \psi)(x) = \lambda \psi(x). \tag{2.39}$$

Now if the magnitude of λ is not 1, repeated application of this formula will give a wavefunction that either blows up at spatial positive infinity or negative infinity. We would like to find wavefunctions that can extend throughout an infinite solid with bounded probability density, and hence require $|\lambda| = 1$. From that it follows that $\lambda = e^{i\theta}$, and we define $k = \theta/a$, where we need to specify an interval of width 2π to uniquely define θ, say $[-\pi, \pi)$. In other words, k is ambiguous by addition of a multiple of $2\pi/a$, as expected. So we have shown

$$\psi_k(x + a) = e^{ika} \psi_k(x). \tag{2.40}$$

The last step is to define $u_k(x) = \psi_k(x) e^{-ikx}$; then (2.40) shows that u_k is periodic with period a, and $\psi_k(x) = e^{ikx} u_k(x)$.[1]

While the energetics of Bloch wavefunctions underlies many properties of solids, there is also Berry phase physics arising from the dependence of u_k on k that was understood only rather recently. Note that, even though this is one-dimensional, there is a nontrivial closed loop in the parameter k that can be defined because of the periodicity of the Brillouin zone $k \in [-\pi/a, \pi/a]$:

$$\gamma = \oint_{-\pi/a}^{\pi/a} \langle u_k | i \partial_k | u_k \rangle dk. \tag{2.41}$$

How are we to interpret this Berry phase physically, and is it even gauge-invariant? We will derive its physical meaning (connected to electrical polarization) in Chapter 4, but an intuitive clue is provided if we make the replacement $i\partial_k$ by x, as

---

[1] Readers interested in more information, such as the inclusion of spin and the three-dimensional case, can consult the solid state text of Ashcroft and Mermin (1976).

would be appropriate if we consider the action on a plane wave. This suggests, correctly, that the Berry phase may have something to do with the spatial location of the electrons, but evaluating the position operator in a Bloch state gives an ill-defined answer; for this real-space approach to work, we would need to introduce localized Wannier orbitals in place of the extended Bloch states.

2.4 The Simplest Tight-Binding Model

The tight-binding approximation is a simple version of a method known as linear combination of atomic orbitals (LCAO). The basic concept is provided here because tight-binding models are often used to demonstrate various kinds of topological behavior, but any solid-state textbook will have considerably more details. In this section we treat a single orbital per unit cell, because the new features that emerge when there are multiple orbitals per unit cell are a major subject of Chapter 4.

The goal is to find a way to deal with electronic wavefunctions on an infinite periodic array without having to consider the enormous Hilbert space of all possible wavefunctions. We look for a variational ground state made as a superposition of *one* orbital wavefunction on each atom. Let atom m have one orbital that we keep, $|m\rangle$. For example, for a chain of hydrogen atoms, we would keep the 1s orbital on every atom, and figure that this is probably a good approximation to the low-energy states of the chain, since the gap to the $n = 2$ orbitals is quite large.

In the simplest case of a diatomic molecule made of two identical atoms, and with one orbital on each atom, suppose that the ground state is made from the even combination of those orbitals, while the first excited state is made from the odd combination. The splitting is determined by the magnitude of the hopping t. Writing $|1\rangle$ and $|2\rangle$ for the orbitals kept on atom 1 and 2, the hopping integral is

$$-t = \langle 1|V_2|2\rangle = \langle 1|V_1|2\rangle, \tag{2.42}$$

where V_1 and V_2 are the atomic potentials centered on atom 1 and 2, respectively. For the energy we find

$$E_\pm = \epsilon_0 + V_{\text{cross}} \pm |t| \tag{2.43}$$

where ϵ_0 is the original orbital energy and $V_{\text{cross}} = \langle 1|V_2|1\rangle = \langle 2|V_1|2\rangle$, where we have assumed that the two atoms are identical.

Generalizing this to an infinite chain gives a Hamiltonian with two terms: a diagonal term with the energy of each orbital V_0 (actually computing this is a little subtle for an infinite chain, so we assume some finite value is provided), and a hopping term $-t$ between nearest-neighbor atoms that is essentially similar to the two-atom case. We seek normalized wavelike solutions of the Hamiltonian

$$|\psi\rangle = \sum_{n=1}^{N} \frac{e^{ikan}|n\rangle}{\sqrt{N}}. \tag{2.44}$$

Here the chain of atoms is taken to form a ring, so that the boundary conditions are periodic, and soon we will take $N \to \infty$.

Acting on this wavefunction with the Hamiltonian, which is a tridiagonal matrix in the basis of atomic orbitals, we find that indeed these waves can be eigenstates with energy satisfying

$$E = V_0 - 2t \cos(ka). \tag{2.45}$$

Here and above we are making the approximation that the overlap between different atomic orbitals $\langle i | j \rangle$ can be neglected; including this effect leads to a generalized eigenvalue problem rather than a conventional eigenvalue problem, but the basic physics of the approximation is unchanged.

While the wavefunction in (2.44) is an energy eigenstate, it may not be obvious that it satisfies Bloch's theorem. Let ϕ be the wavefunction of the atomic orbital being used, so that the wavefunction of $|n\rangle$ is $\phi(x - an)$. Then the explicit wavefunction is

$$\psi(x) = \sum_{n=-\infty}^{\infty} e^{ikna} \phi(x - na), \tag{2.46}$$

and

$$\psi(x + a) = \sum_{n=-\infty}^{\infty} e^{ikna} \phi(x - (n-1)a) = \sum_{m=-\infty}^{\infty} e^{ik(m+1)a} \phi(x - ma) = e^{ika} \psi(x), \tag{2.47}$$

where $m = n - 1$.

An interesting question is about what spatial information remains if we only have access to the tight-binding Hamiltonian and not to the specific properties of the orbitals. This will become important when we later consider multiple orbitals within the unit cell, which allows the spatial distribution of the electronic state to change with crystal momentum. As hinted above, the Berry phase of Bloch wavefunctions contains spatial information and allows numerous important properties to be extracted.

The main subtlety arising in the case of multiple orbitals with different locations in the unit cell is that at least two different conventions for the form of the tight-binding basis and Hamiltonian are widely used in the literature ["Convention I" and "Convention II," in the language of the book (Vanderbilt, 2018), which has a lengthy explanation]. While both conventions give the same physical results if properly implemented, the expressions used for Berry phase quantities such as the electric polarization are simplest in Convention I, although the tight-binding Hamiltonian may seem to be unnecessarily complicated in that convention.

The basic difference is whether phase factors appearing in the finite-dimensional tight-binding "Bloch Hamiltonian," defined in a moment, involve distances between orbitals within the unit cell or only lattice vectors. Let us formalize our understanding of tight-binding models a little, and go to three dimensions in writing formulae. Under some simplifying assumptions, a tight-binding model with N orbitals in the unit cell is related to finding eigenvalues of an $N \times N$ matrix. Suppose the orbitals in the unit cell with lattice vector $\mathbf{R}$ are described by states $|\phi_{\mathbf{R}j}\rangle$, $j = 1, \ldots, N$, with wavefunction

$$\phi_{\mathbf{R}j}(\mathbf{r}) = \varphi_j(\mathbf{r} - \mathbf{R} - \mathbf{r}_j). \tag{2.48}$$

Here φ_j contains information about the local orbital character. Assume again that these states are orthonormal; relaxing this assumption just converts the matrix problem to a generalized eigenvalue problem. Assume also that the position operator is diagonal in this basis with elements $\mathbf{R} + \mathbf{r}_j$. Information such as the values of V_0 and t in the above example comes from the real-space Hamiltonian

$$H_{jk}(\mathbf{R}') = \langle \phi_{0j} | H | \phi_{\mathbf{R}'k} \rangle. \tag{2.49}$$

Here $\mathbf{R}'$ is a (relative) lattice vector, possibly zero. In the example above with one orbital per unit cell in one dimension, $H(\pm a) = -t$ and $H(0) = V_0$.

We can construct states of Bloch form from these orbitals,

$$|\psi_j^{\mathbf{k}}\rangle = \sum_{\mathbf{R}} e^{i\mathbf{k}\cdot(\mathbf{R}+\mathbf{r}_j)} |\phi_{\mathbf{R}j}\rangle, \tag{2.50}$$

where the sum is over all lattice vectors. Note that the form of the exponent means that these wavefunctions *are not* periodic in $\mathbf{k}$ with the reciprocal lattice, assuming at least one $\mathbf{r}_j$ is nonzero, but pick up phase factors. This is Convention I, also known as periodic gauge; Convention II just consists in dropping the $\mathbf{k} \cdot \mathbf{r}_j$ in the exponent.

These are not yet energy eigenstates, but we can form linear combinations of them that are eigenstates and still satisfy Bloch's theorem. The Hamiltonian in this basis is diagonal in $\mathbf{k}$ and has the form

$$H_{jk}^{\mathbf{k}} = \langle \psi_j^{\mathbf{k}} | H | \psi_k^{\mathbf{k}} \rangle = \sum_{\mathbf{R}'} e^{i\,\mathbf{k}\cdot(\mathbf{R}'+\mathbf{r}_k-\mathbf{r}_j)} H_{jk}(\mathbf{R}'). \tag{2.51}$$

Box 4.1 discusses a famous one-dimensional example with two orbitals per unit cell using this approach.

Equation (2.45) is the first of many examples in this book of a band structure of electrons. What happened to all the high-energy states of an electron moving in vacuum? We got rid of them by working only with the variational states made up of superpositions of the atomic orbitals. If we excited an electron to high energy in a solid, it would undoubtedly go into one of the higher orbitals that we threw away. But the reason why we can think of electrons under weak fields as moving in the band, and even see Bloch oscillations that directly probe the periodic nature of crystal momentum, is that there is an *energy gap* between the lowest band and higher bands. Under weak, slowly varying fields, the electron cannot gain enough energy to jump across this gap. It is useful in the theory of metals to divide processes into "intraband," at low energies/frequencies, and "interband," once transitions become allowed, as the underlying physics can be quite different.

2.5 Dirac Band Structure of the Honeycomb Lattice

The honeycomb lattice has turned out to be one of the most popular actors in topological physics, repeatedly putting in prominent appearances. In this book, this includes the physics of graphene, Haldane's Chern insulator in Section 3.2 and Kitaev's honeycomb model in Section 7.4. In each of these, the properties of the energies and wavefunctions of noninteracting electrons with nearest, and at most next nearest neighbor, hopping are central ingredients to their topological properties. This section presents a brief account of this band structure for nearest-neighbor hopping, see, for example, the exposition in Goerbig (2011) in the context of graphene for further details.

The first thing to derive are the famous gapless Dirac points with the linear ("relativistic") dispersion in their vicinity, and the symmetric spectrum of valence and conduction band, evoking particles as well as the Dirac sea filled with negative energy antiparticles. This thus presents an instance of an emergent, and stable, relativistic Lorentz invariance.

All these features are present for the nearest-neighbor hopping problem. The honeycomb lattice has two sublattices, labeled A and B, that is, it can be thought of a Bravais lattice with a two-atom basis (Figure 3.8). The corresponding Bravais lattice is a triangular lattice, which has coordination $z = 6$: each site has six nearest-neighbors, displaced by vectors $\pm\mathbf{n}_1$, $\pm\mathbf{n}_2$ and $\pm\mathbf{n}_3 = \pm(\mathbf{n}_2 - \mathbf{n}_1)$, where $\mathbf{n}_{1,2}$ are chosen to enclose an angle $2\pi/6$. These are therefore the vectors linking each honeycomb lattice site with the sites on the same sublattice in adjacent unit cells, that is, its next-nearest neigbors. Importantly, the honeycomb lattice is bipartite, that is, all nearest-neighbor bonds link sites on opposite sublattices. The

Fourier transform of the hopping matrix is hence a purely off-diagonal $n_s \times n_s = 2 \times 2$ matrix with entries given by $s_{\mathbf{k}} = |s_{\mathbf{k}}| \exp(i\zeta_{\mathbf{k}}) = 1 + \exp(i\mathbf{k}\cdot\mathbf{n}_1) + \exp(i\,\mathbf{k}\cdot\mathbf{n}_2)$

$$\mathcal{H} = t \begin{pmatrix} 0 & s^*(\mathbf{k}) \\ s(\mathbf{k}) & 0 \end{pmatrix} . \tag{2.52}$$

The eigenvalue spectrum of this matrix is

$$\epsilon_{\mathbf{k},\lambda} = \lambda t \sqrt{3 + 2\sum_{i=1}^{3} \cos(\mathbf{k}\cdot\mathbf{n}_i)} , \tag{2.53}$$

where $\lambda = \pm 1$ distinguishes the two bands. The corresponding eigenfunctions reside equally on the two sublattices:

$$\frac{1}{\sqrt{2}} \begin{pmatrix} 1 \\ \lambda\exp(i\zeta_{\mathbf{k}}) \end{pmatrix} . \tag{2.54}$$

2.5.1 Dirac Points and Dirac Equation

The Dirac points correspond to zero energy. This occurs at the six corners of the Brioullin zone, but these form triplets related to each other by addition of reciprocal lattice vectors, so that there are only two inequivalent, but symmetry-related Dirac points. The location is conventionally denoted by $\mathbf{K}$ and $\mathbf{K}'$, where the choice $\mathbf{K}' = -\mathbf{K}$ is possible. This allows the definition of a valley index $\iota = \pm 1$ so that the Dirac points are at $\iota\mathbf{K}$, with $\mathbf{K} = (4\pi/3, 0)$ in Figure 2.1.

One can now carry out a gradient expansion by Taylor expanding $\mathcal{H}$ in Eq. 2.52 in deviations $\mathbf{q}$ from $\iota\mathbf{K}$. This yields the Dirac equation

$$\mathcal{H}^\iota = \iota v_F \left(q_x \sigma^x + q_y \sigma^y \right) . \tag{2.55}$$

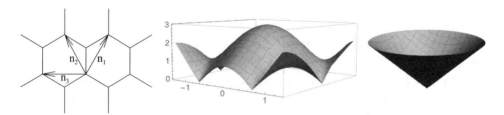

Fig. 2.1 Dirac band structure (middle) of the honeycomb lattice (left); see also Figure 3.8. The arrows denote the basis vectors $\mathbf{n}_{1,2}$ of the underlying triangular Bravais lattice. The Dirac points are visible where the energy vanishes. The Dirac cone is magnified in the right panel; x- and y-axes are the crystal momenta, labeled in units of π. Only the positive energy band is shown.

Here, $\sigma^{x,y}$ are Pauli matrices. In the present convention, the sublattice labels are interchanged between the two valleys, that is, the top spinor components at $\pm\mathbf{K}$ are associated with opposite sublattices; this will be important in the context of strain-induced artificial gauge fields in Chapter 8.

This gradient expansion fixes the effective Fermi velocity in terms of the nearest-neighbor hopping strength t as

$$v_F = \frac{3|t|a}{2\hbar} \, , \tag{2.56}$$

where in the following, as often in this book, we use a convention which sets $\hbar$ to unity; also, we will drop a in the following. This is the nearest-neighbor distance on the honeycomb lattice, which equals about 0.14 nm for graphene, while v_F turns out to be in the ballpark of 10^6 m/s.

The low-energy spectrum near the Dirac point, valid at energies $\epsilon \ll t$ much less than the total bandwidth, exhibits the desired linear dispersion

$$\epsilon_{\mathbf{q}} = \pm v_F q \, . \tag{2.57}$$

2.5.2 Relativistic Landau Levels

A magnetic field can now be added via minimal coupling, in the same way as for the nonrelativistic case discussed following Eq. 2.30, that is, by replacing the momentum operator $\mathbf{p}$ by $\mathbf{p} - q\mathbf{A}$. This leads to the reappearance of the ladder operators a, $a^\dagger$ from Eq. 2.33, along with the magnetic length ℓ. Implementing this starting from Eq. 2.55 yields

$$\mathcal{H}^\iota = \iota\sqrt{2}\frac{v_F}{\ell}\begin{pmatrix} 0 & a \\ a^\dagger & 0 \end{pmatrix}. \tag{2.58}$$

The corresponding eigenvalues are the relativistic Landau level energies

$$\epsilon_{n,\lambda} = \lambda\frac{v_F}{\ell}\sqrt{2n} \, . \tag{2.59}$$

This has the following salient features. First, the Landau level energies are no longer equidistant like in the nonrelativistic case, but now scale as $\sqrt{n}$. The reason they get "closer together" at higher energies is that the density of states grows with energy in the relativistic case. While for the standard case of a parabolic dispersion relation, $\rho(E)$ is constant, the linear dispersion leads to a linear density of states, $\rho(E) \propto E$, so that the integrated density of states up to energy E is $\rho_{\text{tot}}(E) \propto E^2$. The number of Landau levels hosting a fixed number of single particle states thus scales as $n \sim E^2$, that is, $\epsilon_n \sim \sqrt{n}$.[2]

---

[2] For an anisotropic honeycomb lattice, when the sum of hopping integrals in two directions equals that in the third, the two Dirac cones merge, at which point the dispersion in one direction is linear, and quadratic in the other. It is a simple exercise to work out how the Landau level energies scale with n in this case.

Second, there now are Landau levels with positive and negative energies, $\lambda = \pm 1$. In particular, this means that there is no longer a *lowest* Landau level. Instead, and third, there now exists a special *central* Landau level with $n = 0$. To see what is special about this, consider the Landau level wavefunctions, that is, the eigenvectors of $\mathcal{H}$ in Eq. 2.58. These are given by

$$|n = 0\rangle_{\mathrm{D}} = \begin{pmatrix} 0 \\ |n = 0\rangle \end{pmatrix} \tag{2.60}$$

for $n = 0$, and for $n \neq 0$ by

$$|n \neq 0\rangle_{\mathrm{D}}^{\lambda, \iota} = \frac{1}{\sqrt{2}} \begin{pmatrix} |n - 1\rangle \\ \iota\lambda |n\rangle \end{pmatrix}. \tag{2.61}$$

Here, we are using the same wavefunctions as in Eq. 2.35, having suppressed the m-index, which plays the same role in either case. The important feature is that the wavefunction of the central Landau level resides on one sublattice only, A for one of the Dirac points, and B for the other. For all other Landau levels, they reside on both with equal probability.

2.6 Landau Theory of Symmetry-Breaking Phases

Most phases of matter, including solids, magnets, superfluids, and many others, can be understood in terms of "broken symmetry." At high temperature, fluctuations are induced by the requirement to maximize entropy, and these fluctuations tend to destroy order. As temperature is lowered, the energy gain from developing order can overwhelm the entropy cost from lowering disorder. A remarkable fact that only became clear after the solution of the two-dimensional Ising model by Onsager in 1948 is that this energy-entropy competition can lead to a sharp phase transition, described mathematically by a singularity in some derivative of the free energy that emerges in the thermodynamic limit (the limit of an infinite number of degrees of freedom).

For our purposes, we need a way to describe such a breaking of symmetry mathematically. Rather than try to describe every microscopic degree of freedom in a complicated interacting system, we will eventually follow Landau and introduce a classical field theory in terms of some emergent or coarse-grained field that describes the type of order we wish to study. This will lead to a very useful body of ideas collectively known as Landau–Ginzburg theory.

To start, let us first consider an Ising model on a hypercubic lattice (chain, square, cubic lattices in $d = 1, 2$ and 3). Our microscopic description is in terms of a discrete spin variable $s_i = \pm 1$ at each site, with the energy function

$$\mathcal{H} = -J \sum_{\langle ij \rangle} s_i s_j. \tag{2.62}$$

Here $\langle ij \rangle$ denote nearest-neighbor bonds and J is some interaction strength with units of energy in terms of which we measure the temperature, $T = 1/\beta$. The corresponding partition function is over all microstates, weighted by the Boltzmann factor $\exp[-\beta \mathcal{H}(\{s_i\})]$:

$$Z = \sum_{s_i = \pm 1} \exp[-\beta \mathcal{H}(\{s_i\})]. \tag{2.63}$$

At temperature $T = \infty$, the system is equally likely to be in any microstate. At $T = 0$, only two microstates occur: one with all spins up and one with all spins down. The surprise is that, if the lattice of spins is in more than one dimension, there is a nonzero temperature T_c, proportional to J, below which the zero-temperature description is qualitatively but not quantitatively correct.

2.6.1 Mean-Field Theory

As an explicit example, one can construct the lattice mean-field description of the Ising model. One assumes that each spin "sees" a mean field due to the average magnetization, m, of its z neighbors, where z is known as the coordination number of the lattice. The total mean-field H_{eff} then consists of the exchange field zJm and the applied field h. This mean field induces a magnetization for the spin in question. Self-consistency is achieved by demanding that the induced magnetization m_i equals that of its neighbors, m: computing the expectation value of the magnetization of a single spin in a field gives

$$m = \tanh(\beta H_{\text{eff}} m) = \tanh(\beta z J m + \beta h m), \tag{2.64}$$

where k_B is Boltzmann's constant. The behavior of this self-consistent equation changes at a special value T_c of temperature with $k_B T_c = zJ$. The full significance of this temperature will require more explanation, but for now, note that for $h = 0$, $T > T_c$ only has the solution $m = 0$, while $T < T_c$ allows for two solutions with lower free energy, $|m| \neq 0$. Thus the mean-field transition temperature is $T_c = zJ/k_B$ or $\beta_c = (zJ)^{-1}$.

Physicists say that the system breaks symmetry below T_c and picks out a particular sign of the average spin m, where the angle brackets denote thermal averaging. Mathematicians have more satisfactory definitions because the average spin strictly speaking is always zero in the canonical ensemble: one can look at either whether there is a nonzero correlation function $\langle s_i s_j \rangle$ as i and j become infinitely far apart, or look for a singularity in some derivative of the free energy (in a first derivative for a first-order transition, in some higher derivative for a second-, or higher-, order transition).

Even if m is always zero in terms of the Boltzmann sum, physical systems do actually break symmetry, chiefly for dynamical reasons: for example, a bar magnet of iron will in principle explore the whole phase space and flip its north and south poles, but the time it takes to do so may be larger than the age of the universe, let alone the time for which the bar magnet will physically exist. Hence we will mostly be content to discuss broken symmetry as real, for example, $m \neq 0$ in the Ising model, even if that is somewhat sloppy mathematically.

2.6.2 Symmetry Breaking and Universality

The mean-field theory of the previous section appears somewhat ad hoc: Why can one replace all neighbors by a mean-field? How does one guess the nature of candidate symmetry breakings? From here to develop a full description of Landau–Ginzburg theory and the renormalization group is a serious undertaking which we are in large part excused of as this book is concerned with topological physics. However, for two reasons we nonetheless present a conceptual outline here. First, knowledge of Landau–Ginzburg theory is necessary to appreciate how much topological physics is different. Second, it will lead us to encounter Berezinskii–Kosterlitz–Thouless physics, one of the important topics linking local and topological forms of order. The reader interested in a more detailed and comprehensive treatment is referred to Cardy (1996) or other texts on advanced statistical physics.

Our conceptual outline is intended to expose the basic mathematical ingredients and give a couple of examples illustrating in particular how seemingly disparate systems can end up with similar universal properties as suggested in Chapter 1, while seemingly similar systems turn out to behave fundamentally differently. Mathematically, a powerful way to understand the broken symmetry is in terms of two symmetry groups: G, the symmetry group of the high-temperature phase, and H, the residual symmetry group that survives in the low-temperature phase.

For instance, in the case of a magnet on a lattice, such as (2.62), G consists of the full symmetry of the lattice – translations, reflections, rotations, . . . – together with those of the internal (e.g., spin) space such as inversion for Ising spins, $\mathcal{I}_I$: $s \to -s$, or rotations in the plane for XY spins,

$$\mathcal{R}_{XY} : \begin{pmatrix} s_x \\ s_y \end{pmatrix} \to \begin{pmatrix} \cos\theta & -\sin\theta \\ \sin\theta & \cos\theta \end{pmatrix} \begin{pmatrix} s_x \\ s_y \end{pmatrix}. \tag{2.65}$$

The above case, where it was the Ising inversion symmetry, $\mathcal{I}_I$, which was discarded upon ordering, is a particularly simple case leaving all the lattice symmetries intact. (Generally, there will be symmetry operations combining lattice and internal symmetries, e.g., in the presence of spin-orbit coupling.)

One then defines the order parameter manifold as the quotient $M = G/H$, where dividing by H (taking cosets of H in G) means that we identify two elements of G that differ by multiplication by an element of H; note that unlike G and H, M is in general no longer a group. The idea of this coset is that H is still an unbroken symmetry of the low-temperature phase, so we should think of the system in this phase as invariant under H, which is true if the system is described as a set of microscopic states and the set is unchanged under the action of symmetries in H.

The notion of an order parameter is basic in Landau theory: it is what we use to model all the complicated microscopic states in terms of one, or a few, macroscopic variables. The idea of the order parameter manifold is that, for many interesting phenomena, we do not care especially about the magnitude of the order parameter itself. We care instead about the set of distinguishable low-temperature states at an arbitrary temperature in the ordered phase, which is exactly M. Note that M does not have any information about the magnitude of the order parameter: at every temperature in the uniaxial ordered phase of the isotropic (Heisenberg) ferromagnet, for example, the order parameter manifold is the same sphere: the high-temperature symmetry is $G = \mathrm{SO}(3)$ (rotations in 3D), the low-temperature symmetry is $H = \mathrm{SO}(2)$ (rotations around the axis of the order), and the order parameter manifold is $M = \mathrm{SO}(3)/\mathrm{SO}(2) = S^2$, the unit sphere in three dimensions. One reason the set M is important will become clear when we discuss topological defects later in this chapter.

A basic tenet of Landau–Ginzburg (LG) theory states that, if G is a subgroup of H, the transition where the order parameter vanishes can (but does not have to) be continuous. In this case its universal properties are predetermined generically (i.e., with high probability in the mathematical sense: to escape this, one needs to fine-tune parameters so as to end up, e.g., at a multicritical point). If this condition is not met, the transition will generically be first-order. The intuition behind this famous prediction of Landau's, for which there may not be a rigorous proof, is as follows: the two phases have symmetries that are not in a subset/superset relation, which means that we cannot say one has more or less symmetry than the other; hence it would be fine-tuned for the continuous breaking of one symmetry (and appearance of order) to take place at the same point in parameter space as the continuous restoration of the other symmetry (and disappearance of order).

Symmetry considerations enter the construction of the basic quantity underlying the LG analysis, the LG functional $\mathcal{L}_{\mathrm{LG}}$. It starts from the order parameter field, $m(\mathbf{r})$, as a function of spatial location $\mathbf{r}$, and constructs an expression that is fully symmetric under G, the group describing the symmetries of the high-temperature (also known as disordered) phase.

Such a recipe is not very practicable in general, but two approximations are made which make it more tractable. As universality occurs in continuous phase transitions, there is a regime where the order parameter field $m(\mathbf{r})$ is small, so that one can Taylor expand $\mathcal{L}_{LG}$ in powers of m. Also, as one is interested in low-energy/long-wavelength physics, one can use a systematic gradient expansion around uniform solutions, rather than allowing arbitrarily fast spatial variations.

Concretely, in the case of the Ising ferromagnet, the order parameter is a simple scalar, and $\mathcal{I}_I : m \to -m$ forbids terms in $\mathcal{L}_{LG}$ which are odd in m. Then, writing the average value of $m(\mathbf{r})$ as m, and expanding in small fluctuations around it, one obtains

$$\mathcal{L}_{LG} = \int d^d\mathbf{r} \left(a_0 + a_1 m^2 + a_2 m^4 + \ldots + b_0 (\nabla m)^2 + \ldots \right). \qquad (2.66)$$

We have assumed rotational symmetry in the form of the gradient term; more generally, we would omit terms with an odd power of spatial derivatives if the system is statistically invariant under spatial inversion $\mathbf{r} \to -\mathbf{r}$. For the XY ferromagnet, little changes: the main difference arises from the fact that $\mathbf{m} = (m_x, m_y) = m(\cos\theta, \sin\theta) \sim me^{i\theta}$ is now a vector, so that m^2 needs to be understood as the expression $|\mathbf{m}|^2$ invariant under $\mathcal{R}_{XY}$. This innocuous change – the mean-field equation (2.64) remains basically unchanged – turns out to change the critical behavior completely, as we will see below.

By contrast, an entirely different Ising problem turns out to be much more like the XY magnet, and the next few lines describe how this comes about while also illustrating the construction of $\mathcal{L}_{LG}$ in a slightly more complex setting.

Consider an Ising antiferromagnet (2.62), $J < 0$, defined on a triangular lattice. There is generally no simple way rigorously to obtain the nature of the ordered phase which is reached as the temperature is lowered, but a guess is provided by the minimum of the interaction energy (2.62) upon Fourier transforming. Indeed, for the case of the ferromagnet on the hypercubic lattice, one finds $\mathcal{H}(q) = -J \sum_{\alpha=1}^{d} \cos(q_\alpha)$, which is manifestly minimized at $q_\alpha \equiv 0$, indicating a preference for a uniform order with every spin having the same average $s_i \sim m$: this is just the ferromagnetic order parameter.

For the triangular Ising antiferromagnet

$$\mathcal{H}(q) = |J| \left[\cos q_x + 2\cos\left(\frac{q_x}{2}\right) \cos\left(\frac{\sqrt{3}q_y}{2}\right) \right], \qquad (2.67)$$

which is minimized for $(q_x, q_y) = \pm\mathbf{K} = \left(\pm\frac{4\pi}{3}, 0\right)$. There are thus two minima, $s^\pm$, at the corners of the Brillouin zone, which are spatially modulated.

To construct a symmetry-broken spin configuration, one expands $m(\mathbf{r})$ as before, with the requirement of real values for the spin field yielding $m(\mathbf{r}) = m_+(\mathbf{r})s^+ + m_-(\mathbf{r})s^-$ with $m_+ = m_-^* = me^{i\theta}$.

This feeds into the construction of $\mathcal{L}_{\mathrm{LG}}$: inversion again removes odd terms in m. In addition, translations now act nontrivially, with $\mathcal{T}_x : m_\pm \rightarrow \exp\left(\pm\frac{4\pi}{3}i\right) m_\pm$ for a unit translation along the $\hat{x}$-axis. The phase factors can only be cancelled for combinations $m_+m_- = m^2$, or for "cubic" powers $m_\pm^{3\cdot n}$. Here $n = 1$ is forbidden by inversion, so that the lowest spatially uniform terms are

$$\mathcal{L}_{\mathrm{LG}} = \int d^2r \left[a_2 m^2 + a_4 m^4 + a_6 m^6 + b_6 m^6 \cos(6\theta) + \dots\right]. \qquad (2.68)$$

Note that we all of a sudden have an XY degree of freedom $me^{i\theta}$ in an Ising system! Indeed, we have ended up with an XY theory with a six fold clock term, $b_6 m^6 \cos(6\theta)$. This is one example of the origin of universality: the long-wavelength properties do not care much about the microscopic nature of the spin – Ising or XY – but rather about the order parameter symmetry, which does inherit items such as inversion symmetry but can nonetheless exhibit entirely different emergent symmetry properties.

2.6.3 Quantum and Statistical Path Integrals

Landau–Ginzburg Theory

With this in hand, we now turn to the construction of field theories to describe conventional symmetry breaking and its universality. There are two field theories we will deal with. The first is Landau–Ginzburg theory, now treated beyond the mean-field approximation. This uses the $\mathcal{L}_{\mathrm{LG}}$ derived above to define a partition function

$$Z_{\mathrm{LG}} = \int Dm \exp\left[-\mathcal{L}_{\mathrm{LG}}\right]. \qquad (2.69)$$

This is a statistical integral over the order parameter field $m(r)$. You may be familiar with sums over spin configurations in an Ising model, for example. It is often convenient to work in the continuum and integrate over configurations of a continuous field, and such integrals are very similar mathematically to the path integrals appearing in quantum field theory.

Here the measure of the integral can be defined more precisely in Fourier space, where omitting high-wavevector components is typically necessary for a sensible theory. In condensed matter, this makes physical sense as a short-distance cut-off below which the field m is not meaningful: m cannot vary on a scale shorter than the separation between neighboring sites. This is a vastly simplified version of the original problem in at least two ways: considering the integration over the coarse-grained order parameter field $m(\mathbf{r})$, and we are not doing any microscopic calculation of the coefficients that appear in the expansion.

A remarkable fact is that the above Landau–Ginzburg theory can be not just qualitatively correct but actually exact for some properties, such as critical exponents near second-order phase transitions, even without a microscopic calculation of the coefficients. Such properties are referred to as universal: they depend on symmetry and dimensionality but little else. For example, the liquid–gas critical point in the phase diagram of water has the same critical exponents as the Ising phase transition in three dimensions. As a mathematical example of where universality comes from, the terms beyond a_2 in the above energy turn out not to impact critical exponents and selected other properties, as long as the lower-order coefficients have appropriate signs. We will study one or two examples of critical points later, concentrating on examples where topological considerations are important.

Nonlinear σ Model

Landau–Ginzburg theory was motivated as a high temperature expansion taking advantage of the smallness of the order parameter m. An alternate field theory, the nonlinear σ-model, can be viewed as an expansion starting from zero temperature. We will concentrate on systems in which the zero-temperature phase breaks a continuous symmetry, so that the set of inequivalent order parameter states M is a continuous manifold; this includes, for example, Heisenberg and XY magnets, in which $M = S^2$ and $M = S^1$, respectively, but not Ising magnets, where M is a set with two elements. Here S^d is the d-dimensional surface of the unit sphere in $d + 1$-dimensional real space. For an XY magnet[3], we can label a ground state simply by an angle θ between 0 and 2π (the order parameter has a magnitude Δ as well, but all ground states have the same magnitude of the order parameter by symmetry).

When the temperature is slightly increased, fluctuations of the order parameter will take place. The nonlinear σ-model is a theory that ignores fluctuations of the order parameter *magnitude* but captures fluctuations in its *direction*, which are lower in energy or softer. More precisely, the nonlinear σ-model into a symmetric space $M = G/H$ is defined as a path integral over an M-valued field. For the XY case above, this can be written simply in terms of a spatially varying angle $\theta(\mathbf{r})$:

$$Z_{\mathrm{NL}\sigma\mathrm{M}} = \int (D\theta)\, e^{-\int d^d\mathbf{r}\, g(\nabla\theta)^2}, \qquad (2.70)$$

where we have incorporated β into the definition of the coupling g. We will return to this model once we have said a bit more about topological defects in Section 2.8; it will turn out (Box 2.3) that for our XY example in two spatial dimensions, the

---

[3] We choose the example of $M = \mathrm{U}(1) \cong \mathrm{SO}(2)$ here for a reason. It turns out that, for the nonlinear σ-model to include gapless excitations, the form of the theory can become more complicated. Namely, for Lie groups more complicated than U(1), an additional term of topological origin is required, leading to the Wess–Zumino–Witten model that we discuss in Box 4.2.

physics depends crucially on vortices, the simplest kind of topological defect, and in fact shows a phase transition that would not be present if, hypothetically, the field θ were not periodic and Eq. 2.70 became just the massless Gaussian model or the free scalar field theory.

We have written both of the above field theories in a classical or Euclidean representation, where there is a positive weight on each field configuration. A natural question is how the partition function integral in such a theory is related to the quantum path integral that may be familiar from an advanced course in quantum mechanics, which in general has a complex integrand. The easiest example of the analytic continuation to imaginary time that connects the two types of path integrals is for the harmonic oscillator. Its partition function at a finite temperature T is

$$Z_{\text{harmonic}} \approx \int dx(\tau) \exp\left[-\int_0^\beta d\tau \left(\dot{x}^2(\tau)/2m + kx^2/2\right)\right], \qquad (2.71)$$

where there are periodic boundary conditions on $x(\tau)$: $x(\beta) = x(0)$. A worthwhile calculation (hint: simplify the integral by considering Fourier components of $x(\tau)$) leads to the result

$$Z_{\text{harmonic}} = \frac{1}{2\sinh(\beta\hbar\omega/2)} = \sum_{n=0}^{\infty} e^{-\beta\hbar\omega(n+1/2)}, \qquad (2.72)$$

where the last expression is what we would calculate from the spectrum. Now analytically continuing this calculation from imaginary time τ gives a trace of the form appearing in a quantum path integral,

$$Z_{\text{harmonic}} = \text{Tr}\, e^{-\beta H} \to \text{Tr}\, e^{-itH/\hbar} = \int dx_0\, U(x_0, t; x_0, 0), \qquad (2.73)$$

where in the last step we have used the position basis to put the result in terms of matrix elements of the unitary time evolution operator U. Now the divergence of Z at real times $t = 2\pi n/\omega$, for integer n, can be simply interpreted: at these times all the energy eigenstates that appear in an arbitrary initial condition appear with exactly the same phases, so the state is (aside from an overall phase factor) exactly the initial state, the time evolution operator is the identity, and the trace diverges.

A variety of field theories of topological phases will appear in Chapter 6, and several books specifically devoted to the many other applications of field theories in condensed matter physics are listed in the appendix. Although we will not discuss the uses of this approach, it turns out that the Berry phase of a spin-half discussed in Section 2.1 is the crucial ingredient in developing a path integral for quantum spins, which can be used to understand, for example, a key difference between integer-spin and half-integer-spin antiferromagnetic chains (Auerbach, 1994).

2.7 Two Mathematical Approaches to Topology

This book is primarily about the physics of materials in which some property has a topological character, in the sense of being described by mathematical objects that are robust under continuous deformations. This section introduces two commonly used flavors of algebraic topology, which will be used repeatedly in physics contexts in later sections of the book. The first flavor, cohomology, can be motivated intuitively by the question of which integrals over a path (or surfaces, etc.) are invariant of smooth changes of the path but sensitive to its topology. The second flavor, homotopy, was actually the first use of topology in condensed matter physics: it appears naturally in the classification of topological defects in symmetry-breaking phases, which we review.

Homotopy and cohomology are often sensitive to the same topological character of a manifold, and consequently cohomological integrals are often used by physicists to compute the homotopy class of some material or configuration. A particularly important tool in Chapters 3 and 4 will be integrals constructed from the Berry phase of a material's Bloch states. The particular form of these integrals will make a little more sense once we see how in this chapter certain objects are naturally suited to compute topological properties. As a warmup for our topological discussion, Box 2.2 gives an example of how integrals over a local geometrical object, the Gaussian curvature, become topological. Here the emergence of topology may be a little easier to picture than when the local geometrical object emerges from quantum mechanics via the Berry phase.

Box 2.2 Topology from Geometry: The Gauss–Bonnet Theorem

A topologist has been described as "a mathematician who can't tell the difference between a doughnut and a coffee cup." As a prelude to the connections between geometry and topology, we start by discussing an integral that will help us classify two-dimensional compact manifolds (surfaces without boundaries) embedded smoothly in three dimensions. For our purposes, a manifold is a d-dimensional surface that locally "looks like" $\mathbb{R}^d$. The integral we construct is topologically invariant in that if one such surface can be smoothly deformed into another, then the two will have the same value of the integral. The integral can't tell the difference between the surface of a coffee cup and that of a doughnut, but it can tell that the surface of a doughnut (a torus) is different from a sphere. Similar connections between global geometry and topology appear frequently.

We start with a bit of local geometry. Given our $d = 2$ surface embedded in $d = 3$ Euclidean space, we can choose coordinates at any point on the surface so that the $(x, y, z = 0)$ plane is tangent to the surface, which can locally be specified by a single function $z(x, y)$. We choose $(x = 0, y = 0)$ to be the given point, so $z(0, 0) = 0$. The tangency condition is

$$\frac{\partial z}{\partial x}\bigg|_{(0,0)} = \frac{\partial z}{\partial y}\bigg|_{(0,0)} = 0. \tag{2.74}$$

Hence we can approximate z locally from its second derivatives:

$$z(x, y) \approx \frac{1}{2} \begin{pmatrix} x & y \end{pmatrix} \begin{pmatrix} \frac{\partial^2 z}{\partial^2 x} & \frac{\partial^2 z}{\partial x \partial y} \\ \frac{\partial^2 z}{\partial y \partial x} & \frac{\partial^2 z}{\partial^2 y} \end{pmatrix} \begin{pmatrix} x \\ y \end{pmatrix}. \tag{2.75}$$

The Hessian matrix that appears in the above is real and symmetric. It can be diagonalized and has two real eigenvalues λ_1, λ_2, corresponding to two orthogonal eigendirections in the (x, y) plane. The geometric interpretation of these eigenvalues is simple: their magnitude is an inverse radius of curvature, and their sign tells whether the surface is curving toward or away from the positive z direction in our coordinate system. To see why the first is true, suppose that we carried out the same process for a circle of radius r tangent to the x-axis at the origin. Parameterize the circle by an angle θ that is 0 at the origin and traces the circle counterclockwise, that is,

$$x = r \sin\theta, \quad y = r(1 - \cos(\theta)). \tag{2.76}$$

Near the origin, we have

$$y = r(1 - \cos(\sin^{-1}(x/r))) \approx r - r\left(1 - \frac{x^2}{2r^2}\right) = \frac{x^2}{2r}, \tag{2.77}$$

which corresponds to an eigenvalue $\lambda = 1/r$ of the matrix in Eq. 2.75.

Going back to the Hessian, its determinant (the product of its eigenvalues $\lambda_1\lambda_2$) is called the Gaussian curvature and has a remarkable geometric significance. As in the example above, the Gaussian curvature remains the signed product of inverse radii of curvature along the two perpendicular principal directions. First, consider a sphere of radius r, which at every point has $\lambda_1 = \lambda_2 = 1/r$. Then we can integrate the Gaussian curvature over the sphere's surface,

$$\int_{S^2} \lambda_1\lambda_2 \, dA = \frac{4\pi r^2}{r^2} = 4\pi. \tag{2.78}$$

Beyond simply being independent of radius, this integral actually gives the same value for any manifold that can be smoothly deformed to a sphere. (More generally, scale invariance is a necessary but not sufficient for an integral to give a topological invariant.)

However, we can easily find a manifold with a different value for the integral. Consider the torus (Figure 2.2) made by revolving the circle in Eq. 2.76, with $r = 1$, around the axis of symmetry $x = t, y = -1, z = 0$, with $-\infty < t < \infty$. To compute the Gaussian curvature at each point, we sketch the calculation of the eigenvalues of the Hessian as follows. One eigenvalue is around the smaller circle, with radius of

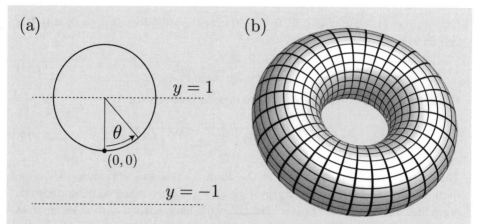

Fig. 2.2 An example of the Gauss–Bonnet theorem. (a) Coordinates of circle of radius 1 as described in text. Revolving this circle around the line $y = 1$ leads to a sphere, while revolution around the line $y = -1$ leads to a torus like that shown in (b). (b) This torus has zero total Gaussian curvature: the negative curvature at points on the inside (i.e., around the hole in the doughnut) compensates the positive curvature on the outside. The circles on the torus's surface indicate the principal directions.

curvature r: $\lambda_1 = 1/r = 1$. Then the second eigenvalue must correspond to the perpendicular direction, which has a radius of curvature that depends on the angle θ around the smaller circle (we keep $\theta = 0$ to indicate the point closest to the axis of symmetry). The distance from the axis of symmetry is $2 - \cos\theta$, so we might have guessed $\lambda_2 = (2 - \cos\theta)^{-1}$, but there is an additional factor of $\cos\theta$ that appears because of the difference in direction between the surface normal and this curvature. So our guess is that

$$\lambda_2 = -\frac{\cos\theta}{2 - \cos\theta}. \tag{2.79}$$

As a check and to understand the sign, note that this predicts a radius of curvature 1 at the origin and other points closest to the symmetry axis, with a negative sign in the eigenvalue indicating that this curvature is in an opposite sense as that described by λ_1. At the top, the radius of curvature is 3 and in the same sense as that described by λ_1, and on the sides, λ_2 vanishes because the direction of curvature is orthogonal to the tangent vector.

Now we compute the curvature integral. With ϕ the angle around the symmetry axis, the curvature integral is

$$\int_{T^2} \lambda_1\lambda_2\, dA = \int_0^{2\pi} d\theta \int_0^{2\pi} (2-\cos\theta)\, d\phi\, \lambda_1\lambda_2 = \int_0^{2\pi} d\theta \int_0^{2\pi} d\phi\, (-\cos\theta) = 0. \tag{2.80}$$

Again this zero answer is generic to any surface that can be smoothly deformed to the torus. The general result (the Gauss–Bonnet formula) of which the above are examples is

$$\int_S \lambda_1 \lambda_2 \, dA = 2\pi \chi = 2\pi (2 - 2g), \tag{2.81}$$

where χ is a topological invariant known as the Euler characteristic and g is the genus, essentially the number of holes in the surface.[a] If we go beyond the sphere and torus to a manifold with boundaries, the Euler characteristic becomes $2 - 2g - b$, where b is the number of boundaries: one can check this by noting that by cutting a torus, one can produce two annuli (by slicing a bagel) or alternately one cylinder with two boundaries (by slicing what one of us calls a bundt cake and the other a Gugelhupf). The Gauss–Bonnet formula and Euler characteristic are examples of a general principle: we will encounter other examples where a topological invariant is expressed as an integral over a local quantity with a geometric significance.

---

[a] A good question is why we write the Euler characteristic as $2 - 2g$ rather than $1 - g$; one way to motivate this is by considering polygonal approximations to the surface. The discrete Euler characteristic $V - E + F$, where V, E, F count vertices, edges, and faces, is equal to χ. For example, the five Platonic solids all have $V - E + F = 2$.

2.7.1 Invariant Integrals along Paths: Exact Forms

As our first example of a topological property, let's ask about making line integrals along paths (not path integrals in the physics sense, where the path itself is integrated over) that are nearly independent of the precise path: they will turn out to depend in some cases on topological properties (homotopy or cohomology). We will assume throughout, unless otherwise specified, that all functions are smooth (i.e., $\mathbb{C}^\infty$, meaning derivatives of all orders exist).

First, suppose that we deal with paths on some open set U in the two-dimensional plane $\mathbb{R}^2$.[4] We consider a smooth path $(u(t), v(t))$, where $0 \le t \le 1$ and the endpoints may be different.[5]

Now let $f(x, y) = (p(x, y), q(x, y))$ be a two-dimensional vector field that lets us compute line integrals of this path:

$$W = \int_0^1 dt \left(p \frac{du}{dt} + q \frac{dv}{dt} \right), \tag{2.82}$$

where p and q are evaluated at $(x(t), y(t))$.

---

[4] Open set: some region of nonzero size around each point in the set is also in the set.

[5] To make these results more precise, we should provide for adding one path to another by requiring only piecewise smooth paths, and require that u and v be smooth in an open set including $t \in [0, 1]$. For additional rigor, see the first few chapters of Fulton (1995).

Mathematical note: in more fancy language, f is a differential form, a 1-form to be precise. All that means is that f is something we can use to form integrals over paths that are linear probes of the tangent vector of the path. Another way to state this is that the tangent vector to a path, which we call a vector, transforms naturally in an opposite way to the gradient of a function, which we call a covector (a linear functional on vectors).[6] We will say a bit more about such forms in a moment.

Our first goal is to show that the following three statements are equivalent: (a) W depends only on the endpoints $(u(0), v(0))$ and $(u(1), v(1))$ (Figure 2.3a); (b) $W = 0$ for any closed path; (c) f is the gradient of a function g: $(p, q) = (\partial_x g, \partial_y g)$. The formal language used for (c) is that f is an *exact form*: $f = dg$ is the differential of a 0-form (a smooth function) g.[7]

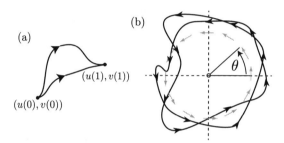

Fig. 2.3 The plane and punctured plane as examples of the cohomology of differential forms. (a) On the plane $\mathbb{R}^2$, the integrals of a gradient along a path depend only on the endpoints, so the two paths shown give the same result. The integral of a gradient along any closed path is zero. (b) On the punctured plane $\mathbb{R}^2 - (0, 0)$, closed paths can give nonzero integrals of the 1-form shown with gray arrows, which is singular at the origin. The path shown circles the origin twice and has winding number 2. As a simple example of continuous versus discrete calculation of topological invariants, the winding number could be computed either using the integral of the form shown, which locally is the gradient of the angle $\theta = \tan^{-1}(y/x)$, or by counting the signed number of crossings of a radius, such as the positive x-axis, by the path.

---

[6] To convince yourself that this is true, think about how both transform under a rotation $x_i' = R_{ij}x_j$ on the underlying space. A tangent vector transforms by matrix multiplication with R, i.e., $\mathbf{v}' = R\mathbf{v}$, while a gradient vector transforms differently: $\frac{\partial f(\mathbf{x}')}{\partial x_i'} = \frac{\partial f}{\partial x_j}\frac{\partial x_j}{\partial x_i'} = \nabla_x f R^{-1}$. This distinction is very similar to that between kets (state vectors) and bras (linear functionals on state vectors) in quantum mechanics.

[7] As an aside, we note that in an undergraduate thermodynamics course, the reader may have come across some statements about the difference between the left-hand and right-hand sides of the equation describing changes of the internal energy U: $dU = dQ - dW$, where the differentials on the right-hand side are sometimes denoted by slightly different symbols than d. The mathematical statement is that while the function of state dU is an exact form, the heat added and work done, dQ and dW, are not. We are not aware of any treatments of topological phenomena in this context.

Note that (c) obviously implies (a) and (b), since then $W = g(u(1), v(1)) - g(u(0), v(0))$. To show that (b) implies (a), suppose (b) is true and (a) is not. Then there are two paths γ_1, γ_2 that have different integrals but the same endpoints. Form a new path γ so that, as t goes from 0 to $\frac{1}{2}$, γ_1 is traced, and then as t goes from $\frac{1}{2}$ to 1, γ_2 is traced opposite its original direction (now you can see why piecewise smooth paths are needed if one wants to be rigorous). Then this integral is nonzero, which contradicts (b).

It remains to show that (a) implies (c). Define $g(x, y)$ as equal to 0 at $(0, 0)$, or some other reference point in U if U does not include the origin. Everywhere else, set g equal to the W obtained by integrating over an arbitrary path from $(0, 0)$ to the final point, which by (a) is path-independent. (If U is not connected, then carry out this process on each connected component.) We will show that $\partial_x g = p$, and the same logic then implies $\partial_y g = q$. We need to compute

$$\partial_x g = \lim_{\Delta x \to 0} \frac{g(x + \Delta x, y) - g(x, y)}{\Delta x}. \tag{2.83}$$

We can obtain g by any path we like, so let's take an arbitrary path to define $g(x, y)$, then add a short horizontal segment to that path to define the path for $g(x + \Delta x, y)$. The value of the integral along this extra horizontal segment converges to $p(x, y)(\Delta x)$, as needed.

It turns out that the above case is simple because the plane we started with is topologically trivial. Before proceeding to look at a nontrivial example, let us state one requirement on f that is satisfied whenever f is exact ($f = dg$). The fact that partial derivatives commute means that, with $f = dg = (p, q)$, $\partial_y p = \partial_x q$. We can come up with an elegant notation for this property by expanding our knowledge of differential forms.

Before, we obtained a 1-form f as the differential of a scalar g by defining

$$f = dg = \partial_x g \, dx + \partial_y g \, dy. \tag{2.84}$$

Note that we now include the differential elements dx, dy in the definition of f, and that 1-forms form a real vector space (spanned by dx, dy): we can add them and multiply them by scalars. To obtain a 2-form as the differential of a 1-form, we repeat the process: writing $f = f_i dx_i$ (with $x_1 = x$, $x_2 = y$, $f_1 = p$, $f_2 = q$)

$$df = \sum_{i,j} \frac{\partial f_i}{\partial x_j} dx_j \wedge dx_i, \tag{2.85}$$

where the $\wedge$ product (the wedge product, or exterior product) between differential forms satisfies the rule $dx_i \wedge dx_j = -dx_j \wedge dx_i$, which implies that if any coordinate appears twice, then we get zero: $dx \wedge dx = 0$. For some intuition about why this anticommutation property is important, note that in our 2D example,

$$df = (\partial_x f_y - \partial_y f_x) dx \wedge dy, \tag{2.86}$$

so that the function appearing in df is just the curl of the 2D vector field represented by f. So our statement about partial derivatives commuting is just the statement that if $f = dg$, then $df = 0$, or that the curl of a gradient is zero. We label any 1-form satisfying $df = 0$ a *closed form*. While every exact form is also closed, we will see that not every closed form is exact, with profound consequences.

2.7.2 Locally Invariant Integrals along Paths: Closed Forms and Cohomology

As an example of nontrivial topology, we would now like to come up with an example where integrals over paths are only path-independent in a limited topological sense: the integral is the same for any two paths that are *homotopic*, one of the fundamental concepts of topology (to be defined in a moment). Basically, two paths are homotopic if one can be smoothly deformed into another. Consider the vector field

$$f = (p, q) = \left(-\frac{y}{x^2 + y^2}, \frac{x}{x^2 + y^2} \right) = \frac{-y dx + x dy}{x^2 + y^2}, \tag{2.87}$$

where in the second step we have written it using our 1-form notation. This vector field is well defined everywhere except the origin (Figure 2.3b). This 1-form looks locally like the gradient of $g = \tan^{-1}(y/x)$ (which just measures the angle in polar coordinates), but that function can only be defined smoothly on some open sets. For example, in a disc around the origin, the 2π ambiguity of the inverse tangent prevents defining g globally.

So if we have a path that lies entirely in a region where g can be defined, then the integral of this 1-form over the path will give the change in angle between the starting point and end point $g(u(1), v(1)) - g(u(0), v(0))$. What about other types of paths, for example, paths in $\mathbb{R}^2 - (0, 0)$, the 2D plane with the origin omitted, that circle the origin and return to the starting point? We can still integrate using the 1-form f, even if it is not the gradient of a scalar function g, and will obtain the value $2\pi n$, where n is the winding number: a signed integer that describes how many times the closed path $(u(t), v(t))$ circled the origin as t went from 0 to 1.

Now this winding number does not change as we make a small change in the closed path, as long as the path remains in $\mathbb{R}^2 - (0, 0)$. What mathematical property of f guarantees this? Above we saw that any exact 1-form (the differential of a scalar function) is also closed. While f is not exact, we can see that it is closed:

$$df = \left(\partial_x \frac{x}{x^2 + y^2}\right) dx \wedge dy + \left(\partial_y \frac{-y}{x^2 + y^2}\right) dy \wedge dx = \frac{2 - 2}{x^2 + y^2} dx \wedge dy = 0.$$
(2.88)

In other words, $(-y, x)/(x^2 + y^2)$ is curl-free (irrotational), while $(-y, x)$ has constant nonzero curl. Now suppose that we are given two paths γ_1 and γ_2 that differ by going in different ways around some small patch dA in which the 1-form remains defined. The difference in the integral of f over these two paths is then the integral of df over the enclosed surface by Stokes's theorem, which is zero if f is a closed form.

So we conclude that if f is a closed form ($df = 0$), then the path integral is path-independent if we move the path continuously through a region where f is always defined. For an exact form ($f = dg$), the integral is completely path-independent. In the case of $\mathbb{R}^2 - (0, 0)$, the 1-form in Eq. 2.87 is locally but not completely path-independent. Both closed forms and exact forms are vector spaces (we can add and multiply by scalars), and typically infinite-dimensional, but their quotient as vector spaces is typically finite-dimensional. (The quotient of a vector space A by a vector space B is the vector space that identifies any two elements of A that differ only by an element of B as discussed in the context of local order parameters earlier in this chapter.) A basic object in cohomology is the first de Rham cohomology group (a vector space is by definition a group under addition),

$$H^1(M) = \frac{\text{closed 1-forms on } M}{\text{exact 1-forms on } M} = \frac{Z^1(M)}{B^1(M)}.$$
(2.89)

If you wonder why the prefix "co-" appears in cohomology, there is a dual theory of linear combinations of curves, surfaces, and so forth, called homology, in which the differential operator in de Rham cohomology is replaced by the boundary operator. However, while arguably more basic mathematically, homology seems to crop up less frequently in physics.

An even simpler object is the zeroth de Rham cohomology group. To understand this, realize that a closed 0-form is one whose gradient is zero, that is, one that is constant on each connected component of U. There are no (-1)-forms and hence no exact 0-forms. So the zeroth group is just $\mathbb{R}^n$, where n is the number of connected components.

Note that there are many different ways to compute the winding number of a path, for example either continuously, by the integral of the 1-form, or discretely, by counting the signed number of crossings of the positive x-axis. Many of the more sophisticated topological invariants appearing later in this book likewise can be computed either continuously through an integral or discretely through a counting process, but often the integral is more fundamental in the sense that it is what arises directly from the microscopic expression for an observable quantity.

We can show that $H^1 = \mathbb{R}$ for the unit circle S^1 using the angle form f in Eq. 2.87, by showing that this form (more precisely, its equivalence class up to exact forms) provides a basis for H^1. Given some other form $\tilde{f}$, we use the unit circle path, parameterized by an angle θ going from zero to 2π, to define

$$c = \frac{\int_0^{2\pi} \tilde{f}}{\int_0^{2\pi} f}. \tag{2.90}$$

Now $\tilde{f} - cf$ integrates to zero. We can define a function g via

$$g(\theta) = \int_0^{\theta} (\tilde{f} - cf). \tag{2.91}$$

Now g is well defined and periodic because of how we defined c, and $\tilde{f} = cf + dg$, which means that $\tilde{f}$ and cf are in the same equivalence class as their difference dg is an exact form. We say that $\tilde{f}$ and cf are cohomologous because they differ by an exact form. So cf, $c \in \mathbb{R}$, generates H^1, and $H^1(S^1)$ is isomorphic to $\mathbb{R}$. With a little more work, one can show that $\mathbb{R}^2 - (0, 0)$ also has $H^1 = \mathbb{R}$.

We close this section with a few more remarks on cohomology that are not essential on a first reading. Actually we can connect the results of this section to the previous one: a general expression for the Euler characteristic is

$$\chi(M) = \sum_i (-1)^i \dim H^i(M) = \sum_i (-1)^i \dim \frac{Z^i(M)}{B_i(M)}. \tag{2.92}$$

The dimension of the ith cohomology group is called the ith Betti number (to be pedantic, the Betti numbers are defined for homology rather than cohomology, but these are related by a simple duality). There is a compact way to express the idea of cohomology and homology that lets us introduce some notation and terminology that comes in useful later. If Ω_r is the vector space of r-forms, and C_r is the dual space of r-chains (here duality means that an r-chain can be viewed as an r-dimensional path on which an r-form gives a number), then the action of the boundary operator ∂ and the differential d (i.e., exterior derivative) is as follows:

$$\begin{aligned} \longleftarrow C_r \underset{\partial_{r+1}}{\longleftarrow} C_{r+1} \underset{\partial_{r+2}}{\longleftarrow} C_{r+2} \longleftarrow \\ \longrightarrow \Omega_r \underset{d_{r+1}}{\longrightarrow} \Omega_{r+1} \underset{d_{r+2}}{\longrightarrow} \Omega_{r+2} \longrightarrow . \end{aligned} \tag{2.93}$$

The kernel of a linear map like the differential is defined as the set of elements in the initial (source) space taken to zero, and the image is the set of elements in the final (target) space that are images under the map of some point in the initial space. Thus our definitions earlier in this section can be summarized by saying that exact forms are the image of the differential map, and closed forms are the kernel. The rth cohomology group is the quotient $\ker d_{r+1}/\operatorname{im} d_r$, and the rth homology group is $\ker \partial_r/\operatorname{im} \partial_{r+1}$.

The duality relationship is provided by Stokes's theorem. Recall that this theorem relates the integral of a form over a boundary to the integral of the differential of the form over the interior. In terms of the linear operator (f, c) that evaluates the form f on the chain c, the theorem has the compact expression

$$(f, \partial c) = (df, c). \tag{2.94}$$

Now we move on to a different type of topology that is perhaps more intuitive and will be useful for our first physics challenge: how to classify defects in ordered systems.

2.7.3 Winding Numbers and Homotopy

What if we did not want to deal with smooth functions and calculus? An even more basic type of topology, homotopy theory, can be defined without reference to calculus, differential forms, and so on (although in physics the assumption of differentiability is usually applicable). Suppose that we are given a continuous map from $[0, 1]$ to a manifold M such that 0 and 1 get mapped to the same point; we can think of this as a closed curve on M. We say that two such curves γ_1, γ_2 are homotopic if there is a continuous function (a homotopy) f from $[0, 1] \times [0, 1]$ to M that satisfies

$$f(x, 0) = \gamma_1(x), \quad f(x, 1) = \gamma_2(x). \tag{2.95}$$

Intuitively, f describes how to smoothly distort γ_1 to γ_2. Now homotopy is an equivalence relation and hence defines equivalence classes: $[\gamma_1]$ is the set of all paths homotopic to γ_1. Furthermore, concatenation of paths (i.e., tracing one after the other) defines a natural group structure on these equivalence classes: the inverse of any path can be obtained by tracing it in the opposite direction. (To be precise, one should define homotopy with reference to a particular point where paths start and end; for a symmetric space where all points are basically equivalent, which is the usual case in physics, this is unnecessary.) We conclude that the equivalence classes of closed paths form a group $\pi_1(M)$, called the fundamental group or first homotopy group. Higher homotopy groups $\pi_n(M)$ are obtained by considering mappings from the n-sphere S^n to M in the same way.

The homotopy groups of a manifold are not totally independent of the cohomology groups: for example, if $\pi_1(M)$ is trivial, then so is the first de Rham cohomology group. The manifolds used as examples above, $\mathbb{R}^2 - (0, 0)$ and S^1, both have $\pi_1(M) = \mathbb{Z}$: thus there is an integer that we can use to classify (equivalence classes of) paths. In fact this integer should be thought of just as winding number and it can be computed by the angle form given above; as in this case, it is frequently useful to use a cohomological integral to compute which homotopy class includes a given path.

So our two-dimensional examples already contains the two types of topology that occur most frequently in physics: de Rham cohomology and homotopy. A powerful relationship between them is given by the Hurewicz theorem: the first (co)homology group is the Abelianization of the first homotopy group for a connected space.[8] We will use homotopy in more detail in the following section, when we explain how it can be used to classify topological defects such as vortices in symmetry-breaking phases. Higher homotopy groups π_n will be defined there and used to classify other kinds of topological defects and configurations, for example, in the case of skyrmions in the quantum Hall effect (Section 3.7).

2.8 Topological Defects in Symmetry-Breaking Phases

Topological defects in an ordered phase can be classified using mappings from spheres in real space to the order parameter manifold M, that is, the homotopy groups $\pi_n(M)$. We will explain this result and see a number of examples. Another reasons the manifold M is useful in practice is that moving from one point in M to another is naively a soft or massless fluctuation, while changing the magnitude of the order parameter is a hard or massive fluctuation. A field-theory description that involves only the degrees of freedom in M, known for historical reasons as a nonlinear σ-model, is frequently useful at low energies for this reason.

Continuous spins also allow a number of important phenomena related to topological defects: the simplest example of this idea is a vortex in an XY model or, equivalently, a neutral superfluid. We will discuss this kind of vortex in 2D systems in the most detail, as the vortex is easily visualized and also controls the phase diagram (Box 2.3). In general, a topological defect refers to a configuration of a continuous model that cannot smoothly relax to a uniform configuration because of a nontrivial topological invariant (usually a winding number or some generalization thereof).

Vortices are the simplest example of topological defects. The local spin variable in the 2D XY model is a unit vector on the circle. Suppose that we are at low temperature so that the spin tilts only slightly from one site to the next. Then, in going around a large circle, we can ask how many times the spin winds around

---

[8] This theorem also gives an isomorphism between higher homology and homotopy groups if the lower homotopy groups are trivial.

the unit circle, and define this as the winding number $n \in \mathbb{Z}$. Note that if the winding number is nonzero, then the continuum limit must break down at some point within the circle, as otherwise we would have the same angular rotation $2\pi n$ around circles of smaller and smaller radius, implying larger and larger gradients and hence infinite energy density, since the energy density is proportional to the squared gradient of the order parameter. In fact the total energy is weakly divergent as computed in Box 2.3.

Many types of topological defects are now known in various condensed matter and particle physics systems. Vortices in the XY model have integer charge or winding number, but frequently topological defects have charge taking values in a finite group like $\mathbb{Z}_2$ (the group with two elements ± 1). Finally, in addition to the thermodynamics of such defects, one can consider their dynamics: many observed properties of superfluid helium are controlled by the motion of vortex loops.

The mathematical classification of topological defects has been carried out for a variety of systems. Vortex-like defects (defects that can be circled by a loop) are related to the group $\pi_1(M)$, where M is the manifold of degenerate values of the order parameter once its magnitude has been set (for example, S^1 for XY and S^2 for Heisenberg, where S^d is the unit sphere in $d + 1$ dimensions). $\pi_1(M)$ is known as the first homotopy group and is the group of equivalence classes of mappings from S^1 to M: for example, the mappings from S^1 to S^1 are characterized by an integer winding number $n \in \mathbb{Z}$, so $\pi_1(S^1) = \mathbb{Z}$, while $\pi_1(S^2) = 0$ (the group with one element) as any loop on the sphere is contractible to a point.[9]

In other words, $\pi_1(M)$ gives the set of equivalence classes up to smooth deformations of closed paths on M. The group operation on equivalence classes in the group is defined by concatenation of paths. An example of what this means physically can be understood from Figure 2.4. The order parameter winds clockwise (counterclockwise) on circling the positive (negative) vortex clockwise. Going around the rectangular path shown would wind and unwind around the order parameter manifold only slightly, with zero net winding, and this path can be continuously deformed to the path far away, which does not wind at all. The second homotopy group $\pi_2(M)$ classifies mappings from S^2 to M, and describes defects circled by a sphere, such as pointlike defects in 3D. For example, $\pi_2(S^2)$ is nonzero, and there are stable point defect configurations in 3D of Heisenberg spins (known descriptively as hedgehogs) but not of XY spins.

There are also topological configurations that do not involve defects; an example is the skyrmion of Heisenberg spins in $d = 2$, which we discuss in more detail in the context of the quantum Hall effect in Section 3.7. Not only does the skyrmion not have a core as it can be everywhere continuous; it also can be uniform at infinity. The use of $\pi_2(S^2) = \mathbb{Z}$ in two dimensions is that the skyrmion is a nontrivial map

---

[9] Intuitively, imagine a basketball with an unknotted rubber band stretched to form a loop on its surface. Then move the rubber band continuously to a small circle around the North Pole.

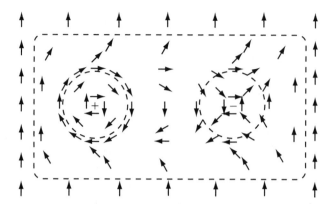

Fig. 2.4 The direction of an XY-type order parameter can be represented as a vector on the unit circle. On going around either dashed circle, the order parameter winds once around a circle, but it winds in opposite directions in going around the positive $(+)$ and negative $(-)$ vortices. The interaction energy between a vortex–antivortex pair is finite as the pair can be embedded in a configuration that is uniform at long distances from the center of the pair, because the topological charges of the two defects cancel: there is no nontrivial winding of the order parameter on going around the dashed rectangle enclosing both vortex cores. This illustrates what it means to say that there is an addition (group operation) in the first homotopy group. Note also that the velocity of a superfluid is given by the gradient of the order parameter, and hence the circulation is opposite around positive and negative vortices.

from the whole plane, compactified at infinity,[10] to the order parameter manifold, rather than just a map from the circle at infinity. Shankar's monopoles and other defect-free configurations in 3D of order parameter manifolds such as SO(3) are related to the group π_3 (cf. Nakahara, 1998).

Vortices in symmetry-broken phases become richer when the order parameter carries electromagnetic charge, as in the case of superconductors (Section 8.5). The most important difference will be that the superconducting vortex is not an infinitely large object like the superfluid vortex. Instead, the magnetic field through the center of the vortex is screened by supercurrents, and ultimately the strength of these currents goes to zero exponentially with radius beyond a certain scale known as the London penetration depth. In the superfluid vortex the currents fall off slowly as a power law in radius, unless an antivortex intervenes (Figure 2.4).

When we move beyond the treatment of just the order parameter and think about the excitations of the system, additional physics results as particles can be required to be present in the vortex core (see Section 8.7), for topological reasons. The richest topological defects of all are in states combining symmetry breaking and topological order, such as topological superconductors (Chapter 9), where

[10] Compactified means that we only consider configurations or functions on the plane that take the same value for all points at spatial infinity. Then the plane becomes topologically equivalent to the sphere, by stereographic projection.

one focus of current research is to find exotic Majorana zero mode quasiparticles trapped in vortices; such quasiparticles offer promise for a topological approach to quantum computation.

Box 2.3 The Berezinskii–Kosterlitz–Thouless Transition

One of the most remarkable examples of a collective effect arising from many topological defects is the superfluid transition in two spatial dimensions. We sketch a theoretical prediction by Kosterlitz and Thouless, anticipated in part in work of Berezinskii, that received spectacular experimental confirmation in work of Bishop and Reppy on ^{4}He films. While the full analysis of Kosterlitz and Thouless is a tour de force of renormalization-group methods, which we will not cover here, the essentials can be obtained from simple physical arguments about topological defects.

Our starting point is the two-dimensional XY model: the local spin variable on each site of a lattice is a unit vector on the circle, with the lattice Hamiltonian

$$ H = -J \sum_{\langle ij \rangle} \mathbf{s}_i \cdot \mathbf{s}_j = -J \sum_{\langle ij \rangle} \cos(\theta_i - \theta_j), \qquad (2.96) $$

where J is an energy and the sum is over nearest-neighbor pairs. In the second equality we have introduced an angle θ via $s_x + i s_y = e^{i\theta}$.

This model has the same symmetry as the superfluid transition in a film of atoms with no low-temperature internal degrees of freedom, by the following argument: the Bose condensation transition means that one quantum state has a macroscopic number of atoms, and the wavefunction of this state $\psi(\mathbf{r})$ can be taken loosely as the order parameter in the ordered phase.[a]

Suppose that we are at low temperature so that the spin moves only slightly from one site to the next. Then, in going around a large circle, we can ask how many times the spin winds around the unit circle, and define this as the winding number $n \in \mathbb{Z}$. As mentioned above, if the winding number is nonzero, then the continuum limit must break down at some point within the circle or else the energy would diverge; in a moment, we will calculate the energy of a vortex and a vortex-antivortex pair in Eq. 2.105 and see this divergence.

This will also let us see from a fairly simple calculation how there can be continuously varying exponents in the power law correlations of the 2D XY model at low temperature. The assumption we'll need to make is that vortices are unimportant at sufficiently low temperature, so that the 2π periodicity of the phase can be ignored (a vortex in 2D is a point where the magnitude of the order parameter vanishes, around which the phase of the order parameter changes by a multiple of 2π). This calculation can be justified by looking at the renormalization group flow in a space of two parameters: the temperature, and the vortex fugacity (essentially a parameter controlling how many vortices there are). An excellent reference for this RG flow is the original paper by Kosterlitz and Thouless (1973). If there are no vortices and the magnitude of the order parameter is constant, then the effective partition function is

$$Z = \int D\theta(r) \, e^{-\frac{K}{2} \int (\nabla \theta)^2 \, d^2 r},$$
(2.97)

where θ is no longer restricted to be periodic. Here K is a dimensionless coupling incorporating temperature that, in a lattice model such as (2.96), can be obtained by linearization. We can define a superfluid stiffness with units of energy $\rho_s = (k_B T) K$ that measures the energy required to create a twist in the superfluid phase. One way to look at this nonlinear sigma model is as describing slow variations of the ordered configuration at low temperature: the magnitude is fixed because fluctuations in magnitude are energetically expensive, but since there are degenerate states with the same magnitude but different θ, slow variation of θ costs little energy.

In the model with no vortices, that is, with θ treated as a real-valued rather than periodic field, the spin correlation function, which we will find goes as a power law, is

$$\langle \mathbf{s}(0) \cdot \mathbf{s}(r) \rangle = \mathrm{Re} \langle e^{i\theta(0)} e^{-i\theta(r)} \rangle.$$
(2.98)

(Actually taking the real part is superfluous if we define the correlator of an odd number of θ fields to be zero.) We choose not to rescale the θ field to make K equal to unity, since such a rescaling would modify the periodicity constraint $\theta = \theta + 2\pi$ in the model once vortices are restored. To obtain this correlation function, we first compute the correlation of the θ fields $G(r) = \langle \theta(0)\theta(r) \rangle$, which we will need to regularize by subtracting the infinite constant $G(0) = \langle \theta(0)^2 \rangle$.

From previous experience with Gaussian theories, we know to write $G(r)$ as an integral over Fourier components:

$$G(r) = \frac{1}{(2\pi)^2 K} \int^{a^{-1}} \frac{e^{i\mathbf{k} \cdot \mathbf{r}}}{k^2} \, d\mathbf{r}.$$
(2.99)

However, this integral is divergent at $k = 0$ (this is called infrared divergent; by contrast, ultraviolet divergent means a divergence at short length scales, at $k = \infty$). In condensed matter we expect that short length scales are cut off by some microscopic length a, as in the above integral, while long length scales are more interesting. We can regularize the short-distance divergence by subtracting out the formally infinite quantity $G(0)$, and then using $\tilde{G}(r) = G(r) - G(0)$ to calculate physical quantities:

$$\tilde{G}(r) = -\frac{1}{(2\pi)^2 K} \int^{a^{-1}} \frac{1 - e^{i\mathbf{k} \cdot \mathbf{r}}}{k^2} \, d\mathbf{r}.$$
(2.100)

Now the integral can only be a function of the ratio r/a (you can check this by changing variables). As $a \to 0$, the leading term in the integration is proportional to

$$\tilde{G}(r) = \frac{1}{(2\pi)^2 K} \int^{a^{-1}} \frac{dk}{k} = -\frac{\log(r/a)}{2\pi K} + \ldots,$$
(2.101)

where one factor of 2π was picked up by the angular integration. Note that at large r, $\tilde{G}$ is divergent, which makes sense since θ is unbounded.

We now want to calculate the resulting spin-spin correlator. To do this we need to use a fact about Gaussian integrals. The Gaussian average

$$\langle e^{iJ\phi} \rangle = (A/\sqrt{2\pi}) \int_{-\infty}^{\infty} d\phi \, e^{iJ\phi} e^{-\frac{1}{2}A\phi^2} = e^{-\frac{1}{2}A^{-1}J^2} = e^{-\frac{1}{2}\langle J^2\phi^2\rangle} \qquad (2.102)$$

generalizes to the continuum limit in the following way:

$$\langle e^{-i\theta(r)+i\theta(0)} \rangle = e^{-\frac{1}{2}\langle(\theta(r)-\theta(0))^2\rangle} = e^{G(r)-G(0)}. \qquad (2.103)$$

So finally,

$$\langle \mathbf{s}(0) \cdot \mathbf{s}(r) \rangle = \mathrm{Re}\langle e^{i\theta(r)-i\theta(0)} \rangle = \frac{1}{(r/a)^{1/2\pi K}}. \qquad (2.104)$$

We expect on physical grounds that this power law correlation function (algebraic long-range order) cannot survive up to arbitrarily high temperatures; above some maximum temperature, there should be a disordered phase with exponentially decaying correlations.

To understand how our physical expectation of exponentially short correlations at high temperature is met, we give a simple picture due to Kosterlitz and Thouless (which is supported by a more serious RG calculation). The picture is that the phase transition results from an unbinding of logarithmically bound vortex-antivortex pairs, which can be viewed as the plasma-gas transition of a two-dimensional Coulomb plasma. The phase with unbound vortices is a plasma phase since it is has unbound positive and negative charges (vortices) as in a plasma. Note, of course, that the logarithmic interaction between vortex charges in 2D is just like that between Coulomb charges in 2D. A more serious calculation constructs an RG flow in terms of vortex fugacity to show that below a critical temperature vortices are irrelevant (their fugacity scales to 0), while above that temperature vortices are relevant (their fugacity increases upon rescaling). Vortex-antivortex pairs (see Figure 2.4) are logarithmically bound because the energy of a single vortex of winding number n goes as, from integrating the energy density $(k_B T)K(\nabla\theta)^2/2$ from (2.97),

$$E = \frac{1}{2}K(k_B T)\int_a^L (n/r)^2 d^2r \sim \pi n^2 K \log(L/a). \qquad (2.105)$$

Here L is the long-distance cutoff (e.g., system size) and a is the short-distance cutoff (e.g., vortex core size). Although the energy of a single vortex in the infinite system diverges, the interaction energy of a vortex-antivortex pair does not; each vortex has energy given by (2.105), but with the system size replaced by the intervortex spacing. Note that changes of order unity in the definition of this spacing or the core size will add constants to the energy but not change the coefficient of the logarithm.

We would like to compare the free energy of two phases: one in which vortices are bound in pairs, and essentially do not modify the Gaussian model, and one in which vortices are numerous and essentially free, although the system is still charge-neutral (total winding number 0). Suppose the vortices in the free phase have typical separation L_0. Then each vortex can be distributed over a region of size L_0^2, and the entropic contribution to the free energy per vortex is $-TS = -T \log \Omega = -2T \log(L_0/a)$, where $\Omega = (L_0/a)^2$ is the number of distinguishable states. The energy cost is $E = \pi n^2 J \log L_0/a$, so there should be a phase transition somewhere near $T_{KT} = \pi J/2k_B$, where we have written $K = J/k_B T$ in order to define a coupling energy scale J.

A bit more work shows that this coupling scale, as we have defined it, is exactly the superfluid stiffness ρ_s that measures the energy induced by a twist in the superfluid phase. More precisely, the Berezinskii–Kosterlitz–Thouless transition occurs when the asymptotic long-distance stiffness ρ_s^∞, including renormalization by bound vortex pairs, satisfies

$$\rho_s^\infty = \frac{2k_B T_{KT}}{\pi}. \qquad (2.106)$$

This prediction of a universal jump at T_{KT} in the superfluid stiffness was beautifully confirmed in experiments (Bishop and Reppy, 1978). Another way to state this result is that a 2D superfluid that starts at short distances with stiffness less than this value allows vortices to proliferate and reduce the long-distance superfluid stiffness to zero. This behavior is rather different from that in higher dimensions, where the superfluid stiffness flows to zero smoothly.

It is remarkable that such relatively straightforward considerations can uncover the KT transition. Equally remarkably, the renormalization group can predict phase diagrams in considerable detail. As a rich illustrative example, we return to the case of the triangular Ising magnet and consider it as a quantum magnet in a transverse field of strength Γ:

$$\mathcal{H} = -J \sum_{\langle ij \rangle} s_i^z s_j^z - \Gamma \sum_i s_i^x \qquad (2.107)$$

Here, the Ising spins s_i of Eq. 2.62 have been replaced by spin-1/2 operators s^α. Via a Trotter–Suzuki transformation this can be transformed into a stacked classical magnet, with the same exchange interactions in the plane but with ferromagnetic couplings between adjacent planes. In this mapping, the temperature of the quantum magnet is encoded by the extent of the additional, third dimension, which is infinite at $T = 0$ but finite otherwise. For details of this mapping, see Box 6.1.

The coupling in the third direction adds a term proportional to $\cos q_z$, with a negative prefactor, to the Fourier transform of the couplings (Eq. 2.67), which remains unchanged otherwise. The candidate ordering thus takes place at $(\pm 4\pi/3, 0, 0)$ now, and the Landau–Ginzburg functional (Eq. 2.68) remains essentially unchanged except for the integration over the additional dimension (and extra gradient terms, which we have suppressed).

The analysis of the XY-model with a clock term was analyzed in great detail by José et al. (1977) in a milestone paper demonstrating the power of the renormalization group in its relatively early days. There, it was found that in $d = 2$ dimensions, the KT transition out of the paramagnet is not changed by the addition of the clock term b_6 (this term is thus said to be irrelevant at the transition). Upon lowering the temperature further, the KT phase with its drifting exponent is encountered, which terminates at another transition where the clock term finally becomes relevant. The order parameter then locks into one of the clock directions. This corresponds to a three-sublattice ordering at wavevector $\pm\mathbf{K}$, as expected.

This contrasts to the quantum phase transition at $T = 0$ in the original quantum model, which occurs when the transverse field strength Γ is tuned. This corresponds to an infinite extent in the third dimension of the effective classical model, and hence a $d = 3$ RG analysis is needed. Here, it is found that the clock term is still irrelevant at the transition, which therefore is in the XY universality class in $d = 3$. However, it immediately becomes relevant inside the ordered phase; such terms are known as dangerously irrelevant, and again lead to locking into one of six symmetry-breaking phases.

The resulting phase diagram is shown in Figure 2.5. The ordered phase is labeled as bond-ordered, on account of its appearance in the dual quantum dimer model on the honeycomb lattice (see Box 6.1). The details of the conventional ordered phases are unimportant in the context of our interest in topological phases; however, looking ahead to Chapter 6, the robust ordering tendency of quantum dimer models visible in this phase diagram is common to all quantum dimer models on bipartite lattices such as the honeycomb, and also responsible for the absence of the desired topologically ordered resonating valence bond liquids on the square lattice.

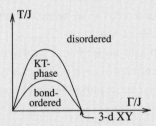

Fig. 2.5 Phase diagram of the transverse field Ising model on the triangular lattice. Upon lowering transverse field and/or temperature, a KT transition to a phase with algebraic correlations takes place. The exponent of the algebraic correlations drifts until long-range order sets in at another KT transition. At $T = 0$, these collapse to a single quantum phase transition in the 3D XY universality class. From Moessner and Sondhi (2001a).

---

[a] More precisely, as we will use in the discussion of superconductivity and the Josephson effect in Chapter 8, the order parameter comes from an expectation value of an operator changing the particle number, and its phase is not an overall wavefunction phase.

3

Integer Topological Phases: The Integer Quantum Hall Effect and Topological Insulators

The first topological phase to be discovered experimentally was the integer quantum Hall effect (IQHE), which remains one of the most simple and dramatic manifestations of macroscopic quantization. Its subsequent theoretical explanation revealed that even the physics of independent electrons can be remarkably subtle. The Hall effect is a standard probe of conducting materials: when a current I_x runs perpendicular to an applied magnetic field B_z, a transverse voltage V_y builds up. When the material is a two-dimensional electron gas and the magnetic field is sufficiently strong, then there are finite ranges of magnetic field over which the voltage and current satisfy

$$G_{xy} \equiv \frac{I_x}{V_y} = \frac{ne^2}{h},\tag{3.1}$$

where n is an integer, e is electron charge and h is Planck's constant. Note that all geometric factors drop out for a transverse measurement in two dimensions, so the conductivity σ_{xy} is equal to the conductance G_{xy}.

The remarkable fact is just how well (3.1) is satisfied. The integer quantum Hall effect is the property that, even at nonzero temperature in an imperfect material, a transport quantity is quantized to remarkable precision: the transverse (aka Hall) conductivity on a plateau is $\sigma_{xy} = ne^2/h$, where n is integral to 1 part in 10^9. Taking this experiment as a measure of a fundamental unit of resistance for metrology purposes,[1] the Klitzing constant, one obtains $R_K = h/e^2 \approx 25812.80745\ \Omega$. As a metrological aside, it is amusing to note that what is being measured has changed with time, even though the physical measurement is fundamentally unchanged. In the words of the discoverer (von Klitzing, 2017): "When the QHE was discovered, the SI ohm was more precisely known than the fine-structure constant, which

---

[1] Until the 1970s, topology seems to have been regarded as entirely without applications. The writer and occasional mathematics teacher Solzhenitsyn (1968) had his surrogate Nerzhin say, "Suddenly he felt sorry for Verenyov. Topology belonged to the stratosphere of human thought. It might conceivably turn out to be of some use in the twenty-fourth century, but for the time being ..."

is why the first QHE publication was entitled 'New method for high-accuracy determination of the fine-structure constant based on quantized Hall resistance.' However, this title will be wrong when the new SI features a fixed value for the von Klitzing constant. When that happens [as it has in 2019], the manuscript's original title, 'Realization of a resistance standard based on natural constants,' rejected at the time by Physical Review Letters, will turn out to be more appropriate."

This precision results because the transport is determined by a topological invariant, as stated most clearly in work of Thouless. Consequently we say that Thouless-type topological order is present in phases where a response function is determined by a topological invariant. The gapless edge or surface states that often exist at the boundary between a Thouless-type phase and an ordinary insulator are a consequence of this topological response. Another type of topological order (Wen-type) is present in phases with fractional particles, such as fractional quantum Hall phases and the quantum dimer model (Chapter 5).

The importance of Thouless-type topological order for condensed matter physics has expanded greatly in the past few years with the discovery of topological phases in which spin-orbit coupling, rather than magnetic field, produces the new phase. These phases exist in both two and three dimensions, unlike the IQHE which is essentially two-dimensional. They have unusual edge and surface states that we use to introduce their physics as measured in experiments, leading into the deeper theoretical analysis in the following chapter. The easiest way to picture the 2D topological insulator phase is as two copies of the IQHE in which spin-up and spin-down electrons (along some axis) feel opposite magnetic fields from the spin-orbit coupling.

This is also known as the quantum spin Hall effect (QSHE) because in this ideal case it supports a spin transport phenomenon analogous to (3.1). However, there are important and subtle differences between the QSHE and a doubled IQHE, and the theoretical explanation of these in terms of a new kind of topological invariant by Kane and Mele (2005b) triggered an avalanche of theoretical developments that continues to the time of this volume's writing. We will explain the basic idea of their work and its relation to the *charge* transport measurement at Würzburg that gave the first experimental evidence for the QSHE (Koenig et al., 2007).

A rather important consequence of Kane and Mele's work was that theorists, in trying to understand the new topological invariant, realized that it had a fully three-dimensional analogue (Fu et al., 2007; Moore and Balents, 2007; Roy, 2009). The term topological insulator was coined to describe this phase (Moore and Balents, 2007) because there is no longer any quantized transport quantity like the Hall effect, at least in the material's native state; what exactly is quantized is a kind of magnetoelectric effect. The three-dimensional topological insulator resulting from this invariant was soon observed experimentally, but not by a transport experiment; instead its distinctive surface state was imaged in a photoemission experiment (Hsieh et al., 2008). The relative ease of making and measuring 3D topological insulators has led to an explosion of experimental work. A summary of the early history of the 2D and 3D topological insulators with examples of key materials is Moore (2010).

In this chapter we seek instead to give physical arguments and examples for the topological properties that give rise to the states we discuss, deferring a more mathematical discussion to the following chapter on wavefunction geometry and Berry phases. The essential aspects of these states can be understood from fairly simple assumptions about magnetic fields, spin-orbit coupling, disorder, and time-reversal symmetry. At a deeper level, the topological properties are described by an invariant resulting from integration of an underlying Berry phase. It turns out that the Berry phase can be rather important even when it is not part of a topological invariant. In crystalline solids, the electrical polarization, the anomalous Hall effect, and part of magnetoelectric polarizability also derive from Berry phases of the Bloch electron states, as described in Chapter 4.

First we give some background for the original quantum Hall discovery by Dorda, Pepper, and von Klitzing in 1980, which triggered a flood of developments continuing to the present day, and for the more recent discovery of topological insulators in two and three dimensions. We cover the classic explanation by Laughlin for the precision of quantization in the IQHE, then in later sections explain the more recently discovered integer topological phases with reference to simple tight-binding models on the honeycomb lattice that were important in their understanding. The chapter closes with an important physical example that combines several concepts from Chapter 2: skyrmions in the quantum Hall effect, which add a dose of quantum mechanics to the theory of topological defects.

3.1 IQHE: Basic Phenomena and Theory

What does it mean to say that electrons move only in two dimensions? Imagine a perfectly planar interface between two insulators so that the electronic eigenstates decompose into products of a wavefunction in the normal direction z and a plane wave in the xy plane: $\psi(x, y, z) = \phi(z)e^{i(k_x x + k_y y)}$. If there is a single bound state for electrons at the interface, and if the temperature is much less than the energy separation between this bound state and other transverse wavefunctions, then electronic motion in the z direction is frozen out: the electron is perfectly two-dimensional with respect to low-energy processes that are unable to promote the electron to a different z state. Even if a finite number of bound states (subbands) are occupied, the essential features remain two-dimensional because the electronic motion in the normal direction is strongly quantized. The technique of modulation doping in molecular beam epitaxy allows the creation of 2D electron gases (2DEGs) with extremely long electron mean free paths, comparable to the sample size (Box 3.2).

The original, nonquantized Hall effect is a valuable probe of materials because, in a simple Drude model of 2D transport with a single relaxation time τ, a Hall ratio R_H can be formed in which the relaxation time drops out and only the density and charge of the carriers survives (Ashcroft and Mermin, 1976):

$$R_H = \frac{V_y}{I_x B} = \frac{1}{n_{2D} e}. \tag{3.2}$$

Here the areal density of carriers is n_{2D}. Then the Hall conductivity is

$$\sigma_{xy} = \frac{n_{2D} e}{B}. \tag{3.3}$$

The quantized values $\sigma_{xy} = ne^2/h$ then correspond to an electron density

$$n_{2D} = \frac{neB}{h} = \frac{n}{2\pi\ell^2}, \tag{3.4}$$

or n electrons per (single-electron) flux quantum $\frac{eB}{h}$. We have introduced the magnetic length $\ell = \sqrt{\hbar/eB}$.[2]

The inverse areal density $2\pi\ell^2$ is known to be special from a classic calculation in nonrelativistic quantum mechanics (Section 2.2). A 2D electron moving in a perpendicular magnetic field has, ignoring the Zeeman coupling, highly degenerate electron energy levels spaced by $\hbar\omega_c$, where $\omega_c = eB/m$ is the classical cyclotron frequency (the angular frequency at which an electron orbits under the Lorentz force of the magnetic field). These Landau levels have a degeneracy which scales with the area of the system; each Landau level contains one electronic state per flux quantum. So the occurrence of features in transport when the chemical potential passes through a Landau level because of a change in the magnetic field is not very surprising.

What requires explanation is that the plateaux have nonzero width, that is, that σ_{xy} is flat on a plateau to a tremendously good approximation. It is as though the state of the system is not changing at all, even as the magnetic field changes by a significant percentage. The existence of plateaux is even more surprising when one realizes that in the calculation above of a perfectly clean system, the electrical field E_y associated with the transverse voltage can be made zero by a Lorentz transformation to the frame moving with velocity E_y/B. In this frame the electrons are

---

[2] This is one of several places where our use of SI units, following most experimental papers, may cause confusion when comparing to other sources using Gaussian-CGS units. Conversions are given in Table A.1 of the appendix.

stationary, which implies that the Hall ratio in the original frame is exactly proportional to $1/B$ with no plateaux (Eq. 8.10)! The current theoretical understanding is that the IQHE system is in fact changing on a plateau, but that the only changes involve localized electronic states that do not contribute to transport (see Chapter 8). In a moment we will repeat a classic argument of Laughlin for what happens on a quantum Hall plateau.

In order to make the following argument more concrete, it is worthwhile to introduce a basis for wavefunctions in the lowest Landau level. For problems with rotational symmetry, a nice basis is ($z = x + iy$)

$$\psi_m(z) = \frac{\left(\frac{z}{\sqrt{2}\ell}\right)^m \exp(-|z|^2/4\ell^2)}{\sqrt{2\pi\ell^2 m!}}, \tag{3.5}$$

the $n = 0$ case of the form given in Section 2.2, where $m = 0, 1, \ldots$. The exponential factor keeps the electrons confined in the vicinity of the origin; more precisely, a quick calculation shows that the magnitude $|\psi_m|^2$ is maximized at radius $r = \ell\sqrt{2m}$, and the mean squared radius is

$$\langle r^2 \rangle_m = 2(m+1)\ell^2. \tag{3.6}$$

As a check, the number of states within a ring of large radius R is approximately $R^2/(2\ell^2) = \pi R^2/(2\pi\ell^2)$, confirming that the areal density is $1/(2\pi\ell^2)$. We will want to study in a moment how the filled Landau level made up of these states responds to an applied electric field, but it may be useful first to review the simpler problem of a particle moving on a ring pierced by a magnetic flux in Box 3.1. The key feature is that an integer number of flux quanta can be gauged away and do not modify the spectrum. The connection is that we will find it useful to generate electric fields through a time-varying flux through a hole in the Landau level, and there must be a relationship between the original state and the final state once an integer number of flux quanta have been added.

Box 3.1 One Particle on a Ring Pierced by Magnetic Flux

A famous consequence of the way electromagnetic fields appear in quantum mechanics is the Aharonov-Bohm effect: even if there is no magnetic field in some region of space, a particle moving in that region can notice the effects of a magnetic field elsewhere. In the simplest case, this effect is a consequence of the fact that the vector potential **A** rather than the magnetic field **B** appears in the Schrödinger equation. Considering a particle moving on an idealized one-dimensional loop (Figure 3.1a) allows us to see, however, that at certain values of the magnetic flux through the loop, the Aharonov–Bohm effect on physical quantities disappears.

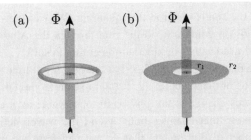

Fig. 3.1 (a) Insertion of magnetic flux Φ through a solenoid passing through a ring modifies the wavefunctions and energies as described in Eq. 3.11. (b) Geometry of Laughlin pumping. The flux is now inserted through a solenoid inside the inner radius r_1 of a two-dimensional electron gas confined to an annulus or Corbino disk. If the electrons are in an integer quantum Hall state, the effect of adiabatically pumping flux $\Phi_0 = h/e$ will be to shuttle an integer number of electrons from the inner edge at radius r_1 to the outer edge at r_2 or vice versa, via the process shown in Figure 3.2.

Without flux, the Hamiltonian on the ring is just

$$H = -\frac{\hbar^2}{2m}(\partial_x \psi)^2, \tag{3.7}$$

where x is a coordinate along the ring: x and $x + 2\pi R$ therefore represent the same physical point, where R is the radius of the ring. So the energy eigenstates are plane waves e^{ikx}, with energies $\hbar^2 k^2/(2m)$, but we must restrict the values of k to

$$k_n = \frac{2\pi n}{2\pi R} = \frac{n}{R}, \quad n \in \mathbb{Z} \tag{3.8}$$

in order for the periodicity condition to be satisfied.

How is this modified when a magnetic flux Φ pierces the ring, for example in an infinitesimally narrow solenoid through the ring's center? The line integral of the vector potential around the ring is the magnetic flux:

$$\oint A_x \, dx = \int B \, dA = \Phi. \tag{3.9}$$

The Hamiltonian now takes the form, supposing that the particle is an electron with charge $-e$,

$$H = \frac{1}{2m}(-i\hbar\partial_x + eA_x)^2, \tag{3.10}$$

which implies that the energies are modified to

$$E_n = \frac{\hbar^2}{2m}\left(\frac{n}{R} + \frac{e\Phi}{2\pi R\hbar}\right)^2 = \frac{\hbar^2}{2mR^2}\left(n + \frac{\Phi}{\Phi_0}\right)^2. \tag{3.11}$$

Here $\Phi_0 = h/e$ is the single-electron flux quantum. We can view the flux as having induced a phase shift on going around the ring through the Aharonov–Bohm effect, which causes the allowed values of kinetic momentum to shift.

However, we see that if the applied flux Φ is an integer multiple of the flux quantum Φ_0, the spectrum has not changed at all. It might seem as though the wavefunctions have changed, but this change is not physically observable as it can be removed by a gauge transformation that changes both A and the wavefunction phase. We say that an integer number of flux quanta that do not penetrate the regions where electrons are present can be gauged away, as they do not affect the spectrum or other observable quantities; it is only the fractional part of fluxes that generate observable Aharonov–Bohm phases. The concept of Laughlin pumping is essentially that, even if the spectrum with one flux quantum is the same as that at zero flux, there could be a nontrivial evolution called spectral flow as the flux is changed. Even in this ring example we can see, starting with the $n = 0$ state, that its energy would smoothly increase as flux was inserted until at one flux quantum, it had the energy of what was originally the $n = 1$ state.

The other remark we make about this model is that at half-integer fluxes $\Phi = \pm\frac{\Phi_0}{2}, \pm\frac{3\Phi_0}{2}, \ldots$ the spectrum has a twofold degeneracy. We can view this degeneracy as reflecting a kind of time-reversal symmetry: without changing the flux through the ring, there are always right-moving and left-moving states at the same energy. At these fluxes, the phase factor induced on moving around the circle is just -1, and hence real, unlike the complex phases at other noninteger flux values. A consequence is that if we require that the spectrum be invariant under taking the complex conjugate of the wavefunctions, which transforms k to $-k$, without changing the flux, then there are only two fluxes that qualify: integer flux and half-integer flux (note that all integer fluxes are indistinguishable from each other by the above argument, and likewise for half-integer fluxes). A way to make this argument more abstract is that the set of integer fluxes is invariant under $\Phi \rightarrow -\Phi$, but so is the set of half-integer fluxes. The idea that a system with periodic variable (here Φ) that flips sign under some symmetry might permit exactly two inequivalent classes turns out to be useful later on, for example in the discussion of electrical polarization and the quantized magnetoelectric effect in Section 4.5.

Consider a Corbino disk geometry where a material is present in an annulus $r_1 < r < r_2$. We wish to consider the Hall current in this material between the inner and outer edges when an electromotive force is applied by changing the flux through a solenoid located at the origin (Figure 3.1b). We assume that the flux Φ is increased from zero adiabatically, that is, at sufficiently low frequency that there are no excitations across the gap. The gauge invariance of the Schrödinger equation implies that when the flux through the solenoid reaches one flux quantum, the quantum-mechanical eigenstates are equivalent via a gauge transformation to those at zero flux.

Laughlin argued that the net result of the adiabatic flux increase must have been to transfer an integer number of electrons, possibly zero, from one edge of the annulus to the other. This is physically plausible if no electronic states span the annulus, so that the question of how many electrons are in the annulus is well defined up to edge contributions; indeed the quantization of the IQHE is expected to break down at sample widths comparable to a magnetic length, when wavefunctions from one edge tunnel appreciably to the other. If an integer n electrons are transferred by an increase of one flux quantum h/e, then the Hall conductance averaged over the period of the flux increase can be computed by computing the current flow across a ring of radius r anywhere in the annulus. Writing Φ for the flux through the annulus, and using Maxwell's equations to relate the flux change to the induced electric field, we obtain

$$\frac{\int_0^T j_r \, dt}{T E_\theta} = \frac{ne/2\pi r}{(T/2\pi r)\frac{d\Phi}{dt}} = \frac{ne}{h/e} = \frac{ne^2}{h}. \tag{3.12}$$

The strength of this argument is that it applies as long as the Fermi level lies in a mobility gap, that is, localized states at the Fermi level do not affect the premise that an integer number of charges be transferred. The actual behavior of electrons in a magnetic field and disorder potential is discussed in some detail in Chapter 8.

To understand concretely how the flux induces charge transport, let us consider how the solenoid modifies the wavefunctions (3.5). Note that if we really took these wavefunctions to model our system exactly, then the annulus of occupied states has simply shifted outward, that is, the system has moved. If the final state is to be the same as the initial state, then there must be reservoirs of charge at the inner and outer edges, able to transfer electrons into and out of the annulus that is in the QHE regime. These must be gapless because the electron transfer occurs adiabatically. The importance of this gapless edge state was pointed out in an important paper by Halperin (1982). A model of how the edge state arises is to think about the evolution of states shown in Figure 3.2, and note the gapless spectrum near the edge; one can think about these discrete modes as arising from quantizing edge ripples on an incompressible droplet. We will have much more to say about these edge states later, as the primary means of learning about the order in integer and fractional quantum Hall states is via edge measurements.

Transport in the Laughlin picture occurs by a very counterintuitive process. Normally transport of independent electrons in a solid is described in terms of

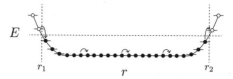

Fig. 3.2 Adiabatic transport by pumping in the quantum Hall effect. Suppose that a weak confining potential is added to the lowest Landau level so that the electrons are trapped in an annulus $r_1 < r < r_2$. The Fermi level (horizontal dotted line) separates occupied states (solid circles) from empty states (open circles). For a weak potential, it is sufficient to picture the symmetric-gauge eigenfunctions as unchanged and localized at different values of the radius as shown. As a flux through the inner circle of the annulus is adiabatically increased from zero to one flux quantum $\Phi_0 = h/e$, the eigenstates smoothly shift over as indicated by the arrows. The effect of the pumping cycle is to transport one electron from the inner edge at r_1 to the outer edge at r_2.

the occupancy of single-electron states changing; an electron hops between different states in roughly the same way as a person climbing stairs moves between stairs. The adiabatic transport taking place in the Laughlin argument as the flux is increased is like an escalator or baggage carousel: the states are continuously evolving and carrying the electrons with them (Figure 3.2). Whether adiabatic or ordinary transport is taking place in a given quantum Hall experiment is a subtle issue because of the gapless edge states described above. Transport between two reservoirs connected by an edge state can be understood in the conventional picture of electronic occupancies, while transport between reservoirs connected by a gapped region, as in the Laughlin argument, requires that the states evolve, not just their occupancies.

In the above discussion of the quantum Hall effect, we ignored entirely the effects of band structure in the underlying material. This is a reasonable approximation for traditional semiconductor 2DEGs in the following sense. For realistic laboratory magnetic fields, the magnetic flux through one unit cell is much smaller than a flux quantum, so a description in terms of Bloch wavepackets is acceptable (Ashcroft and Mermin, 1976). For typical 2DEG densities, only the bottom or top of the 2D band is appreciably occupied and the band physics can be captured simply by an effective mass. Some additional physics and history of ultraclean 2DEGs, including the key advance of modulation doping, is discussed in Box 3.2. These assumptions do not apply when the magnetic field is replaced by the spin-dependent forces that give rise to topological insulators, as those forces vary significantly on the length scale of a unit cell. Hence we will need to develop an alternate picture of the IQHE even to describe the simplest model of the 2D topological insulator, which is as two oppositely directed copies of the IQHE.

Box 3.2 Modulation Doping

Progress in sample growth and measurement techniques are as crucial for pushing back the boundaries of knowledge as ideas and theories. One notable instance of the former is the invention of modulation doping, which lowered the level of electrostatic disorder in the two-dimensional electron liquid to a point where a many-body phenomenon as delicate as the FQHE could be observed. The impact of that invention is hard to overstate – the resulting high-mobility semiconductor heterostructures have found their place in innumerable applications including semiconductor lasers and high-electron-mobility transistors (HEMTs).[a]

The need for modulation doping arises because semiconductors such as gallium arsenide (GaAs) need to be supplied with carriers (electrons or holes) in order to have interesting physics at temperatures well below their intrinsic band gap. The dopant ions such as aluminum that are used to donate these carriers serve as potential scatterers as their ionic potentials are necessarily different from those of the atoms they substitute. (In practice the randomness in the locations of these dopant atoms cannot entirely be

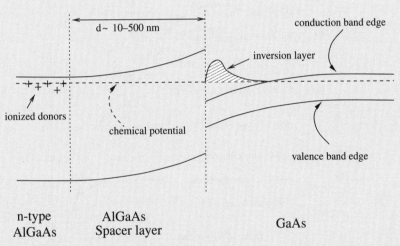

Fig. 3.3 Modulation-doped GaAs heterostructure. The structure is two-dimensional, and the horizontal axis denotes the layer-by-layer growth direction in real space. The vertical direction shows the energies of conduction (top) and valence (bottom) bands. The mobile carriers (electrons, in this case) exist in a thin inversion layer at an interface, where the conduction band dips below the chemical potential. The dopant ions, whose ionization contributes these carriers, are deposited in layers some distance away, so the disorder potential they contribute is rather suppressed by the intervening charge-neutral spacer layer. From Moessner (1997).

eliminated.) At low temperatures where electron-phonon scattering is irrelevant, these dopant potentials are the primary source of resistivity.

Modulation doping takes advantage of the flexibility of layer-by-layer deposition of atoms using techniques such as molecular beam epitaxy. This allows separating the donor layer spatially from the 2DEG, which decreases the disorder potential *exponentially* in the layer separation.

A modulation-doped quantum well (Figure 3.3) is grown layer by layer via molecular beam epitaxy, with a spacer layer separating the dopants from the carriers in the quantum well. That this is useful in suppressing the disorder potential follows from Poisson's equation $\nabla^2 \phi(\mathbf{r}) = \rho_c$, which connects the electric potential ϕ with the charge density ρ_c. In the spacer layer, $\rho_c = 0$, so that the Fourier transform of the potential satisfies

$$\mathbf{k}^2 \phi_{\mathbf{k}} = (k_\perp^2 + k_\parallel^2)\phi_{\mathbf{k}} = 0 \implies k_\perp^2 = -k_\parallel^2. \tag{3.13}$$

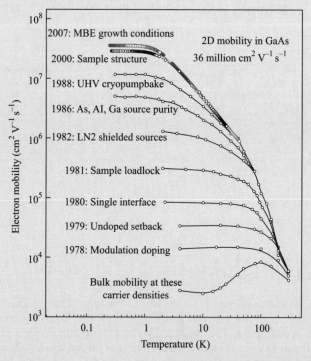

Fig. 3.4 Increase of low-temperature mobility in two-dimensional electron liquids with time. The "bulk" curve denotes a uniformly doped GaAs sample, while "undoped setback" precedes the introduction of modulation doping. Increased mobility was a primary cause of successive discoveries in quantum Hall physics, including the first odd-denominator fraction in 1982 and the first even-denominator fraction in 1989. From Schlom and Pfeiffer (2010). Reprinted with permission by Nature Publishing Group.

This means that a variation of the potential with wavevector $k_\parallel$ in the plane must decay exponentially, with decay constant $\kappa_\perp = k_\parallel$: short-wavelength fluctuations decay faster than long-wavelength ones.

The carriers in the quantum well can thus have extremely long mean free paths, leading to the high mobilities shown in Figure 3.4. Major physics discoveries have occurred as a consequence of these increases in mobility. Graphene, a quite different type of two-dimensional electron gas, has similarly seen new correlated states emerge as a result of increases in mobility. Here the mobility increases were achieved using two techniques: by finding good substrates, such as boron nitride, and by suspending the graphene (i.e., not using a substrate at all).

---

[a] The textbook Ashcroft and Mermin (1976) is a good reference for the basic physics of such heterostructures.

3.2 Two Lattice Models of the IQHE, and Chern Number

Someone new to solid-state physics might be skeptical that the Landau level wavefunctions described in (3.5), for an electron moving in a constant magnetic field in free space, remain relevant when the electron also feels a periodic potential from the crystalline background. The Hofstadter model is possibly the simplest example of how to incorporate the orbital effect of a magnetic field in a model of a crystal. Its full behavior is quite rich, but we will concentrate here on its topological aspects and on recovering Landau level physics when the magnetic field is weak.

Consider free electrons moving on the square lattice with bond length a, and ignore spin for now. The tight-binding Hamiltonian with nearest-neighbor interactions is, in zero magnetic field,

$$H = -\sum_{\langle ij \rangle} t(c_i^\dagger c_j + c_j^\dagger c_i), \tag{3.14}$$

where $t > 0$ is the hopping matrix element and the sum is over nearest-neighbor sites i and j. With periodic boundary conditions on an $N \times N$ lattice, introduce the Fourier-transformed plane wave operators

$$c_{\mathbf{k}}^\dagger = \frac{1}{\sqrt{N}} \sum_i c_i^\dagger e^{-i\mathbf{k} \cdot \mathbf{r}_i}, \tag{3.15}$$

where the components of $\mathbf{k} = (k_x, k_y)$ are integer multiples of $\frac{2\pi}{Na}$, and $\mathbf{r}_i$ is the location of site i. Then the Hamiltonian is diagonalized as

$$H = \sum_{\mathbf{k}} \epsilon_{\mathbf{k}} c_{\mathbf{k}}^\dagger c_{\mathbf{k}} \tag{3.16}$$

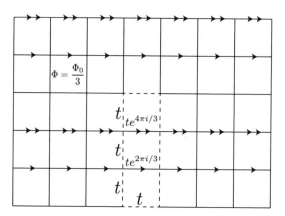

Fig. 3.5 The Hofstadter model: one particle hopping on a square lattice with a magnetic flux Φ through each plaquette. Here $\Phi = \Phi_0/3$, and we have chosen Landau gauge with the vector potential along $\hat{\mathbf{x}}$ and varying with y. All vertical bonds have just the original zero-field hopping t. Horizontal bonds take one of three possible values, denoted by zero arrows, one arrow, or two arrows. A unit cell or supercell of the system at this flux (dotted lines) contains three original unit cells.

with the eigenenergies $\epsilon_{\mathbf{k}} = -2t(\cos k_x a + \cos k_y a)$. This system has one site per unit cell and hence one band in the first Brillouin zone.[3]

The Hofstadter model (Hofstadter, 1976) is obtained by adding a uniform perpendicular magnetic field $\mathbf{B} = B\hat{\mathbf{z}}$ to this tight-binding model, with flux Φ per unit cell (Figure 3.5). How do we incorporate such a magnetic field into the tight-binding model? Moving an electron around any plaquette should pick up a phase Φ through the Aharonov-Bohm effect. That is accomplished by changing the Hamiltonian in (3.14) to incorporate a phase into the hopping amplitudes,

$$ H = -\sum_{\langle ij \rangle} \left[t \exp\left(i\frac{2\pi}{\Phi_0} \int_{\mathbf{r}_j}^{\mathbf{r}_i} \mathbf{A} \cdot \mathbf{dr} \right) c_i^\dagger c_j + \text{h.c.} \right]. \qquad (3.17) $$

Here $\mathbf{A}$ is the electromagnetic vector potential created by the magnetic field. We remark in passing that this phase factor resulting from integrating a gauge potential along a line in real space is known as a Wilson loop and has some important applications in practical methods for calculating topological invariants.

For general values of the flux, the Hofstadter Hamiltonian (3.17) is quite challenging to solve. When the flux is a rational multiple of the single-electron flux quantum,

$$ \Phi = \frac{p}{q}\Phi_0 = \frac{p}{q}\frac{h}{e}, \quad p, q \in \mathbb{Z}, \qquad (3.18) $$

---

[3] A puzzle for students, of relevance to the high-temperature cuprate superconductors, which exist on this lattice: what is the shape of the Fermi surface at half-filling?

then the system still has a finite unit cell and Bloch eigenstates. Note that $p \ll q$ for achievable magnetic fields in crystals, since one flux quantum through a typical unit cell of a few angstroms on a side would be achieved only at fields of order 10^5 Tesla. The importance of the commensurability condition (3.18) can be seen by working in the Landau gauge, with vector potential

$$\mathbf{A} = (-By, 0, 0), \quad \nabla \times \mathbf{A} = B\hat{\mathbf{z}}, \quad B = \frac{\Phi}{a^2}. \tag{3.19}$$

This leads to phase factors on horizontal bonds that indeed satisfy the expected Aharonov-Bohm relation. Let us take $\Phi = \Phi_0/3$ for simplicity as shown in Figure 3.5. The hopping amplitudes on horizontal bonds are arranged in a repeating sequence of period 3 with values $(-t, -te^{2\pi i/3}, -te^{4\pi i/3})$, with the amplitude to hop from right to left being the complex conjugate of the amplitude to hop from left to right. Because the new unit cell or supercell is three times as large as before, and contains three sites, we can expect three bands in a Brillouin zone that is one-third as large as in the case of zero field.

By the same process, if we looked at smaller commensurate fluxes $\Phi = \Phi_0/n$, we would have n bands to consider. Calculating the energies of the Bloch eigenstates is not too difficult, but these energies are not the most important facts about the model. Let us suppose that the bands occupy disjoint regions in energy, so that when some bands are filled but not others, we have a band insulator. Is this band insulator different from the standard insulators one learns about in a first course?

The answer is yes, and in fact these are our first examples of topologically nontrivial bands, in this case bands of nonzero Chern number, which is a fancy way of saying that they have something in common with the Landau levels we found for a free electron. Indeed, as n becomes very large, the bands of the Hofstadter model become flat in energy, and the lowest bands can be viewed as the first few Landau levels. This makes sense if we think about the lowest bands as made out of the lowest-energy states of the zero-field model, which are close to free electron states, for example, they have a similar quadratic dispersion obtained by expanding around the minimum of the band structure.

Even at $n = 3$, there is something topological about the bands obtained by solving the Hofstadter model, which is our first example of a topological band structure. The eigenstates and eigenenergies can be found by diagonalizing the three-by-three Bloch Hamiltonian. The detailed values of the energies are not very important, but for illustration, the lowest band's energy is found to be, setting the lattice spacing $a = 1$:

$$E(\mathbf{k}) = 2\sqrt{2}\cos[\phi(\mathbf{k}) + 2\pi/3], \tag{3.20}$$

where

$$\phi(\mathbf{k}) = \frac{1}{3}\cos^{-1}\left[-\frac{\cos(3k_x) + \cos(3k_y)}{2\sqrt{2}}\right], \tag{3.21}$$

in which the symmetry between x and y is manifest, even if it was broken in the wavefunctions by our gauge choice. More important for our purposes is the fact that this lowest band, for example, responds to an inserted flux in the same way as a Landau level.

More generally, we can associate an integer-valued topological invariant called the Chern number or TKNN integer with each band in the Hofstadter model. That integer has a physical meaning – it describes how much that band, if it is filled with electrons, contributes to the IQHE:[4]

$$\sigma_{yx} = \frac{e^2}{h}\sum_{\text{occ } i} C_i, \quad C_i = \frac{1}{2\pi}\int \mathcal{F}^i \, d^2k. \tag{3.22}$$

Here $C_i \in \mathbb{Z}$ is the Chern number of band i. The integrand $\mathcal{F}^i$ is known as the Berry curvature or Berry flux of the Bloch states in band i and takes the form (cf. Eq. 2.9)

$$\mathcal{F}^i = (\nabla_k \times \boldsymbol{\mathcal{A}}^i)_z, \quad \mathcal{A}^i_\lambda = i\langle u^i | \partial_\lambda u^i \rangle. \tag{3.23}$$

Here we have broken our rule of not introducing microscopic formulae until Chapter 4, in addition to trusting the reader to distinguish between band index i and $i = \sqrt{-1}$, but the historical impact of Eq. 3.22 is sufficient to warrant an exception. At a conceptual level, suffice it to say for now that something about how the wavefunctions wind with momentum leads to an effective gauge field $\mathcal{A}$ in the two-dimensional Brillouin zone, and a topological integer characterizing each band.

The Hofstadter problem with $n = 3$ discussed above has Chern numbers $C_1 = 1, C_2 = -2, C_3 = 1$ for the three bands. In general, for n odd this pattern still obtains: the total Chern number is zero, and all bands have Chern number 1 except for the band in the middle. For even n, only the two bands in the middle have Chern numbers different from one; their Chern numbers are equal and make the total of Chern numbers for the entire band structure equal to 0.

The remarkable expression (3.22), found by Thouless, Kohmoto, den Nijs, and Nightgale in a seminal 1982 paper (Thouless et al., 1982), connects the topology of wavefunctions to a physical observable. Part of their motivation was to explain how the quantization predicted by Laughlin's pumping argument can be recovered

---

[4] The σ_{yx} instead of σ_{xy} in this formula is not a typo but a deliberate minus sign; it originates in the existence of two sign conventions for the Berry curvature and hence C_i. TKNN defined the Berry connection with a minus sign relative to the contemporaneous definition by Berry, which we have followed here. Our convention, which is also that of Vanderbilt (2018), makes natural the expression for polarization in Chapter 4.

by relatively standard means, namely, calculation of σ_{xy} starting from the general Kubo linear-response expression for σ_{xy}. We will derive (3.22) and give an expression for C_i in terms of the Bloch wavefunctions in the following chapter.

For now, we state some basic facts without proof. Each isolated band has a Chern number, and when two bands touch, their total Chern number remains a well-defined integer even though each individual band may not have its own Chern number. We have discussed Chern number here for a two-dimensional band structure, but each band in a three-dimensional band structure has three Chern numbers, related to xy, yz, and xz planes in the Brillouin zone. Finally, the Chern number and the IQHE it predicts remain well-defined concepts even when interactions are added between the electrons, as long as the ground state remains nondegenerate. Without a magnetic field or magnetism in a material, there is a symmetry known as time-reversal symmetry, discussed in the following section, that forces all $C_i = 0$.

The importance of time-reversal symmetry will start to be evident in the second lattice model we discuss, which was introduced by Haldane as a model of the IQHE and Chern bands without a macroscopic magnetic field. As in the Hofstadter problem, we start with a band structure that is not topological. This is the nearest-neighbor hopping model on the honeycomb lattice, a well-known model for graphene, which we introduced in Section 2.5. An illustration of a two-dimensional linear dispersion characteristic of a Dirac point like the ones in graphene is also provided in Figure 3.10; in this case, it arises at the surface of a three-dimensional topological insulator. Topological aspects of such points in three dimensions are discussed in detail in Chapter 7.

One way to view the point made by Haldane is that there are inequivalent ways that a gap could open up at these band touchings, even if we preserve the size of unit cell. Let us first consider a trivial way to open up a band gap. If instead of graphene, which is a single layer of carbon atoms, we had a monolayer honeycomb material with different atoms on the two inequivalent sites (*A* and *B* in Figure 3.8), what would happen? Nature furnishes an example in boron nitride, BN. The boron and nitrogen orbitals would be expected to have different on-site energies, which means that the orbital energy is not just an overall constant that we can neglect. Instead the difference in orbital energies between *A* and *B* sites, $\Delta = \epsilon_A - \epsilon_B$, is an important parameter that affects the band structure. An energy gap opens at the **K** and **K$'$** points. This is not surprising and very similar in spirit to the opening of a gap in a one-dimensional tight-binding chain when odd and even atoms become inequivalent.

Haldane pointed out that a gap also opens when we add a kind of spatially variable magnetic field, even if the field averages to zero, and that the nature of the bands in this case is more topological. Suppose we add second-neighbor hopping, and also a magnetic field that gives phases to second-neighbor hopping elements as shown in Figure 3.8. The precise form of this term is given below in Eq. 3.38, where a spin dependence is added in order to create a new phase. The total magnetic flux through each hexagon turns out to be zero, unlike for the Hofstadter model in Figure 3.5, and the original hoppings of strength t along nearest-neighbor bonds can remain unchanged. However, if there are additional processes such as second-neighbor hopping through the interior of the hexagons, these hoppings might be sensitive to the flux.

In the Haldane model, when a gap is present, the lower and upper bands have equal and opposite Chern numbers ± 1. If this material had the lower band filled and the upper band empty, it would support an integer quantum Hall effect, just as in the Hofstadter model with one band filled. The term Chern insulator is often used for this kind of IQHE phase appearing on a lattice in zero average field, as in the Haldane model with one band filled.

Even though the total field is zero, we can see that the Chern insulator phase in this model has some features reminiscent of the Hofstadter model. The second-neighbor hoppings used to open the gap have imaginary parts, so the Hamiltonian matrix is not purely real as it was for the simple band insulator case representing BN. The appearance of imaginary matrix elements is a sign that the Haldane model breaks the same kind of symmetry as an external magnetic field. We study this time-reversal symmetry in the next section by taking a step backward and starting with some simple pictures from classical physics.

3.3 Time-Reversal Symmetry in Classical and Quantum Physics

While symmetry is the bread and butter of large swaths of physics, time-reversal symmetry seems not to receive the attention it deserves in undergraduate classes. Here we use classical concepts to understand how standard observables transform under change of the sign of time, and then explain how its antiunitary nature makes time-reversal symmetry quite different from standard unitary symmetries at the quantum level.

To visualize time-reversal symmetry intuitively, consider running the slides of a film in reversed order. Then spatial directions are preserved but the sign of time has been flipped ($t \rightarrow -t$), and as a result, objects appear to move backward. If we think about the velocity $\frac{dx}{dt}$, then we would expect this to flip sign under either spatial inversion ($\mathbf{r} \rightarrow -\mathbf{r}$) or time-reversal. We say that the velocity is odd under each of these transformations, by analogy with an odd function on the line, which flips under $x \rightarrow -x$. The acceleration is odd under inversion but unchanged (even) under time-reversal, because it has two powers of inverse time.

According to Newton, accelerations are induced by forces, so now let us think about the Lorentz force law

$$m\frac{d^2\mathbf{r}}{dt^2} = q(\mathbf{E} + \mathbf{v} \times \mathbf{B}).$$ (3.24)

Then, assuming mass and charge are unchanged under both symmetries, we conclude that electric field $\mathbf{E}$ is odd under inversion and even under time-reversal, while $\mathbf{B}$ behaves oppositely: magnetic fields are odd under time-reversal but even under inversion. (As an exercise, consider a loop of current producing a magnetic dipole, and verify that the direction of that dipole has the expected behavior under both symmetries.) It might seem unusual that magnetic materials break time-reversal symmetry spontaneously and hence can produce a magnetic moment, but recall that many other symmetries such as translational and rotational symmetries are routinely broken in the solid state.

We remark in passing that it can be dangerous to assume that both sides of an equation describing macroscopic behavior should transform identically under time-reversal, even if the underlying microscopic equations are time-reversal-invariant. Macroscopic systems in which we only study a subset of degrees of freedom have an entropic arrow of time: for example, we see entropy increase in our daily lives, even though the microscopic equations of motion are reversible. We think of this arrow of time as resulting from coarse graining, which means not being able to observe exactly what every particle of a large system is doing. Also, a remarkable discovery in high-energy physics is that the microscopic equations of our universe are actually not reversible (they break time-reversal symmetry), which one can see in the properties of unusual particles called kaons. But the equations for the particles that appear in solids do have this symmetry, while in magnetic solids, there is spontaneous breaking of this symmetry even though the underlying equations are time-reversal-invariant.

Time-reversal has an interesting effect in quantum mechanics when we think about spin-half particles like electrons. For such a particle, in a system with time-reversal symmetry like an atom in zero magnetic field, there are always degenerate eigenstates (Kramers pairs) for the following reason: every one-electron state is degenerate with, and different from, its time-reversed version. For integer-spin particles, a state can be the same as its time-reversed version, but this never happens for fermions (spin-half particles). Time-reversal symmetry is special in quantum mechanics because it is represented as an antiunitary operation (the product of complex conjugation and a unitary operation), while ordinary symmetries like rotations are unitary operations. Some intuition for why time-reversal must include a complex conjugation, and hence be antiunitary, comes from considering how complex conjugation, acting on the time-dependent Schrödinger equation, gives a solution with the opposite sense of time, or noting that the time-reversed version of the wavefunction of a plane wave moving to the right, e^{ikx}, is just e^{-ikx}, which is also its complex conjugate. A 1931 theorem of Wigner implies that symmetries preserving all observables must be implemented in quantum mechanics by either unitary or antiunitary operators.

This fact might seem esoteric, but the time-reversal symmetry of electrons has been extremely important in some new discoveries in solids in the last decade or so. When an electric field is applied to an atom, or spin-orbit coupling is added, these twofold degeneracies do not split, but they do split when a magnetic field is added, which is a clue that magnetic fields behave differently and break time-reversal symmetry. The next section gives an intuitive picture of how the behavior of electrons under time-reversal, including the existence of Kramers pairs, gives rise to new states with directly measurable experimental consequences.

3.4 Topological Insulators in 2D: Basic Phenomena and Theory

Topological insulators in 2D and 3D result from the spin-orbit coupling intrinsic to all materials, rather than from external magnetic fields. For simplicity, consider an example of spin-orbit coupling familiar from atomic physics:

$$H_{SO} = \lambda \mathbf{L} \cdot \mathbf{S}, \tag{3.25}$$

where $\mathbf{L}$ is the orbital angular momentum of an electron, $\mathbf{S}$ is spin angular momentum, and λ measures of the spin-orbit coupling strength. This coupling arises because the electric field of the nucleus, when viewed in the frame of a relativistically moving electron, acquires a magnetic component that couples to the electron spin via the Zeeman effect. Because it is a relativistic effect, strong spin-orbit coupling requires fast-moving electrons and hence strongly charged nuclei: $\lambda \propto Z^4$ where Z is the atomic number of the nucleus.

In general, crystals have less symmetry than isolated atoms, and the form of spin-orbit coupling is less constrained. However, the spin-orbit force on an electron in a crystal depends on the electron's motion through space and in this sense is loosely similar to a magnetic field, which also generates a velocity-dependent force. More explicitly, think of a 2D electron moving in a constant magnetic field $B\hat{\mathbf{z}}$. Representing the magnetic field in the rotational gauge $\mathbf{A} = (-By/2, Bx/2, 0)$, the linear coupling to the electron is

$$\mathbf{A} \cdot \mathbf{p} = \frac{B}{2}(\mathbf{z} \times \mathbf{r}) \cdot \mathbf{p} = \frac{B}{2}\mathbf{L}_z, \tag{3.26}$$

so a magnetic field is also a coupling to orbital angular momentum. However, this similarity masks some clear differences, for example, the spin-orbit force changes sign if the electron spin changes sign. This is a signal of a more fundamental difference: spin-orbit coupling preserves time-reversal symmetry (is even under time-reversal), unlike a magnetic field, because spin and orbital angular momentum are both odd under time-reversal. We will discuss time-reversal more formally in a moment because it is the key to the topological protection in topological insulators.

Another difference between spin-orbit coupling and a constant magnetic field is that spin-orbit forces in a crystal are expected to vary significantly on the scale of a unit cell. Before developing a theory of the IQHE in a periodic potential, let us state some of the basic phenomenology of the QSHE. If we could make two oppositely directed copies of the IQHE for up and down spins, for example, an $n = 1$ state for $S_z = +\hbar/2$ and an $n = -1$ state (i.e., oppositely directed effective magnetic field) for $S_z = -\hbar/2$, then this overall combination would have time-reversal symmetry, as time-reversal flips both the spin and the magnetic field. The resulting quantum Hall charge currents would cancel out, but there would be a quantized spin current (a quantum spin Hall effect) (Murakami et al., 2004):

$$\mathcal{J}^i_j = \sigma^s_H \epsilon_{ijk} E_k. \tag{3.27}$$

The spin current $\mathcal{J}$ carries two indices, one a spin direction and one a spatial direction, and σ^s_H is a quantized spin conductivity. We will not worry for the moment about the various subtleties of defining a spin current in a solid with spin-orbit coupling, but we will argue in a moment that there is in general no quantized spin transport of the form (3.27) along any spin axis.[5] The appearance of the electric field rather than the magnetic field in the quantum spin Hall equation results from the goal of having a potentially dissipationless current equation. If dissipation provides no arrow of time, then both sides should transform in the same way under the time-reversal operation, which fixes the field on the right side to be **E** rather than **B**.

In real materials, there is no conserved direction of spin like $\hat{\mathbf{z}}$ in this example, and it is an unphysical simplification to separate electrons into two independent species. Looking at the edge, one would expect naively that the two oppositely directed edge modes would mix with each other under realistic spin-orbit coupling and disorder, leading to localization and no propagating edge mode. This simple expectation turns out to be incorrect. Under some circumstances, a propagating edge mode does survive and distinguishes ordinary and topological 2D insulators. The key in understanding when spin-orbit leads to a protected edge state is to consider the action of time-reversal symmetry on electrons; this step is what makes the

---

[5] In mathematical terms, in realistic solids not only is spin rotational symmetry (SU(2)) broken, but there are no unbroken U(1) subgroups as crystals do not have continuous rotations around any axis. So there is no absolute conservation of spin because there is no continuous symmetry left to protect it, although spin can be approximately conserved for a relatively long time scale (greater than picoseconds) in some solids.

theory of the 2D topological insulator a highly nontrivial advance on the theory of the IQHE. One result is that, in the more topological language introduced below, 2D time-reversal-symmetric insulators are not characterized by an integer, but by a $\mathbb{Z}_2$ invariant that only takes two possible values: even, in ordinary insulators, and odd, in topological insulators.

We wish to explain the surprising fact that the quantum spin Hall phase survives, with interesting modifications, once we allow more realistic spin-orbit coupling, as long as time-reversal symmetry remains unbroken. The time-reversal operator $\mathcal{T}$ acts differently in Fermi and Bose systems, or more precisely in half-integer versus integer spin systems. Kramers showed that the square of the time-reversal operator is connected to a 2π rotation, which implies that

$$\mathcal{T}^2 = (-\mathbf{1})^{2S}, \tag{3.28}$$

where S is the total spin quantum number of a state: half-integer-spin systems pick up a minus sign under two time-reversal operations. For example, for a scalar wavefunction ($S = 0$) time-reversal is normally implemented simply as complex conjugation, which squares to the identity. For a single spin-half, a commonly used form is

$$\mathcal{T} = i\sigma^y K = \begin{pmatrix} 0 & 1 \\ -1 & 0 \end{pmatrix} K, \tag{3.29}$$

where σ^y is the Pauli matrix and K is complex conjugation. Then

$$\mathcal{T}^2 = \begin{pmatrix} 0 & 1 \\ -1 & 0 \end{pmatrix} K \begin{pmatrix} 0 & 1 \\ -1 & 0 \end{pmatrix} K = K^2 \begin{pmatrix} -1 & 0 \\ 0 & -1 \end{pmatrix} = -\mathbf{1}. \tag{3.30}$$

Note that the inverse of the time-reversal operator is thus $\mathcal{T}$ or $-\mathcal{T}$, depending on the spin.

A consequence of this is the existence of Kramers pairs: every eigenstate of a time-reversal-invariant spin-half system is at least twofold degenerate. In other words, every state of a spin-half particle is degenerate with *and distinct from* its time-reversed version, while an eigenstate of an integer-spin particle can be invariant under time reversal. We will show that a time-reversal invariant Hermitian operator H', which could be the system's Hamiltonian, is zero between the members of a Kramers pair, that is, a state ψ and its time-reversal conjugate $\phi = \mathcal{T}\psi$.

First, let us give the proper definition of an antiunitary operator. A unitary operator U is one that preserves inner products: $\langle U\phi | U\psi \rangle = \langle \phi | \psi \rangle$ for every pair of states ψ and ϕ, which yields the compact result $U^\dagger U = \mathbf{1}$ or $U^\dagger = U^{-1}$. To say that a Hamiltonian is invariant under U means that its matrix elements satisfy

$$\langle U\phi | H | U\psi \rangle = \langle \phi | H | \psi \rangle \Rightarrow U^\dagger H U = H. \tag{3.31}$$

An antiunitary operator K involves complex conjugation, so we modify the above definition so that the effect of K on inner products includes a complex conjugation or reversal of order:

$$\langle K\phi | K\psi \rangle = \langle \phi | \psi \rangle^* = \langle \psi | \phi \rangle. \tag{3.32}$$

To check that this makes sense, take the Hilbert space to be scalar wavefunctions and allow K to be just complex conjugation. Now the statement that antiunitary operator K is a symmetry of Hamiltonian H is similarly modified to

$$\langle K\phi | H | K\psi \rangle = \langle \psi | H | \phi \rangle. \tag{3.33}$$

A useful alternate form of the statement that H has K as a symmetry comes from rewriting

$$\langle K\phi | H | K\psi \rangle = \langle K\phi | HK\psi \rangle = \langle K\phi | (KK^{-1})HK\psi \rangle = \langle K^{-1}HK\psi | \phi \rangle, \tag{3.34}$$

which becomes the above statement of time-reversal invariance if $K^{-1}HK = H$, that is, if H and K commute.

We now apply this rule with $K = \mathcal{T}$ to the matrix element between an eigenstate ψ and its time-reversal conjugate of a time-reversal-symmetric Hamiltonian H'. This leads to the first equality in

$$\langle \mathcal{T}\psi | H' | \psi \rangle = \langle \mathcal{T}\psi | H' | \mathcal{T}^2\psi \rangle = -\langle \mathcal{T}\psi | H' | \psi \rangle = 0. \tag{3.35}$$

The second step is just the fact that $\mathcal{T}^2 = -1$, and the last step is to note that the first and third entries are opposite each other, and if $x = -x$, then $x = 0$. Since H' is time-reversal invariant, $\mathcal{T}|\psi\rangle$ must be an eigenstate with the same energy as $|\psi\rangle$, and from the above it is a distinct, and in fact orthogonal, state.

We used the notation H' to stress that this result applies both to the original Hamiltonian and any perturbation respecting time-reversal symmetry. We immediately see that no mixing or scattering is induced between the two states of a Kramers pair; for example, the pair does not split into even and odd combinations with a gap, as one might have expected. Combining Kramers pairs with what is known about the edge state, we can say a bit about why an odd-even or $\mathbb{Z}_2$ invariant might be physical here. If there is only a single Kramers pair of edge states and we consider low-energy elastic scattering, then a right-moving excitation can only backscatter into its time-reversal conjugate, which is forbidden by the Kramers result above if the perturbation inducing scattering is time-reversal invariant. However, if we have two Kramers pairs of edge modes, then a right-mover can back-scatter to the left-mover that is *not* its time-reversal conjugate. This process will, in general, eliminate these two Kramers pairs from the low-energy theory.

Our general belief based on this argument is that a system with an even number of Kramers pairs at the edge will, under time-reversal-invariant backscattering,

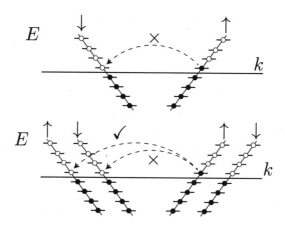

Fig. 3.6 Low-energy scattering processes at the edge of a 2D topological insulator (top), compared to those in an ordinary quantum wire (bottom). What is shown is the energy-momentum relation along the edge. Solid circles denote occupied states and open circles empty states. The lowest-energy current-carrying state (one extra electron moving right) has only a single decay process allowed by energy conservation, and Kramers's theorem prevents any time-reversal-symmetric perturbation from inducing this process because the two states involved are time-reversal conjugate, i.e., members of a Kramers pair. In the simplest case, they can be viewed as opposite spin, but the absence of scattering persists even with spin-orbit coupling as long as time-reversal is unbroken. In the ordinary wire (below), with an even number of Kramers pairs, there are two possible decay processes for the current-carrying state shown, and while one is forbidden by time-reversal symmetry, the other is allowed (check mark) and ultimately leads to localization.

localize in pairs down to zero Kramers pairs, while a system with an odd number of Kramers pairs will wind up with a single stable Kramers pair. In other words (Figure 3.6), an ordinary quantum wire with an even number of Kramers pairs is susceptible to localization, while the edge of a 2D insulator is "half of a quantum wire," and stable as long as $\mathcal{T}$ remains a good symmetry. It is as though we took a single quantum wire and separated it into two halves at two opposite edges of the 2D insulator, thereby making it stable to time-reversal-symmetric perturbations. The same concept is helpful in understanding the 3D topological insulator introduced below.

Additional support for this odd-even argument will be provided by our next approach. We would like, rather than just trying to understand whether the edge is stable, to predict from bulk properties whether the edge will have an even or odd number of Kramers pairs. Since deriving the bulk-edge correspondence directly is quite difficult, what we will show is that starting from the bulk T-invariant system, there are two topological classes. These correspond in the example above (of separated up- and down-spins) to paired IQHE states with even or odd Chern number for one spin. Then the known connection between IQHE integer and number of edge states is good evidence for the statements above about Kramers pairs of edge modes.

Before turning to the bulk, we sketch a graphical version of the original argument by Kane and Mele for why Kramers degeneracy means that there are two inequivalent possibilities for edge structure. Imagine a crystalline edge of a 2D insulator, that is, one at which periodicity is unbroken along the edge and Bloch's theorem applies in this direction (Figure 3.7). We can then view a system that is gapped in bulk as a one-dimensional band structure with a bandgap except possibly at the edge. A lattice system in an IQHE state, also known as a Chern insulator, will have at least one state in the gap at each edge reaching from the conduction band to the valence band, with a preferred direction.

Now consider what edge structures are allowed by time-reversal symmetry. The action of time-reversal symmetry on the Bloch Hamiltonians $H(\mathbf{k})$, whose eigenvalues are the energies of Bloch states at crystal momentum $\mathbf{k}$, in a

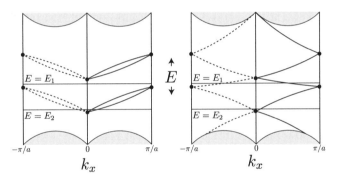

Fig. 3.7 Consider an edge of a semi-infinite two-dimensional insulator preserving lattice translational symmetry by distance a in the $\hat{\mathbf{x}}$ direction. The diagrams show two possible band structures in the resulting one-dimensional Brillouin zone. There are bulk states (shaded) from occupied (conduction) and unoccupied (valence) bands above and below the bulk bandgap. For states inside the bulk bandgap, there are two possible arrangements consistent with time-reversal symmetry, which requires Kramers degeneracies at momentum $k = 0$ and $k = \pm \pi/a$. The time-reversal symmetry means that $E(k) = E(-k)$ and drawing the $k \geq 0$ half is sufficient: the dashed bands are just reflected versions of the solid bands, so we could view the right half as an effective Brillouin zone. In an ordinary insulator (left), there can be edge states but their bands form loops, so that at some energies, such as energy E_1 in the diagram, there may not be any edge states, while at other energies such as E_2 in the diagram, there are an even number of Kramers pairs. Kane and Mele (2005b) pointed out another possibility: in the topological insulator (right), the two bands coming into one Kramers pair at $k_x = 0$ go into different Kramers points at $k_x = \pi/a$. This forces a zigzag pattern and an odd number of Kramers pairs at every energy in the bulk bandgap.

time-reversal-invariant system is as follows: with Θ the representation of $\mathcal{T}$ in the Bloch Hilbert space,

$$\Theta H(\mathbf{k})\Theta^{-1} = H(-\mathbf{k}). \tag{3.36}$$

In words, the time-reversal operator acts both on the Bloch Hamiltonian and also on crystal momentum $\mathbf{k}$. In particular, only the time-reversal invariant momenta at which $\mathbf{k} = -\mathbf{k}$ are forced to have Kramers pairs. In one dimension there are just two such points, $k = 0$ and $k = \pm\pi/a$, recalling that crystal momenta are only defined modulo a reciprocal lattice vector. This combined action is one reason why more than two decades elapsed between the discovery of the integer-valued topological invariant that underlies the IQHE and the discovery of the $\mathbb{Z}_2$ invariants in time-reversal-invariant systems. The mathematical expression of those invariants in two and three dimensions is described starting in Section 4.4. Figure 3.7 gives the original picture of Kane and Mele for how this requirement leads to two distinct possibilities consistent with time-reversal invariance.

3.5 A Lattice Model of the 2D Topological Insulator

Returning to the idea of two copies of the IQHE generated by spin-orbit coupling, consider the model of graphene introduced in Kane and Mele (2005a). This is a tight-binding model for independent electrons on the honeycomb lattice (Figure 3.8). The spin-independent part of the Hamiltonian consists of a nearest-neighbor hopping, which alone would give a semimetallic spectrum with Dirac nodes at certain points in the 2D Brillouin zone, plus a staggered sublattice potential whose effect is to introduce a gap:

$$H_0 = t \sum_{\langle ij \rangle \sigma} c_{i\sigma}^\dagger c_{j\sigma} + \lambda_v \sum_{i\sigma} \xi_i c_{i\sigma}^\dagger c_{i\sigma}. \tag{3.37}$$

Here $\langle ij \rangle$ denotes nearest-neighbor pairs of sites, σ is a spin index, ξ_i alternates sign between sublattices of the honeycomb, and t and λ_v are free parameters.

The insulator created by increasing λ_v is an unremarkable, spin-independent band insulator. However, the symmetries of graphene also permit an intrinsic spin-orbit coupling of the form

$$H_{SO} = i\lambda_{SO} \sum_{\langle\langle ij \rangle\rangle \sigma_1 \sigma_2} \nu_{ij} c_{i\sigma_1}^\dagger s_{\sigma_1 \sigma_2}^z c_{j\sigma_2}. \tag{3.38}$$

Here $\nu_{ij} = [(2/\sqrt{3})\hat{\boldsymbol{d}}_1 \times \hat{\boldsymbol{d}}_2]_z = \pm 1$, where i and j are next nearest neighbors and $\hat{\boldsymbol{d}}_1$ and $\hat{\boldsymbol{d}}_2$ are unit vectors along the two bonds that connect i to j. Here s^z is the matrix representation of the spin operator, that is, $\hbar\sigma^z/2$. Including this type of spin-orbit coupling alone would not be a realistic model. For example, the

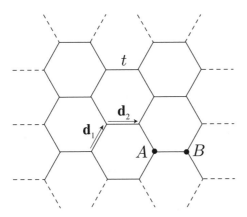

Fig. 3.8 The honeycomb lattice used to model graphene (see also Section 2.5) and to define the tight-binding models of Haldane (1988) and of Kane and Mele (2005a). With just nearest-neighbor hopping t, this provides a commonly used model of graphene, with gapless Dirac points at two inequivalent points in the Brillouin zone. A simple band gap respecting time-reversal and spin symmetries would be created if we allowed different on-site potentials on the A and B sublattices; one site on each sublattice is indicated. Adding second-neighbor hoppings gives the possibility of opening a gap between topologically nontrivial bands. For the two sites connected by the sum of the vectors $\mathbf{d}_1$ and $\mathbf{d}_2$, the factor ν_{ij} of Eq. 3.38 is $\nu_{ij} = [(2/\sqrt{3})\hat{d}_1 \times \hat{d}_2]_z = -1$.

Hamiltonian $H_0 + H_{SO}$ conserves S_z, the distinguished component of electron spin, and reduces for fixed spin (up or down) to the model introduced by Haldane for a lattice quantum Hall effect (Haldane, 1988). Generic spin-orbit coupling in solids should not conserve any component of electron spin.

As a test of what happens when S_z is not conserved, consider the addition of Rashba-type spin-orbit coupling:

$$H_R = i\lambda_R \sum_{\langle ij \rangle \sigma_1 \sigma_2} c_{i\sigma_1}^\dagger \left(\mathbf{s}_{\sigma_1 \sigma_2} \times \hat{\mathbf{d}}_{ij} \right)_z c_{j\sigma_2}, \qquad (3.39)$$

with $\mathbf{d}_{ij}$ the vector from $i \rightarrow j$ and $\hat{\mathbf{d}}_{ij}$ the corresponding unit vector, and the spin operators $\mathbf{s}$ now involve all three of the Pauli matrices. Note that Rashba spin-orbit coupling is not intrinsic to graphene but generated by an inversion-symmetry breaking field, such as an electric field, in the out-of-plane direction (Bychkov and Rashba, 1984). (Inversion means the symmetry $\mathbf{r} \rightarrow -\mathbf{r}$ introduced in Section 3.3.) The Rashba coupling is a standard form, discussed in more detail around Eq. 3.40, that is believed to be a reasonable model for the dominant spin-orbit coupling in adsorbed graphene.

A topological phase survives, but is strongly modified by the Rashba term, consistent with the physical picture of even and odd classes above. For a system with

S_z conserved, there are many phases labeled by an integer n, as in the IQHE: if spin-up electrons are in the $\nu = n$ state, then spin-down electrons must be in the $\nu = -n$ state by time-reversal symmetry, where the sign indicates that the direction of the effective magnetic field is reversed. Once S_z is not conserved because $\lambda_R \neq 0$, there are only two categories, the ordinary and topological insulators. The decoupled $\nu = \pm n$ cases with S_z conserved can be adiabatically connected (i.e., without closing the gap), once S_z is not conserved, to the ordinary insulator for even n and to the topological insulator for odd n, at least once disorder is allowed. This continuity is described using properties of band structures in Section 4.4.

This completes our outline of two-dimensional insulating systems. This kind of topological insulator was observed by a transport measurement in (Hg,Cd)Te quantum wells (Koenig et al., 2007), building on an analysis in terms of the Bernevig-Hughes-Zhang tight-binding model (Bernevig et al., 2006). A simplified description of this experiment is that it observed, in zero magnetic field, a two-terminal conductance $2e^2/h$, consistent with the expected conductance e^2/h for each edge if each edge has a single mode, with no spin degeneracy. More recent work has observed some of the predicted spin transport signatures as well, although as expected the amount of spin transported for a given applied voltage is not quantized, unlike the amount of charge. Some intuition for why this particular material shows the quantum spin Hall effect is provided by noting that strained HgTe is an inverted band gap semiconductor: one can imagine that as spin-orbit coupling increases a band moves from the conduction band to the valence band. This is not alone sufficient for the topological phase, as an even number of band inversions would return the system to the ordinary class (as is believed to happen in PbTe), but in HgTe calculations of the type described in Chapter 4 show that material is indeed topological.

3.6 3D Topological Insulators: Basic Phenomena

The notion of a 2D topological insulator as having half of a quantum wire at each edge leads naturally to a 3D generalization, and this 3D topological insulator phase has seen an explosion of interest since its experimental observation in 2008. The existence of a 3D topological phase came as something of a surprise, as the integer quantum Hall effect was known not to have a genuinely 3D analogue: a three-dimensional system in a magnetic field can in principle be put into layered quantum Hall states, with chiral metals on their sides (Chalker and Dohmen, 1995). In this section we sketch a picture of the 3D topological phase and some of its key properties, particularly its surface state, deferring a full theoretical analysis of the bulk until the following chapter. For readers preferring a tight-binding model like the above, it is possible to imitate the two-dimensional case by starting from a bulk semimetal, namely, the tight-binding model on the diamond lattice, and adding an appropriate spin-orbit term to open a topological gap (Fu et al., 2007).

Let us start by trying to create a surface state that is half of an ordinary 2D metal, in the same way as we previously were led to consider half of a spin-independent quantum wire. In the wire, we can count branches of the energy-momentum relation as in Figure 3.6. At a surface, we instead count sheets of the Fermi surface. To understand what this means, start by thinking about an ordinary 2D metal with time-reversal symmetry. Consider a Hamiltonian consisting of a quadratic spin-independent part and a Rashba spin-orbit part:

$$H = \frac{p^2}{2m} + \lambda(\mathbf{p} \times \boldsymbol{\sigma})_z. \tag{3.40}$$

The effect of the Rashba term is to spin-split the Fermi surface; the energy-momentum plot along a line of momenta in the 2D Brillouin zone now has the form of two displaced parabolas (Figure 3.9). However, at every energy the Fermi surface still consists of an even number of sheets (either two or zero), which in this case are circles. We could equally well, for the purposes of this discussion, consider Dresselhaus spin-orbit coupling of the form $\mathbf{p} \times \boldsymbol{\sigma}$, which would also lead to two sheets of the Fermi surface, now with inward-directed and outward-directed spins.

It is believed that any strictly two-dimensional band structure with time-reversal symmetry will have this property: including spin, the Fermi surface contains an even number of sheets. We could break time-reversal symmetry by applying

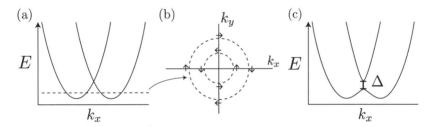

Fig. 3.9 Band structure of a two-dimensional electron gas with Rashba spin-orbit coupling (Eq. 3.40). (a) Energy-momentum relation along a cut with $k_y = 0$: the bands consist of two displaced parabolas describing opposite spin states. Note that opposite momenta will have opposite spin states as required by time-reversal symmetry. (b) At a generic value of Fermi energy (dotted line) above the band minimum, the Fermi surface consists of two sheets, which for this model are circles around which the spin direction precesses. (c) Applying a Zeeman magnetic field $h\sigma_z$ is one way to break time-reversal symmetry: it induces a gap Δ at $\mathbf{k} = 0$ of magnitude proportional to h, and for energies in this gap, there is only a single Fermi surface sheet.

a Zeeman magnetic field, as shown in Figure 3.9b. Now consider instead the linear-in-momentum Hamiltonian

$$H = \lambda(\mathbf{p} \times \boldsymbol{\sigma})_z. \tag{3.41}$$

This Hamiltonian is not bounded below, so we would not accept it as the ultimate Hamiltonian of a system, but it is nevertheless a reasonable description of the surface states in a bulk band gap of a topological insulator (Figure 3.10). The relativistic dispersion relation with zero effective mass is similar to that in graphene, which is a strictly 2D material (a layer of carbon atoms, whose low-energy electronic structure is well described by a nearest-neighbor tight-binding model on the honeycomb lattice discussed in this and the previous chapter).

However, graphene has such a relativistic dispersion relation around the *two* inequivalent Dirac points, $\pm\mathbf{K}$, in the Brillouin zone, Section 2.5.1. It also has a twofold spin degeneracy at each value of momentum if spin-orbit coupling is neglected, and weak spin-orbit coupling splits this degeneracy but keeps the number of sheets of the Fermi surface even. Having an odd number of sheets of the Fermi surface with time-reversal symmetry is the defining feature of the 3D TI surface state, which appears at the boundary between two topologically inequivalent bulk band structures (e.g., a TI material and vacuum). Another way this difference is stated is that in a 3D TI surface state, the Fermi surface encloses an odd number of Dirac points, while in a generic 2D system it encloses an even number. One reason that one- or two-dimensional systems with an odd number of Fermi surface sheets are interesting, whether arising either via topology or via an applied magnetic field, is that they can serve as precursors for topological superconductivity as discussed in Chapter 9.

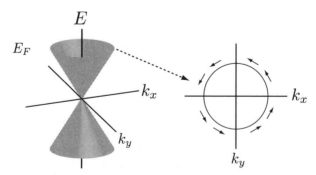

Fig. 3.10 Simplified band structure of a 3D topological insulator. The Dirac cone of surface states (left) terminates above and below in bulk electronic states. Only the surface states associated with one surface are shown, and there is a single spin direction at each value of momentum. There is a single Fermi surface sheet in the bandgap. Angle-resolved photoemission spectroscopy of Bi_2Se_3, showing this structure within the bandgap, is shown in Figure 1.2.

Just as in the two-dimensional case, there is an odd-even effect. One topological insulator surface in isolation cannot be localized by disorder that respects time-reversal; it can be viewed as an ordinary two-dimensional metal which was stabilized by being split into two halves. However, two surfaces hybridized together, as in a thin film, can localize or develop a gap and are not very different from a normal two-dimensional metal with strong spin-orbit coupling. The "factor of 2" difference between a topological insulator surface state and a standard 2D metal, in the sense of having one Dirac cone or two, respectively, has remarkably far-reaching consequences beyond the stability to disorder. Three examples that will appear later in this book are the quantized magnetoelectric effect, unusual magnetotransport phenomena, and novel superconducting proximity effects.

As in the two-dimensional case, these properties emerge from a bulk topological invariant of $\mathbb{Z}_2$ type. Strictly speaking there are four bulk invariants of a 3D time-reversal-invariant band structure (Moore and Balents, 2007; Fu et al., 2007; Roy, 2009), of which three correspond to layered versions of the 2D topological insulator and the fourth is genuinely three-dimensional and responsible for the surface state described here. We postpone the analysis of the bulk invariants to Chapter 4 because they are substantially more complicated than the two-dimensional case. In the two-dimensional case, we could get into the topologically nontrivial phase by essentially separating electrons into up spin and down spin, with S_z commuting with the Hamiltonian. Then two oppositely directed copies of the IQHE provided a simple realization of the quantum spin Hall state or 2D topological insulator, and the challenge was to show that this state is stable to more realistic spin-orbit coupling. In three dimensions, getting into the fully 3D topological phase requires starting with the vector nature of spin-orbit coupling from the very beginning, as no model with S_z conservation realizes the nontrivial value of the invariant (Chapter 4).

Since time-reversal symmetry is required for the topological insulator phase to be well defined, it is natural to ask how the odd-even distinction is violated by breaking time-reversal. Adding a small Zeeman magnetic field $h\sigma_z$ to the Hamiltonian in (3.40) opens up a gap at the Γ point ($\mathbf{p} = 0$) but not elsewhere, meaning that there is now a range of energies with a single sheet of the Fermi surface. This effect is visible in transport measurements (Quay et al., 2010) and has become very important as a means to replicate some of the physics of topological insulator surface states in the quest to create unconventional superconductors with Majorana excitations (Chapter 9).

What about adding a magnetic field to the model topological insulator surface state Hamiltonian (3.41)? Remarkably, starting with one Dirac cone and applying a strong magnetic field gives a half-integer quantum Hall effect

$$\sigma_{xy} = (n + 1/2)e^2/h, \quad n \in \mathbb{Z} \tag{3.42}$$

rather than an integer quantum Hall effect, so the odd numbers of Dirac fermions at the surface of a 3D topological insulator mean that the overall surface Hall effect is half-integer (Fu and Kane, 2007). This might seem of minor interest as the two surfaces measured in parallel in an experiment on a slab will still give an integer result overall. However, understanding the deeper significance of this surface half-integer led to both a more general picture of the topological insulator phase and a better understanding of magnetoelectric effects in all materials.

While the quantum Hall effect was defined in terms of an observable physical response (the Hall effect) that is independent of whether the underlying system is made up of independent or interacting electrons, the band-structure definitions of the topological insulator are based on an independent-electron picture. However, the half-quantized surface Hall effect is a valid characterization of the phase by observable quantities. It is nevertheless more subtle than the quantum Hall effect, which is a property of the unperturbed system, because here a time-reversal-breaking perturbation was required to put the surface state in the quantum Hall regime.

As motivation for the reader to invest in learning the theoretical underpinnings covered in Chapter 4, here are a few questions raised by the above physical pictures of the surface. We emphasized that the edge states and surface states that were the focus of this chapter are independent of surface details, so they must be forced to exist by some bulk property. What bulk property determines whether a given material falls in the topological or ordinary category? The answer in every case is a topological invariant, which can be expressed as an integral of the Berry phase of the electron wavefunctions. That raises the question of what physical significance these integrals might have when they are not quantized, for example, for the partially filled bands of a metal. Often that physical significance turns out to be related to the basic question of how the topological phases discussed in this chapter, and even ordinary nontopological materials, respond to applied electromagnetic fields.

3.7 Skyrmions in the Quantum Hall Effect

The physics of skyrmions in spin-polarized quantum Hall states lies at the intersection of several threads of this book. They appear as excitations of a phase which displays at the same time topological and conventional order; they generate a Berry flux through a twist of an internal order parameter; and they are themselves topological objects. Hence their study combines many of the concepts introduced in Chapter 2. They are also unusual in that they were theoretically predicted to exist (Sondhi et al., 1993) before their experimental discovery (Barrett et al., 1995), a situation which was (and is) not all that common in quantum Hall physics, for the reasons discussed in Section 5.1.

3.7.1 Multicomponent Quantum Hall Systems and Flat-Band Ferromagnetism

In the case of electrons with an internal degree of freedom, such as its spin, there is a copy of each Landau level for each internal state. Even in the case of vanishing Zeeman coupling, when there are separate identical Landau levels for each spin direction, a filled Landau level ($\nu = 1$) is a ferromagnet and hence spontaneously breaks spin rotational symmetry, for reasons which we discuss in the following paragraphs. The SU(2) invariance of a vanishing Zeeman splitting despite the presence of a strong applied field in the quantum Hall regime is not as outlandish as it sounds. This can happen because the Zeeman coupling $g^*\mu_B B$ can vanish as a result of the vanishing effective electronic g-factor g^*. Often quite small at the outset, it can even be arranged to vanish exactly in some semiconductor heterostructures by applying external pressure, which changes g^* as a result of spin-orbit coupling.

The origin of this ferromagnetism is conceptually very simple, following from Pauli's principle that the a many-body wavefunction of fermions must be antisymmetric under particle exchange. Consider a pair wavefunction, written as a product of spatial and spin wavefunctions:

$$\Psi_{\sigma_1,\sigma_2}(\mathbf{r}_1, \mathbf{r}_2) = \psi_{\sigma_1,\sigma_2} \otimes f(\mathbf{r}_1 - \mathbf{r}_2) \otimes F(\mathbf{r}_1 + \mathbf{r}_2), \qquad (3.43)$$

where the center of mass wavefunction F is unimportant for interaction effects. Choosing an antisymmetric relative spatial wavefunction f suppresses the probability for two electrons to be near each other, as it vanishes for vanishing separation $\mathbf{r}_1 - \mathbf{r}_2$. In the presence of a repulsive Coulomb interaction, this is energetically favorable. To keep the total wavefunction antisymmetric then requires a symmetric spin wavefunction ψ_{σ_1,σ_2}. This is just the triplet, that is, a wavefunction corresponding to total spin $S_{\text{tot}} = 1$, while the singlet $S_{\text{tot}} = 0$ is antisymmetric under particle exchange.

The reason this is called flat band ferromagnetism is that putting a node into a spatial wavefunction normally exacts a *kinetic* energy cost – just recall a particle in a box, the energy of an eigenfunction of which grows with the number of its nodes. For this reason, the majority of magnetic compounds are in fact antiferromagnets.

Here, this is different as the special feature of quantum Hall systems is their Landau level structure – all wavefunctions in a Landau level have the same kinetic energy, and hence the node can be put in without a kinetic energy penalty. Another familiar example, incidentally, of Coulomb-driven ferromagnetism is encoded in one of Hund's rules, stating that electrons in a partially filled shell of an atom, that is, all of which occupy symmetry-related orbitals with the same kinetic energy, have their spins aligned.

3.7.2 Exchange-Enhanced Zeeman Splitting

A more detailed discussion of the two-particle problem is given in the section of the fractional quantum Hall effect (Section 5.1). In particular, the above argument would of course demand the pair wavefunction not just to vanish linearly but with a power as high as possible. For $\nu = 1$, the highest possible power, however, turns out to be 1, hence the ferromagnetism is captured correctly by this picture. In the absence of an internal degree of freedom, a filled Landau level of noninteracting electron gives rise to an incompressible state, that is, a state at which the chemical potential jumps as a function of filling on account of the cyclotron gap $\hbar\omega_c$. By contrast, for electrons with an internal degree of freedom, such as a spin, a priori no such gap exists. This happens only once the Landau level for each flavor (say, up and down spin) are filled. Thus, in the presence of flat-band ferromagnetism for interacting electrons, the system will be gapless at $\nu = 1$ for SU(2) spin symmetry (i.e., $E_Z = 0$) on account of the Goldstone mode corresponding to global rotations of the orientation of the ferromagnetic order.

The long-wavelength description of such a ferromagnet in a field is provided by a nonlinear σ-model (see also Eq. 2.70) based on the order parameter manifold, as introduced in Chapter 2. This is supplemented by a Zeeman energy term due to the applied field:

$$E = E_\sigma + E_Z = \frac{\rho_s}{2} \int d^2r |\nabla \mathbf{m}|^2 - g^* \mu_B B \nu \int d^2r\, m^z. \qquad (3.44)$$

Here, the stiffness arises exclusively from the electron interactions. Analogously to the steps in Section 5.1.3 one finds in the lowest Landau level that

$$\rho_s = \sum_{q \neq 0} V(q)|q^2| \exp(-|q|^2/2) \qquad (3.45)$$

is given solely in terms of the Fourier transform $V(q)$ of the electrons' real space interaction potential, $\mathcal{V}(r)$.

In the presence of a symmetry-breaking field, $g^* \neq 0$, the Zeeman splitting, $g^* \mu_B B$ per electron, and therefore the energy gap between spin-split Landau levels, is therefore enhanced by the exchange field, $\propto \rho_s$, of the neighboring aligned spins.

The nonlinear σ-model describes ferromagnetic spin wave excitations, including a gapless Goldstone mode for $g^* = 0$: the universal properties are in the first instance those of an ordinary ferromagnet. However, things will turn out to be much richer as the topological structures of the quantum Hall problem stabilize an excitation topologically distinct from spin waves.

3.7.3 Skyrmions: Charged Topological Spin Textures

An interaction-generated gap in common parlance is taken as the feature distinguishing the FQHE from the IQHE. In that sense, for a multicomponent system, $\nu = 1$ can be thought of as a FQHE as the gap is eventually an interaction effect: in the presence of spontaneous symmetry breaking, otherwise negligible anisotropies become important. This has led to the statement that, as far as the QHE at $\nu = 1$ is concerned, "one is a fraction, too."

So, what is the nature of the charged excitations? These can take the form of skyrmions, topologically stable spin textures whose name was imported from Skyrme's model in nuclear physics.

The starting observation explaining the appearance of skyrmions is that the quantum Hall effects tend to occur whenever flux and charge are commensurate, that is to say when ν is an integer or a simple rational number (this notion will be made more concrete in the framework of Chern–Simons theory in Chapter 6). Therefore, starting from a quantum Hall state, adding or removing a charge interferes with this commensurability, and the system is left with an excitation.

As we discuss next, the topological charge of the skyrmion can be used to reestablish the commensurability in an elegant way. To see this, let us first present a topological classification of excitations of a ferromagnet in $d = 2$, which can be done using the tools developed in Section 2.7.3. Imposing the "compactifying" boundary condition that, infinitely far from the origin, all the spins point along a given direction, amounts to identifying all points "at infinity," or alternatively turns the infinite plane R^2 into the two-dimensional surface of a sphere, S^2. In Figure 3.11, the spin direction "at infinity" is chosen to be along the applied field if one is present, in order to maintain a finite Zeeman energy.

The order parameter manifold – the set of directions in which the magnetization vector $\mathbf{m}$ can point – itself can be described as the surface of a sphere as in Section 2.6. Hence, homotopy theory tells us that the possible order parameter configurations can be classified by an integer topological invariant n_Q of mapping from the surface of a sphere to itself:

$$\pi_2(S^2) = \mathbb{Z}. \tag{3.46}$$

A change in the local orientation of $\mathbf{m}$ induces a Berry flux in the following way. Imagine an electron moving in the plane, with its spin adiabatically following the local magnetization direction. The spin direction of the electron thus describes a trajectory on the sphere, just as described in Section 2.1, and the electron is thus subject to a Berry connection $\mathcal{A}$, generated by the spin texture, in addition to the vector potential A of the applied magnetic field. The Berry connection from the spin texture and the vector potential of the applied magnetic field are not just

Fig. 3.11 A magnetic skyrmion is a texture in two space dimensions of a three-dimensional magnetization field. The magnetization at the center of the skyrmion points in a direction opposite to the distant points "at infinity." In between, the direction changes smoothly, tilting away from the vertical axis along radial directions, and winding like an XY vortex in the horizontal plane in the azimuthal direction.

analogous but combine to affect the electron in effectively the same way, since rotating the direction of spin around a closed path produces exactly the same U(1) phase effect on the wavefunction as moving the electron around a real-space path in a field.

The quantum Hall ferromagnet thus has the option of reinstating flux-charge commensurability by generating a texture in **m** so that the total charge precisely equals

$$Q = \nu \int d^2r \, \nabla \times (\mathcal{A} + A). \tag{3.47}$$

(Note that in $2d$, a magnetic field, being the curl of a two-component vector potential, is a pseudoscalar and can thus be identified with a charge.) The topological charge described above is thence simply proportional to the total electronic charge of the spin texture. In particular, (anti-)skyrmions have topological charge ± 1; skyrmions can reinstate the commensurability required by the topological nature of the QHE thanks to the – conceptually entirely separate – topological stability of their charge.

None of these topological considerations address issues of energetics, that is, the question which charged excitation is actually the cheapest. For quantum Hall skyrmions, there are three contributions which need to be considered in the first instance: the cost of the twist E_σ, and the Zeeman anisotropy cost E_Z from Eq. 3.44, as well as the Coulomb energy E_Q of the charge distribution $Q(r) = \nu \nabla \times \mathcal{A}(r)$.

Of all the possible topologically equivalent spin textures, an entire family gives a minimal $E_\sigma = 4\pi\rho_s$, namely, those which can be parameterized by spinors c

(such that $\mathbf{m} = c^{\dagger}\boldsymbol{\sigma}c$, $\boldsymbol{\sigma}$ being the vector of Pauli matrices), whose entries are holomorphic functions:

$$c(z) = \begin{pmatrix} f(z) \\ g(z) \end{pmatrix},$$ (3.48)

where $f(z)$ and $g(z)$ have n_Q zeroes each, and $\lim_{z\to\infty} f(z)/g(z) = 0$ imposed by our boundary condition. Placing the zero of $f(z)$ at the origin, and that of $g(z)$ at $z = \infty$, yields a charge $n_Q = 1$ skyrmion sketched in Figure 3.11:

$$c = \begin{pmatrix} \lambda \\ z \end{pmatrix},$$ (3.49)

where λ parameterizes the "size" of the skyrmion in the sense that the spins point in the plane perpendicular to the quantization direction at a distance of λ from the origin.

While the Coulomb repulsion is optimized for a charge which is maximally spread out, the Zeeman energy favors compact skyrmions, as these involve flipping spins over a smaller area. The trade-off between these yields a texture which is no longer perfectly holomorphic.

In experiment, the net polarization of the electron system can be probed by nuclear magnetic resonance (NMR). For the GaAs quantum wells on which the first such experiments were done, the Knight shift of a ^{71}Ga nucleus is sensitive to the electronic spin polarization via a hyperfine interaction. It can thus probe the

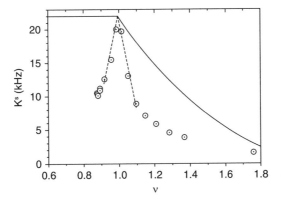

Fig. 3.12 The Knight shift K^s as a function of filling factor ν in a GaAs quantum well (Barrett et al., 1995). The fitted dashed line corresponds to a particles with 3.6 times the spin of a noninteracting electronic quasiparticle, denoted by the solid line. This indicates the presence of extended topological spin textures, skyrmions, as low-energy excitations near $\nu = 1$. Reprinted with permission by *Physical Review*.

reduction of the magnetization density due to the excitations forced into the system by deviating from filling $\nu = 1$.

Thus, it was inferred that the effective spin of an excitation is about 3.6 times that of a noninteracting electron, as depicted in Figure 3.12. This confirmed the prediction that skyrmions are indeed the relevant low-energy excitations in this regime.

4

Geometry and Topology of Wavefunctions in Crystals

One of the highlights of a first course in solid-state physics is the demonstration that important transport and optical properties of a solid can be obtained using the concept of band structure. When all energy bands are either completely filled or completely empty, the solid is insulating; when at least one band is partially filled, the solid is metallic. This description is remarkably successful, although it fails to capture many-electron phenomena such as the Mott insulator.

The point of this chapter, speaking in generalities, is to start explaining how simple band theory is incomplete even at the one-electron level, at a more microscopic level than in the previous chapter. Many important properties of solids do not follow simply from counting electrons and knowledge of $E_i(\mathbf{k})$, the dependence of the energy in band i on wavevector $\mathbf{k}$. The Bloch wavefunctions, not just the energies, are well known to be important for optical transition matrix elements and atomic-scale properties, but one might think that a first description of low-frequency, long-wavelength processes would not need knowledge of these details. For example, the semiclassical dynamics for metallic electrons, at least in the form often presented in textbooks, does not involve the wavefunctions at all, only $E_i(\mathbf{k})$.

It is not very clear on a first exposure to topological phases, such as that in the preceding chapter, how they are connected to a student's first semester in the theory of solids. The rewards for addressing that connection by reconsidering how wavefunctions enter into the physics of a solid are considerable. Through nearly four decades of developments since the seminal papers of Thouless and others in the early 1980s, it has become clear that many important properties of solids crucially depend on wavefunction geometry.

By wavefunction geometry, we mean the geometrical quantities such as Berry phases that result from how the electronic states evolve with momentum. Even familiar ground-state properties such as electrical polarization turn out to be determined by the Berry phases (see Section 2.1) of the Bloch wavefunctions. Indeed the Berry phase contribution is not just a correction or detail, but often the whole story.

The first parts of this chapter try to give an accessible picture of why wavefunction geometry must be important for some simple properties of solids, while the later parts, which can certainly be skipped by readers not interested in explicit formulas, explain in more detail how the states described in Chapter 3, and some generalizations, arise microscopically.

The first example we discuss is how the breaking of inversion symmetry in many materials, and consequent phenomena such as electrical polarization, is manifest in the electronic wavefunctions but not the band energies. This leads to the modern theory of polarization, and we note that a recent book by Vanderbilt discusses that theory and some other topics of this chapter in greater detail (Vanderbilt, 2018). We then return to the integer quantum Hall effect in Section 4.2 and give a microscopic picture of how topology appears: geometrical quantities such as the Berry phase can give rise to topological invariants when integrated over closed manifolds.

Section 4.3 discusses a first example of how many-electron systems can be described in terms of topological invariants very similar to those in the one-electron theory, and in passing indicates how disorder might be treated as well. Topological insulators and their associated $\mathbb{Z}_2$ topological invariants are the subject of Section 4.4. Again the approach taken to topological quantities is to express them as integrals over the Brillouin zone, or parts of it, which provides a unified framework that is closely connected to physical observables via expressions appearing in standard perturbation theory for responses.

We close the discussion of topological aspects of electron wavefunctions in insulators with the striking result of how the non-Abelian Berry phase in multiband systems determines part, and in some materials all, of the magnetoelectric response, that is, the electrical polarization induced by a magnetic field, or the magnetic moment induced by an electric field. The present chapter is focused on band insulators (and in some cases their conducting surfaces), while the role of Berry phases and other geometrical quantities in bulk metals and semimetals forms a major part of Chapter 7. The gauge structure of wavefunctions in a crystal and of electromagnetic fields turn out not just to be mathematically similar; in multiple cases, in both insulators and metals, the microscopic origin of electromagnetic response at low energy is determined by the gauge fields and related structures in the Brillouin zone.

The appearance of geometric quantities in the theory of solid-state properties like polarization and Hall effect can be viewed in mathematical terms as showing that remarkable simplifications of perturbation theory occur in certain limits. Symmetry remains a good guide to which effects are allowed to be nonzero in a crystal, but the emergence in many problems of geometric quantities such as the Berry curvature of Bloch states turns out to give new insights into the source of these effects. The most remarkable feature is that, in a subset of these problems, measured quantities become not just geometric but topological, in the sense of being robust to small perturbations. We explain the origin of topology in this section in noninteracting crystals, for the most part, and then discuss new features from interactions and disorder in later chapters; for disordered systems (Chapter 8), more abstract mathematical frameworks (Kitaev, 2009; Ryu et al., 2010; Freed and Moore, 2013; Morimoto and Furusaki, 2013) are a powerful alternative to the relatively concrete homotopy approach taken in this chapter.

4.1 Inversion Symmetry, Electrical Polarization, and Thouless Pumping

First we introduce inversion symmetry as an example of how important properties of a material may not be reflected in its energy band structure $E_i(\mathbf{k})$. Suppose we want to construct a material with a bulk photovoltaic effect, that is, a material that supports a bulk current normal to the direction of incident unpolarized light. For example, sunlight incident on a thin film of such a material would produce a current that could be used for power generation. What are the symmetry requirements on such a material? A clue is provided by how *two* materials are used in most existing photovoltaic devices: two differently doped versions of silicon, for example, are put next to each other to form a p-n junction. The direction of the current produced in this device is determined by how spatial symmetries were broken through the arrangement of the two materials.

Ordinary silicon by itself cannot support a bulk photovoltaic effect (BPVE): there are too many symmetries in its crystal structure, which mean there is not a unique direction in which current can flow under unpolarized light. There are other crystals, however, where the arrangement of atoms allows a preferred direction along which the current can flow. Think, for example, of an arrangement of dipolar molecules with all dipoles pointing in the same direction. We refer to a crystal as noncentrosymmetric or inversion-breaking if it does not have a center of inversion, and polar if furthermore its symmetries allow a nonzero vector. Gallium arsenide is an example of a crystal that is inversion-breaking but not polar, as it has too many rotational symmetries to allow one direction (vector) to be distinguished uniquely.

The analysis of crystalline symmetries such as inversions, rotations, and translations is a major subject in any solid-state text. It may be worth reminding the reader of the distinction between the point group, which contains symmetries that fix at least one spatial point, and the space group, which also includes transformations with a translational aspect that do not fix any point. A crystal in which not every symmetry can be decomposed into the product of a point group transformation and a translation is called nonsymmorphic. Perhaps the simplest way to picture a nonsymmorphic symmetry is to imagine a line of footprints and realize that the line has a glide reflection symmetry (the product of a mirror and a translation), even though there is no mirror symmetry. A key difference between spatial symmetries and the time-reversal symmetry discussed in Chapter 3 is that spatial symmetries are represented by unitary operators, while the time-reversal operation is antiunitary (i.e., is the product of a unitary symmetry and charge conjugation). We will see the interaction between symmetry and topology recur many times in this book, and an abstract analysis beyond the single-particle picture is in Chapter 11.

Returning to inversion, we would like to present a puzzle to explain why there must be a key ingredient other than band structure involved in determining whether or not a material breaks inversion. Consider a nonmagnetic material, that is, one which does not break time-reversal symmetry. It turns out that any such material will have an apparent inversion symmetry in its band structure, because if there is a Bloch state of energy E and wavevector $\mathbf{k}$, then applying time-reversal leads to another state with the same energy and wavevector $-\mathbf{k}$. In other words, the energy band diagram will appear to be even, meaning symmetric under $\mathbf{k} \leftrightarrow -\mathbf{k}$, whether or not the microscopic crystal breaks inversion.

Electrical polarization, at least classically, depends on broken inversion symmetry. So there must be some ingredient other than the energy bands which can tell us whether the system has a polarization. One can guess that the answer must involve the detailed structure of the wavefunctions. For comparison, it is useful to think of optical transitions within an atom, which in the electric dipole approximation depend on a change in inversion symmetry between the initial and final electronic states. The matrix elements for the transition are determined by the wavefunctions, and it makes sense that something about the wavefunction would also control electrical polarization.

We will show in a one-dimensional example that, rather than involving differences in the wavefunction between bands at fixed k as in optical transitions, the polarization of an occupied band is determined by how the wavefunctions of that band evolve smoothly with k. In fact, the same mathematical structures will emerge as in the example of the time-dependent wavefunction under adiabatic nondegenerate evolution in Section 2.1. So, while the energetics of Bloch wavefunctions underlies many properties of solids, there are also properties arising from the variation of a wavefunction's periodic part u_k with k, and these were understood only rather recently. This understanding works without modification in higher dimensions and has become the standard way to understand and compute polarization of crystals.

Our approach is to start with the simplest case of a one-dimensional crystal and present a mathematical object made from the u_k. We then show that this object, which is an example of the Berry phases discussed in Chapter 2, has some properties that allow it to be identified as the electrical polarization. The precise sense in which this quantity governs physical measurements of polarization is a subtle topic, for which we encourage the reader to look at Vanderbilt (2018).

We start from the observation that, even for a one-dimensional crystal, there is a nontrivial closed loop in momentum k because of the periodicity of the first Brillouin zone $k \in [-\pi/a, \pi/a]$. We can use this loop to define a Berry phase (Eq. 2.9) made from the variation of the Bloch wavefunctions with momentum:

$$\gamma = \oint_{-\pi/a}^{\pi/a} \langle u_k | i \partial_k | u_k \rangle dk = \oint_{-\pi/a}^{\pi/a} \mathcal{A}_k. \tag{4.1}$$

How are we to interpret this Berry phase physically, and is it even gauge-invariant? We will discuss its meaning from ordinary perturbation theory below, but an intuitive clue is provided if we make the replacement $i\partial_k$ by x, as would be appropriate if we consider the action on a plane wave. This suggests, correctly, that this Berry phase may have something to do with the spatial location of the electrons, but evaluating the position operator in a Bloch state gives an ill-defined answer; for this real-space approach to work, we would need to introduce localized Wannier orbitals in place of the extended Bloch states.

Another clue to what the phase γ might mean physically is provided by asking if it is gauge-invariant. Before, gauge invariance resulted from assuming that the wavefunction could be continuously defined on the interior of the closed path. Here we have a closed path on a noncontractible manifold; the path in the integral winds around the Brillouin zone, which has the topology of the circle, but no interior. What happens to the Berry phase if we introduce a phase change $\phi(k)$ in the wavefunctions, $|u_k\rangle \to e^{-i\phi(k)} |u_k\rangle$, with $\phi(\pi/a) = \phi(-\pi/a) + 2\pi n$, $n \in \mathbb{Z}$? Under this transformation, the integral shifts as

$$\gamma \to \gamma + \oint_{-\pi/a}^{\pi/a} (\partial_k \phi) \, dk = \gamma + 2\pi n. \tag{4.2}$$

So redefinition of the wavefunctions shifts the Berry phase: the Berry phase is not fully gauge-invariant because it changes under large (non-null-homotopic) gauge transformations, that is, those that cannot be smoothly changed into the identity transformation. This will correspond to changing the polarization $P = (-e)\frac{\gamma}{2\pi}$ by a multiple of the polarization quantum, which in one dimension is just the electron charge. (In higher dimensions, the polarization quantum is one electron charge per transverse unit cell area.)

Physically this quantized ambiguity of polarization corresponds to the following idea: given a system with a certain bulk unit cell, there is an ambiguity in how that system is terminated and how much surface charge is at the boundary; adding an integer number of charges to one allowed termination gives another allowed termination (Resta, 1992). The Berry phase is not gauge-invariant, but any fractional part it had in units of the polarization quantum *is* gauge-invariant. Note that the above calculation suggests that, to obtain a gauge-invariant quantity, we need to consider a two-dimensional crystal rather than a one-dimensional one. Then integrating the Berry curvature, rather than the Berry connection, has to give a well-defined gauge-invariant quantity; indeed this is just the Chern number discussed in Chapter 3.

We now consider more microscopically the one-dimensional polarization. More precisely, we attempt to compute the *change* in polarization by computing the integral of current through a bulk unit cell under an adiabatic change of the bulk Hamiltonian described by some parameter λ:

$$\Delta P = \int_0^1 d\lambda \frac{dP}{d\lambda} = \int_{t_0}^{t_1} dt \frac{dP}{d\lambda}\frac{d\lambda}{dt} = \int_{t_0}^{t_1} j(t)\, dt. \qquad (4.3)$$

In writing this formula, we are assuming implicitly that there will be some definition of the infinitesimal dP induced by a small change of parameter. The insight is that such a polarization change can be computed not by reference to the surfaces but via the adiabatic current through every unit cell of the bulk. We write q for one-dimensional momentum and k_x, k_y for two-dimensional momenta in the following. We will use Bloch's theorem in the following form: the periodic single-particle orbitals $u_n(q, r)$ are eigenstates of

$$H(q, \lambda) = \frac{1}{2m}(p + \hbar q)^2 + V^{(\lambda)}(r). \qquad (4.4)$$

The current operator is

$$j(q) = (-e)v(q) = \frac{-ie}{\hbar}[H(q,\lambda), r] = \frac{-e}{m}(p + \hbar q) = \frac{-e}{\hbar}\partial_q H(q,\lambda). \quad (4.5)$$

The current at any fixed λ in the ground state is zero, but changing λ adiabatically in time drives a current that generates the change in polarization. To compute this current, we need to use the first correction to the adiabatic theorem; for this and additional physical context on the following derivation, see Vanderbilt (2018). Following Thouless, we choose locally a gauge in which the Berry phase is zero (this can only be done locally and is only meaningful if we obtain a gauge-invariant answer for the instantaneous current), and write for the many-body wavefunction

$$|\psi(t)\rangle = \exp\left(-(i/\hbar)\int_0^t E_0(t')\,dt'\right) \qquad (4.6)$$

$$\times \left[|\psi_0(t)\rangle + i\hbar \sum_{j\neq 0}|\psi_j(t)\rangle(E_j - E_0)^{-1}\langle\psi_j(t)|\dot\psi_0(t)\rangle\right]. \quad (4.7)$$

Here $E_i(t)$ are the local eigenvalues and $|\psi_j(t)\rangle$ a local basis of reference states. The first term is just the adiabatic expression we derived before, but with the Berry phase eliminated with a phase rotation to ensure $\langle \psi_0(t)|\dot{\psi}_0(t)\rangle = 0$.

We want to use the above expression to write the expectation value of the current. The ground state must differ from the excited state by a single action of the (one-body) current operator, which promotes one valence electron (i.e., an electron in an occupied state) to a conduction electron. Using the one-particle states, we get

$$\frac{dP}{d\lambda} = 2\hbar(-e) \operatorname{Im} \sum_{v,c} \int \frac{dq}{2\pi} \frac{\langle u_v(q)|v(q)|u_c(q)\rangle \langle u_c(q)|\partial_\lambda u_v(q)\rangle}{E_c(q) - E_v(q)}. \tag{4.8}$$

For example, we wrote

$$\langle \psi_j(t)|\dot{\psi}_0(t)\rangle = \sum_{v,c} \langle u_c|\partial_\lambda u_v\rangle \frac{d\lambda}{dt}. \tag{4.9}$$

This sum involves both valence and conduction states. For simplicity we assume a single valence state in the following. We can rewrite the sum simply in terms of the valence state using the first-order time-independent perturbation theory expression for the wavefunction change under a perturbation Hamiltonian $H' = dq \, \partial_q H$:

$$|\partial_q u_j(q)\rangle = \sum_{j \neq j'} |u_{j'}(q)\rangle \frac{\langle u_{j'}(q)|\partial_q H(q,\lambda)|u_j(q)\rangle}{E_j(q) - E_{j'}(q)}. \tag{4.10}$$

Using this and $v(q) = \frac{1}{\hbar}\partial_q H(q,\lambda)$ we obtain

$$\frac{dP}{d\lambda} = 2\hbar(-e) \operatorname{Im} \sum_c \int \frac{dq}{2\pi} \frac{\langle u_v(q)|v(q)|u_c(q)\rangle \langle u_c(q)|\partial_\lambda u_v(q)\rangle}{E_c(q) - E_v(q)}$$

$$= 2(-e) \operatorname{Im} \int \frac{dq}{2\pi} \langle \partial_q u_v(q)|\partial_\lambda u_v(q)\rangle. \tag{4.11}$$

We can convert this to a change in polarization under a finite change in parameter λ:

$$\Delta P = 2(-e) \operatorname{Im} \int_0^1 d\lambda \int \frac{dq}{2\pi} \langle \partial_q u_v(q)|\partial_\lambda u_v(q)\rangle. \tag{4.12}$$

The last expression is in two dimensions and involves the same type of integrand (a Berry flux) as in the 2D TKNN formula (Eq. 3.22). However, in the polarization case there does not need to be any periodicity in the parameter λ. If this parameter is periodic, so that $\lambda = 0$ and $\lambda = 1$ describe the same system, then the total current run in a closed cycle that returns to the original Hamiltonian must be an integer number of charges, consistent with quantization of the TKNN integer in the IQHE. This Thouless pumping is another example of how charge can be transported adiabatically and hence without dissipation, as in Laughlin pumping in the IQHE (Figure 3.2). Indeed, when we describe how to think about band structure in terms of fluxes through loops on a torus in Section 4.3, we could derive the Chern number's appearance in the quantum Hall effect as Thouless pumping of charge around a torus as flux is varied.

If we define polarization via the Berry connection (Eq. 2.9),

$$P = i(-e) \oint \frac{dq}{2\pi} \langle u_v(q)|\partial_q u_v(q)\rangle = \frac{-e}{2\pi} \oint dq\, \mathcal{A}, \qquad (4.13)$$

so that its derivative with respect to λ will give the result above with the Berry flux, we note that a change of gauge changes P by an integer multiple of the charge e. Only the fractional part of P is gauge-independent. The physical interpretation of this ambiguity is that it had to be there precisely because a closed cycle in parameter space can pump an integer number of charges through the unit cell as described above. Hence, if we know the precise history of the system, we can compute how much charge was pumped over that history. But if we are only given initial and final states, how much charge was pumped in going from one to the other is ambiguous by an integer multiple of a quantized amount. An example of a famous model of two bands where this can be computed as a useful exercise is in Box 4.1.

The relationship between polarization in 1D, which has an integer ambiguity, and the IQHE in 2D, which has an integer quantization, is the simplest example of the relationship between Chern–Simons forms in odd dimension and Chern forms in even dimension. The shift of the polarization by a quantized amount under large (non-null-homotopic) gauge transformations is related to quantization in one higher dimension.

Box 4.1 Tight-Binding Chain with Two Orbitals per Unit Cell

The canonical tight-binding model with nonzero polarization has two orbitals per unit cell along a chain, with unit cell of size a. We take these two orbitals to have on-site energies $E \pm \Delta$, and taking $E = 0$ henceforth simply amounts to shifting the origin of energy. We need to specify the real-space tight-binding Hamiltonian at Bravais lattice vectors, following our discussion in Section 2.4:

$$H(0) = \begin{pmatrix} \Delta & t+\delta \\ t+\delta & -\Delta \end{pmatrix}$$

$$H(a) = \begin{pmatrix} 0 & 0 \\ t-\delta & 0 \end{pmatrix}, \quad H(-a) = \begin{pmatrix} 0 & t-\delta \\ 0 & 0 \end{pmatrix}. \qquad (4.14)$$

The resulting k-space Hamiltonian is

$$H_k = \begin{pmatrix} \Delta & 2t\cos(ka/2) + 2i\delta\sin(ka/2) \\ \text{h.c.} & -\Delta \end{pmatrix}, \qquad (4.15)$$

with eigenvalues

$$E = \pm\sqrt{4t\cos^2(ka/2) + 4\delta^2\sin^2(ka/2) + \Delta^2}. \qquad (4.16)$$

To understand why the eigenvalues have this form, it is useful to think of expanding the Hamiltonian over Pauli matrices as $H_k = \mathbf{n}_k \cdot \boldsymbol{\sigma}$, and then noting that the eigenvalues are $\pm|\mathbf{n}_k|$.

As we have described the model, it is usually known as the Rice-Mele model (Rice and Mele, 1982). It is a useful exercise to work through the calculation of polarization in this model and compare the results with those in the seminal work of King-Smith and Vanderbilt (1993). If the on-site potentials are the same, the model (with electron-phonon interactions added) was studied in the context of polyacetylene, and is known as the Su–Schrieffer–Heeger (SSH) model (Su et al., 1979). There are two inequivalent polarization states in that inversion-symmetric model, depending on whether odd or even bonds are stronger. They are distinguished in a finite geometry with an even number of sites by one having a bound state at the ends; this is possibly the simplest example of an edge state.

We will say more in the next chapter about the mathematics of the Chern form, which can be viewed as underlying both changes in polarization and the integer quantum Hall effect. The Chern form in the case above (and others to be discussed) came originally from the Berry phases of a band structure, but we will later see how to reinvent it for interacting or disordered systems, for which band structure is not a well-defined concept. Now is probably a good time to note that the Berry curvature is not the only geometric property with physical significance that can be made from Bloch wavefunctions. Very similar to the Berry curvature is the orbital magnetic moment, which also involves the Bloch Hamiltonian and makes an appearance in Chapter 7.

Another physically relevant quantity is the quantum metric or Fubini–Study metric. Define the (Abelian) quantum geometric tensor for band α as

$$T_{\alpha ij} = \langle \partial_i u_\alpha | 1 - \mathcal{P}_\alpha | \partial_j u_\alpha \rangle. \qquad (4.17)$$

The imaginary part of this tensor is antisymmetric and just a constant times the Berry curvature. The real part is symmetric and can be viewed as a metric tensor measuring how much the overlap of Bloch states is reduced from unity when they are at slightly different momenta. This quantity turns out to be related to the spatial spread of Wannier orbitals in similar fashion to how the Berry phase is connected to their center of mass, and additional physical consequences of wavefunction geome-try continue to emerge. We now turn to the most established example, that of Chern number in the integer quantum Hall effect.

4.2 The Integer Quantum Hall Effect and Topological Invariants of Energy Bands

In this section we will give several different ways to understand the TKNN integer or Chern number described in Chapter 3. First, a useful trick for many purposes is to define the Berry flux and first Chern number in a manifestly gauge-invariant way, using projection operators. This is also helpful conceptually: if the Chern number is physical and gauge-independent, that is, independent of what phase factor we choose (or, in the case of a set of bands, independent of what basis we choose), why should we have to introduce a gauge or basis choice to calculate it? For the case of a single nondegenerate band, define $P_j = |u_j\rangle\langle u_j|$ at each point of the Brillouin zone. This projection operator is clearly invariant under U(1) transformations of u_j. The Chern number can be obtained as (Avron et al., 1983)

$$n_j = \frac{i}{2\pi} \int \mathrm{Tr}\left[(\partial_{k_x} P_j) P_j (\partial_{k_y} P_j) - (k_x \leftrightarrow k_y)\right] dk_x\, dk_y. \qquad (4.18)$$

It is a straightforward exercise to verify that this reproduces the TKNN definition (Eq. 3.22).

Then the generalization to degenerate bands, for example, is naturally studied by using the gauge- and basis-invariant projection operator $P_{ij} = |u_i\rangle\langle u_i| + |u_j\rangle\langle u_j|$ onto the subspace spanned by $|u_i\rangle$ and $|u_j\rangle$: the index of this operator gives the total Chern number of bands i and j. In general, when two bands come together, only their total Chern number is defined. The total Chern number of all bands in a finite-dimensional band structure (i.e., a finite number of bands) is argued to be zero below. Often one is interested in the total Chern number of all occupied bands because this describes the integer quantum Hall effect through the TKNN formula; because of this zero sum rule, the total Chern number of all *unoccupied* bands must be equal and opposite.

A similar trick is used to calculate polarization (up to its ambiguity, which is physical) without having to introduce gauge-invariant quantities. Suppose that we discretize the Brillouin zone and calculate the Berry phase integral appearing in the polarization through a series of overlaps of the Bloch states obtained at each momentum, as in Eq. 2.16. Then the logarithm in that equation is independent of the particular phase factors or basis chosen, except for shifts by 2π, which correspond to shifts by the polarization quantum. For some of the non-Abelian Berry phase quantities defined later, it is much more challenging to find efficient computational methods to compute them in practical electronic structure methods, although there has been recent progress with methods such as hybrid Wannier functions (Vanderbilt, 2018).

In the remainder of this section, we describe for mathematically inclined readers a powerful homotopy argument of Avron, Seiler, and Simon (Avron et al., 1983) to show indirectly that there is one Chern number per band, but with a zero sum rule that all the Chern numbers add up to zero. The zero sum rule makes sense because in a finite-dimensional tight-binding model the projector onto all bands is just the identity, and hence independent of wavevector. We will not calculate the Chern number directly, but rather the homotopy groups of Bloch Hamiltonians, and then recognize the Chern numbers in the answer.

To get some intuition for the result, we first consider the example of a non-degenerate two-band band structure, then give the general result, which is an application of a powerful tool in algebraic geometry, the exact sequence of a fibration (Mermin, 1979). The two-band example will let us connect homotopy invariants (which can be viewed intuitively as generalizing the notion of winding number) to cohomological invariants of differential forms, as initially discussed in Chapter 2.

The Bloch Hamiltonian for a two-band nondegenerate band structure can be written in terms of the Pauli matrices and the two-by-two identity as

$$H(k_x, k_y) = a_0(k_x, k_y)\mathbf{1} + a_1(k_x, k_y)\sigma_x + a_2(k_x, k_y)\sigma_y + a_3(k_x, k_y)\sigma_z. \quad (4.19)$$

The nondegeneracy constraint is that a_1, a_2, and a_3 are not all simultaneously zero. Now we first argue that a_0 is only a shift in the energy levels and has no topological significance, that is, it can be smoothly taken to zero without a phase transition. Similarly we can deform the other a functions to describe a unit vector on $\mathbb{Z}_2$: just as the punctured plane $\mathbb{R}^2 - (0, 0)$ can be taken to the circle, we are taking punctured 3-space to the 2-sphere via

$$(a_1, a_2, a_3) \rightarrow \frac{(a_1, a_2, a_3)}{\sqrt{a_1^2 + a_2^2 + a_3^2}} \quad (4.20)$$

at each point in k-space.

Now we have a map from T^2 to S^2. We need to use one somewhat deep fact: under some assumptions, if $\pi_1(M) = 0$ for some target space M, then maps from the torus $T^2 \rightarrow M$ are contractible to maps from the sphere $S^2 \rightarrow M$. Intuitively this is because the images of the noncontractible circles of the torus, which make it different from the sphere, can be contracted on M. By this logic, the two-band non-degenerate band structure in two dimensions is characterized by a single integer, which can be viewed as the Chern number of the occupied band.

What is the Chern number, intuitively? For simplicity let us consider maps from S^2 to the nondegenerate two-band Hamiltonians described above. One picture is in terms of $\pi_2(S^2)$, which is $\mathbb{Z}$: a map from the sphere to the sphere is labeled by the integer number of times one sphere covers the other. So we could view the integer as either a Chern number of the lower band, say, which must be opposite that of the upper band because of a zero sum rule, or instead as coming from the homotopy of the two-by-two Hamiltonians.

An alternate picture is that a nonzero Chern number is an obstruction to globally defining wavefunctions, in the following sense. $\mathcal{F}$, the first Chern form, is a two-form. Let us consider a constant nonzero $\mathcal{F}$, which for the case $S^2 \to S^2$ can be viewed as the field of a monopole located at the center of the target sphere. *Locally*, it is possible to find wavefunctions giving a vector potential $\mathcal{A}$ with $\mathcal{F} = d\mathcal{A}$, but not *globally*. (There has to be a Dirac string passing through the surface of the sphere somewhere.) In other words, states with nonzero Chern number have Chern forms that are nontrivial elements of the second cohomology class: they are closed two-forms that are not globally exact. Note that obtaining integer invariants from these integrals depended on our 1-form coming from some periodic, normalized wavefunction, so that the differential form can't be arbitrarily rescaled.

One subtle thing about this two-band model is that there is a nontrivial invariant in *three* spatial dimensions, since $\pi_3(S^2) = \mathbb{Z}$ (the Hopf invariant). In other words, even if the Chern numbers for the three two-dimensional planes in this three-dimensional structure are zero, there still can be an integer-valued invariant [1]. This map is familiar to physicists from the fact that the Pauli matrices can be used to map a normalized complex two-component spinor, that is, an element of S^3, to a real unit vector, that is, an element of S^2: $n^i = \mathbf{z}^\dagger \sigma^i \mathbf{z}$. This Hopf map is an example of a map that cannot be deformed to the trivial (constant) map, and one can construct tight-binding insulators realizing the nonzero invariant (Moore et al., 2008). The Hopf invariant does not generalize naturally without added symmetries to more than two bands, and hence is not protected except under the somewhat artificial assumption of a fixed-dimensional Hilbert space, but the Chern number does as we see next.

Consider the case of a nondegenerate two-dimensional band structure with multiple bands, which we study using the method of Avron et al. (1983). By the same argument as in the two-band case, we would like to understand π_1 and π_2 of the target space $H_{n \times n}$, nondegenerate $n \times n$ Hermitian matrices. As before, we will

---

[1] The nature of this fourth invariant changes when the Chern numbers are nonzero, as shown by Pontryagin in 1941: it becomes an element of a finite group rather than of the integers.

find that π_1 is zero so that maps from T^2 are equivalent to maps from S^2, but the latter will be quite nontrivial. We first diagonalize H at each point in k-space:

$$H(k) = U(k)D(k)U^{-1}(k). \tag{4.21}$$

Here $U(k)$ is unitary and $D(k)$ is real diagonal and nondegenerate. We can smoothly distort D everywhere in the Brillouin zone to a reference matrix with eigenvalues $1, 2, \ldots$ because of the nondegeneracy: if we plot the jth eigenvalue of D as a function of k_x and k_y, then this distortion ("spectral flattening") corresponds to smoothing out ripples in this plot to obtain a constant plane.

The nontrivial topology is contained in $U(k)$. The key is to note that $U(k)$ in the above is ambiguous: right multiplication by any diagonal unitary matrix, an element of $\mathrm{DU}(N)$, will give the same $H(k)$. So we need to understand the topology of $M = \mathrm{U}(N)/\mathrm{DU}(N) = \mathrm{SU}(N)/\mathrm{SDU}(N)$, where $\mathrm{SDU}(N)$ means diagonal unitary matrices with determinant 1. We can compute π_2 of this quotient by using the exact sequence of a fibration (Mermin, 1979)[2] and the following facts: $\pi_2(\mathrm{SU}(N)) = \pi_1(\mathrm{SU}(N)) = 0$ for $N \geq 2$. These imply that $\pi_2(M) \cong \pi_1(\mathrm{SDU}(N)) = \mathbb{Z}^{n-1}$, that is, $n - 1$ copies of the integers. This follows from viewing $\mathrm{SDU}(N)$ as N circles connected only by the requirement that the determinant be 1. Similarly we obtain $\pi_1(M) = 0$. We interpret these $n - 1$ integers that arise in homotopy theory as just the Chern numbers of the bands, together with a constraint that the Chern numbers sum to zero.

4.3 Many-Particle Interpretation of Topological Invariants

Until this point, we have developed methods in this chapter based on the assumption of a perfect crystal in which electrons move independently. The experimental observation of topological phases in real systems suggests that the topological quantities thus obtained should be physically meaningful even when those assumptions are relaxed, as does the generality of the Laughlin pumping argument discussed in Chapter 3. It turns out that for the integer Hall effect and the electrical polarization, there is an elegant approach, sometimes called the flux trick, to show how the same mathematics extends straightforwardly to the case of an interacting system with disorder. Note that extends straightforwardly means that the concepts and phases remain well defined beyond the single-electron crystalline case; certainly interactions and disorder can modify whether a material is in the topological or trivial class, for example.

---

[2] An exact sequence is a sequence of mappings with the property that the image of one mapping is the kernel of the next. The exact sequence of a fibration, in this context, relates the homotopy groups of two groups G and H and of the coset G/H: $\ldots \pi_n(H) \to \pi_n(B) \to \pi_n(G/H) \to \pi_{n-1}(H) \ldots$.

Let us start with a different picture of what the Bloch wavefunctions of a material describe. The key is to recall how one solves for the Bloch states in a tight-binding model. The crystal momentum vector $\mathbf{k}$ modifies the hoppings through phase factors $\exp[i\mathbf{k} \cdot (\mathbf{r}_1 - \mathbf{r}_2)]$ whenever an electron hop changes spatial location. But those phase factors resemble those that would be created if the unit cell lived on a torus with magnetic fluxes through its noncontractible loops, as those fluxes would generate similar phase factors through the Aharonov-Bohm effect (Box 3.1).

This realization lets us assuage any worry about whether the TKNN integer defined in Eq. 3.22 is specific to noninteracting electrons in perfect crystals. An elegant way to generalize the definition physically, while keeping the same mathematical structure, was developed in (Niu et al., 1985). This definition also makes somewhat clearer, together with our polarization calculation above, why this invariant should describe the Hall conductivity and conductance. First, note that from the formula for the Bloch Hamiltonian in the polarization calculation above, we can reinterpret the crystal momentum q as a parameter describing a flux threaded through a unit cell of size a: the boundary conditions are periodic up to a phase $e^{iqa} = e^{ie\Phi/\hbar}$. We will start by reinterpreting the noninteracting case in terms of such fluxes, then move to the interacting case.

The setup is loosely similar to the Laughlin argument for quantization in the IQHE. Consider adiabatically pumping a flux Φ_x though one circle of a toroidal system, in the direction associated with the periodicity $x \to x + L_x, y \to y$. The change in this flux in time generates an electric field pointing in the $\hat{\mathbf{x}}$ direction. Treating this flux as a parameter of the crystal Hamiltonian, we compute the resulting change in $\hat{\mathbf{y}}$ polarization, which is related to the y current density:

$$\frac{dP_y}{dt} = j_y = \frac{dP_y}{d\Phi_x}\frac{d\Phi_x}{dt} = \frac{dP_y}{d\Phi_x}(E_x L_x). \tag{4.22}$$

We are going to treat the polarization P_y as an integral over y flux but keep Φ_x as a parameter. Then (Ortiz and Martin, 1994)

$$P_y(\Phi_x) = \frac{ie}{2\pi} \int d\Phi_y \, \langle u | \partial_{\Phi_y} u \rangle \tag{4.23}$$

and we see that polarization now has units of charge per length, as expected. In particular, the polarization quantum in the y direction is now one electronic charge per L_x. The last step to obtain the quantization is to assume that we are justified in averaging j_y over the flux:

$$\langle j_y \rangle = \langle \frac{dP_y}{d\Phi_x} \rangle (E_x L_x) \to \frac{\Delta P_y}{\Delta \Phi_x}(E_x L_x), \tag{4.24}$$

where Δ means the change over a single flux quantum: $\Delta\Phi_x = h/e$. So the averaged current is determined by how many y polarization quanta change in the periodic adiabatic process of increasing the x flux by h/e

$$\langle j_y \rangle = \frac{e}{h} \frac{ne}{L_x} (E_x L_x) = \frac{ne^2}{h} E_x. \tag{4.25}$$

The integer n follows from noting that computing $dP_y/d\Phi_x$ and then integrating $d\Phi_x$ gives just the expression for the TKNN integer (Eq. 3.22), now in terms of fluxes. Not every topological property can be generalized to the interacting or disordered case as straightforwardly as the Chern number, however, and we will return to this question in the discussion of disorder in Chapter 8.

4.4 Time-Reversal Invariance and $\mathbb{Z}_2$ Invariants

4.4.1 2D $\mathbb{Z}_2$ Invariants from Chern Number

In Chapter 3, we presented the basic idea of $\mathbb{Z}_2$ topological insulators in two dimensions: a system with an even number of Kramers pairs will, under time-reversal-invariant backscattering, localize in pairs down to zero Kramers pairs, while a system with an odd number of Kramers pairs will wind up with a single stable Kramers pair. Now we would like to build on our improved understanding of Chern number to give a direct Abelian Berry phase approach for the 2D $\mathbb{Z}_2$ invariant, along with a physical interpretation of the invariant in terms of pumping cycles. There are at this point many different ways to look at the $\mathbb{Z}_2$ invariant, but an advantage of the approach given here is that it allows the IQHE and topological insulators to be presented in a single constructive framework. In particular, we can answer the question of what kind of integral over band structure might compute a discrete parity counting such as in the original Kane-Mele argument (Figure 3.6).

We also provide in Box 4.2 an introduction to Wess-Zumino terms in 1+1-dimensional field theory, for interested readers. The common aspect between these two is that in both cases the physical manifold (either the 2-sphere in the Wess-Zumino case, or the 2-torus in the QSHE case) is extended in a certain way, with the proviso that the resulting physics must be independent of the precise nature of the extension. When we then go to three dimensions, it turns out that there is a very nice 3D non-Abelian Berry phase expression for the 3D $\mathbb{Z}_2$ invariant; while in practice it is frequently harder to compute than expressions based on applying the 2D invariant, it is more straightforward mathematically. Indeed it is a higher-dimensional analogue of the relationship between polarization in one dimension and Chern number in two dimensions. For actual calculations, a very important simplification for the case of inversion symmetry (in both $d = 2$ and $d = 3$) was made in (Fu and Kane, 2007): the topological invariant is determined by the product of eigenvalues of the inversion operator at the 2^d time-reversal symmetric points of the Brillouin zone.

Box 4.2 The Wess–Zumino–Witten Model

The Wess–Zumino term is a beautiful example of how it can be useful to think about physical variables, such as fields or wavefunctions, defined on an enlarged space. Consider as a starting point the nonlinear sigma model (NLSM) for the XY model introduced in Eq. 2.70. That action is equivalent to that of a free boson,

$$S_0 = \frac{K}{2} \int_{\mathbb{R}^2} d^2x (\nabla \phi)^2, \qquad (4.26)$$

but for a compact boson field ϕ. This is the nonlinear sigma model into the circle S^1, which is the manifold of the Lie group U(1).

It can be written in a different way if we think about the order parameter manifold (the circle) as the Lie group U(1). Writing $g = e^{i\theta}$, we note that $\partial_i \theta = g^{-1} \partial_i g$, so

$$Z_{\text{NLSM}} = \int \mathcal{D}\theta(x) \exp\left[-\beta \int d^2x \sum_i (g^{-1}\partial_i g)^2/2\right]. \qquad (4.27)$$

In writing the action using g here, we are looking ahead to a generalization. There is not a Lie group structure on the sphere, but we might be tempted to generalize this model to other Lie groups than U(1), for example by taking $g \in$ U(N) or SU(N). Then $g^{-1}\partial_i g$ is an element of the Lie algebra, which has an inner product known as the Killing form; for SU(N), $\mathcal{K}(X, Y) = 2N\text{Tr}(XY)$. Generalizing the kinetic term that is the only term in the action above to the Lie algebra is straightforward.

However, it turns out that the low-energy physics of this generalization with just the resulting term is quite different than the U(1) case. As for the NLSM into the sphere, the fact that the manifolds of unitary groups are curved once we go beyond the circle leads to interactions that result in a mass gap. If we want instead to obtain a gapless model with Lie group symmetry, we must add an additional topological term first written down by Wess and Zumino. This term is quite unusual in that it requires extending the manifold on which the theory lives into an extra dimension. Assume $N >$ 1 in what follows, and pick $g \in G =$ SU(N) for definiteness. Let us compactify the two-dimensional space into S^2 by assuming configurations that are uniform at infinity. Given a configuration of the Lie-group field g on the surface of a sphere, we can always find a way to smoothly deform that configuration to the constant configuration since $\pi_2(\text{SU}(N))$ is trivial.

We will keep writing the generalized model in Euclidean space although their primary relevance is to quantum models in one spatial dimension. To fix this problem, Wess and Zumino wrote a term

$$S_{WZ} = -\frac{2\pi k}{48\pi^2} \int_{B^3} \epsilon_{\mu\nu\lambda} \mathcal{K} \left(g^{-1}\partial_\mu g, \left[g^{-1}\partial_\nu g, g^{-1}\partial_\lambda g \right] \right) \qquad (4.28)$$

that is quite remarkable: even writing this term depends on being able to take an original configuration of g on the sphere S^2 and extend it in to the sphere's interior B^3. (We will not show here that this term accomplishes the desired purpose, just that it is topologically well defined.) We said that one contraction into the ball exists because $\pi_2(G) = 0$. Different contractions exist because $\pi_3(G) = \mathbb{Z}$, and the coefficient of the second term is chosen so that, if k (the level of the resulting Wess–Zumino–Witten theory) is an integer, the different topological classes differ by a multiple of $2\pi i$ in the exponent in the path integral, so that observables are independent of what contraction is chosen. The reason that $\pi_3(G)$ is relevant here is that two different contractions into the interior B^3 can be joined together at their common boundary to form a 3-sphere, in the same way as two disks with the same boundary can be joined together to form the top and bottom hemispheres of a 2-sphere.

The full action of the Wess–Zumino–Witten (WZW) model in the usual notation is then

$$\begin{aligned} S = &-\frac{k}{8\pi} \int_{S^2} d^2x\, \mathcal{K}(g^{-1}\partial^\mu g, g^{-1}\partial_\mu g) \\ &-\frac{k}{24\pi} \int_{B^3} d^3y\, \epsilon^{ijk} \mathcal{K}(g^{-1}\partial_i g, [g^{-1}\partial_j g, g^{-1}\partial_k g]). \end{aligned} \qquad (4.29)$$

The meaning of upper and lower indices in the first term is that the metric of spacetime appears. In the second term, in contrast, the ϵ term appears instead of the metric, a sign that the term is topological in the sense of being metric-independent, like some of the field theories in Chapter 6. In the second term, we have chosen a particular continuation of the field g into the interior B^3 of the sphere S^2. While as mentioned above those continuations certainly exist, we should check to make sure that the physics is independent of precisely which continuation we chose.

This independence is related to another topological fact about SU(N). Consider two different continuations from S^2 into B^3. Actually, as a simpler example, consider two different continuations from S^1 into B^2. We could combine those into a field configuration on S^2, where one continuation gives the northern hemisphere and the other gives the southern hemisphere. In the same way, combining our two continuations from S^2 to B^3 gives a field configuration on S^3. Since $\pi_3(\text{SU}(N)) = \mathbb{Z}$, there are integer-valued classes of such configurations, and in fact the Wess–Zumino term is defined so as to compute this topological invariant Z: more precisely, the difference of the integral above for two different continuations into the bulk is k times $2\pi n$, where $n \in \mathbb{Z}$ measures the topological invariant of the map $S^3 \to S^3$ resulting from combining the two continuations as described above. The ambiguity that appears here related to the nontriviality of $\pi_3(\text{SU}(N))$ will also appear when we discuss magnetic polarizability in a few pages.

When we put this action into a quantum path integral, it therefore leads to a quantization of the level k to *integers*. $\text{SU}(2)_k$ with $k = 2$ can be viewed as a different representation of the same symmetry as the $\text{SU}(2)_1$ realized in the spin-half Heisenberg chain, in the same way as the spins on one site are in different representations of ordinary $\text{SU}(2)$. The full demonstration that the model is gapless is beyond our present scope, but at least we have a topological understanding of why the Wess–Zumino term is a natural quantity to consider. One way to tell apart the gapless points associated with different levels or Lie groups is by computing the central charge c, a measure of how many degrees of freedom are gapless at the critical point, in units where one free boson gives $c = 1$. The WZW model for Lie group g at level k has central charge

$$c = \frac{k \dim \text{SU}(N)}{k + n} \tag{4.30}$$

where $\dim \text{SU}(N) = N(N-1)$. Hence $\text{SU}(3)_1$ has central charge 2, $\text{SU}(2)_2$ has $c = 3/2$, and $\text{SU}(2)_1$ has $c = 1$, consistent with its bosonized representation as a single boson. The WZW models are the simplest examples of conformal field theories (DiFrancesco et al., 1997) that include a Lie group symmetry in addition to the conformal symmetry, and chiral versions can arise as the edge states of non-Abelian quantum Hall states (Chapter 9).

The main subtlety in finding a topological invariant for time-reversal-invariant band structures will be in keeping track of the time-reversal requirements. We introduce $\mathcal{Q}$ as the space of time-reversal-invariant Bloch Hamiltonians. This is a subset of the space of Bloch Hamiltonians with at most pairwise degeneracies (the generalization of the nondegenerate case we described above; we need to allow pairwise degeneracies because bands come in Kramers-degenerate pairs). In general, a $\mathcal{T}$-invariant system need not have Bloch Hamiltonians in $\mathcal{Q}$ except at the four special points (the time-reversal invariant momenta) where $k = -k$. It may be useful to refer back to Figure 3.6 and recall the difference between $k = 0$ and $k = \pm\pi$, which needed to have Kramers degeneracies, and other k values, which did not. The homotopy groups of $\mathcal{Q}$ follow from similar methods to those used above: $\pi_1 = \pi_2 = \pi_3 = 0$, $\pi_4 = \mathbb{Z}$. $\mathcal{T}$ invariance requires an even number of bands $2n$, so $\mathcal{Q}$ consists of $2n \times 2n$ Hermitian matrices for which H commutes with Θ, the representation of $\mathcal{T}$ in the Bloch Hilbert space. In mathematical terms, at general k the implication of time-reversal on the Bloch Hamiltonians is only between k and $-k$,

$$\Theta H(k)\Theta^{-1} = H(-k), \tag{4.31}$$

but at the time-reversal invariant momenta with $k = -k$, the action of time-reversal is more restrictive, leading to Kramers degeneracies for example.

Our goal is now to give a geometric derivation of a formula, first obtained independently via a different approach (Fu and Kane, 2006), for the $\mathbb{Z}_2$ topological invariant in terms of the Berry phase of Bloch functions:

$$D = \frac{1}{2\pi} \left[\oint_{\partial(EBZ)} d\boldsymbol{k} \cdot \boldsymbol{\mathcal{A}} - \int_{EBZ} d^2k\, \mathcal{F} \right] \quad \text{mod } 2. \qquad (4.32)$$

The notation EBZ stands for effective Brillouin zone (Moore and Balents, 2007), which describes one-half of the Brillouin zone together with appropriate boundary conditions. Since the BZ is a torus, the EBZ can be viewed as a cylinder, and its boundary $\partial(EBZ)$ as two circles, as in Figure 4.1b. While $\mathcal{F}$ is gauge-invariant, $\boldsymbol{\mathcal{A}}$ is not, and different (time-reversal-invariant) gauges, in a sense made precise below, can change the boundary integral by an even amount. The formula (4.32) was not the first definition of the two-dimensional $\mathbb{Z}_2$ invariant, as the original Kane–Mele paper (Kane and Mele, 2005b) gave a definition based on counting of zeros of the Pfaffian bundle of wavefunctions. However, (4.32) is easier to connect to the IQHE and sometimes easier to implement numerically, although easiest of all is, if the system has inversion symmetry, to use the method of parity eigenvalues of Fu and Kane (2007).

The way to understand this integral is as follows. Since the EBZ has boundaries, unlike the torus, there is no obvious way to define Chern integers for it; put another way, the $\mathcal{F}$ integral above is not guaranteed to be an integer. However, given a map from the EBZ to Bloch Hamiltonians, we can imitate the Wess–Zumino approach

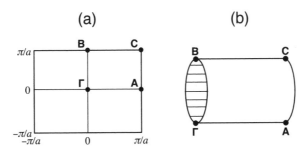

Fig. 4.1 Effective Brillouin zone (EBZ) for time-reversal-symmetric 2D system. (a) A two-dimensional Brillouin zone; note that any such Brillouin zone, including that for graphene, can be smoothly deformed to a torus. The labeled points are time-reversal-invariant momenta. (b) The EBZ. The horizontal lines on the boundary circles $\partial(EBZ)$ connect time-reversal-conjugate points, where the Hamiltonians are related by time reversal and so cannot be specified independently. Figure from Moore and Balents (2007). Reprinted with permission by *Physical Review*.

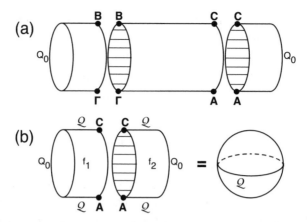

Fig. 4.2 (a) Contracting the effective Brillouin zone to a sphere. (b) Two contractions can be combined to give a mapping from the sphere, but this sphere has a special property: points in the northern hemisphere are conjugate under T to those in the southern, in such a way that overall every band pair's Chern number must be even. Figure from Moore and Balents (2007). Reprinted with permission by *Physical Review*.

above and consider extending the map to be one defined on the sphere (Figure 4.2), by finding a smooth way to take all elements on the boundary to some constant element $Q_0 \in Q$. The geometric interpretation of the line integrals of $\mathcal{A}$ in (4.32) is that these are the integrals of $\mathcal{F}$ over the boundaries, and the requirement on the gauge used to define the two $\mathcal{A}$ integrals is that each extends smoothly in the associated cap. The condition on the cap is that each vertical slice satisfy the same time-reversal invariance condition as an EBZ boundary; this means that a cap can alternately be viewed as a way to smoothly deform the boundary to a constant, while maintaining the time-reversal condition at each step.

The two mathematical steps, as in the Wess–Zumino term, are showing that such contractions always exist and that the invariant D in (4.32) is independent of which contraction we choose. The first is rather straightforward and follows from $\pi_1(\mathcal{H}) = \pi_1(Q) = 0$. The second step is more subtle and gives an understanding of why only a $\mathbb{Z}_2$ invariant or Chern parity survives, rather than an integer-valued invariant as in the IQHE. We can combine two different contractions of the same boundary into a sphere, and the Chern number of each band pair on this sphere gives the difference between the Chern numbers of the band pair obtained using the two contractions (Figure 4.2).

The next step is to show that the Chern number of any band pair on the sphere is even. To accomplish this, we note that Chern number is a homotopy invariant and that it is possible to deform the Bloch Hamiltonians on the sphere so that the equator is the constant element Q_0 (here the equator came from the time-reversal-invariant elements at the top and bottom of each allowed boundary circle). The possibility of deforming the equator follows from $\pi_1(\mathcal{Q}) = 0$, and the equivalence of different ways of deforming the equator follows from $\pi_2(\mathcal{Q}) = 0$. Then the sphere can be separated into two spheres, related by time-reversal, and the Chern numbers of the two spheres are equal so that the total Chern number is zero.

The above argument establishes that the two values of the $\mathbb{Z}_2$ invariant are related to even or odd Chern number of a band pair on half the Brillouin zone. Note that the lack of an integer-valued invariant means, for example, that we can smoothly go from an S_z-conserved model with $\nu = 1$ for spin $\uparrow$, $\nu = -1$ for spin $\downarrow$ to a model with $\nu = \pm 3$ by breaking S_z conservation in between. This can be viewed as justification for the physical argument given above in terms of edge states annihilating in pairs, once we define a $\mathbb{Z}_2$ invariant for disordered systems later.

4.4.2 Pumping Interpretation of $\mathbb{Z}_2$ Invariant

We expect that, as for the IQHE, it should be possible to reinterpret the $\mathbb{Z}_2$ invariant as an invariant that describes the response of a finite toroidal system to some perturbation. In the IQHE, the response is the amount of charge that is pumped around one circle of the torus as a unit flux (i.e., a flux h/e) is pumped adiabatically through the other circle.[3] Here, the response will again be a pumped charge, but the cyclic process that pumps the charge is more subtle.

Instead of inserting a 2π flux through a circle of the toroidal system, we insert a π flux, adiabatically; this is consistent with the part of D in (4.32) that is obtained by integration over half the Brillouin zone. However, while a π flux is compatible with T invariance, it is physically distinct from zero flux, and hence this process is not a closed cycle. We need to find some way to return the system to its initial conditions. We allow this return process to be anything that does not close the gap, but require that the Hamiltonians in the return process *not* break time-reversal. Since the forward process, insertion of a π flux, definitely breaks time-reversal, this means that the whole closed cycle is a nontrivial loop in Hamiltonian space (Figure 4.3). The $\mathbb{Z}_2$ invariant then describes whether the charge pumped by this closed cycle through the other circle of the torus is an odd or even multiple of the electron charge; while the precise charge pumped depends on how the cycle is closed, the parity of the pumped charge (i.e., whether it is odd or even) does not.

This time-reversal-invariant closure is one way to understand the physical origin of the $\mathcal{A}$ integrals in (4.32), although here, by requiring a closed cycle, we have

---

[3] A previous pumping definition that involves a π-flux but considers pumping of $\mathbb{Z}_2$ from one boundary to another of a large cylinder was used in Fu and Kane (2006) and will appear in Chapter 8.

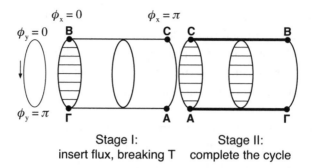

Stage I: Stage II:
insert flux, breaking T   complete the cycle

Fig. 4.3 Graphical representation of charge pumping cycle for $\mathbb{Z}_2$ invariant or Chern parity, from Essin and Moore (2007). The first stage takes place as the flux ϕ_x increases adiabatically from 0 to π. In the second stage the Hamiltonian at $(\phi_x = \pi, \phi_y)$ is adiabatically transported through the space of Hamiltonians to return to the Hamiltonian at $(\phi_x = 0, \phi_y)$. The difference between the second stage and the first is that at every step of the second stage, the Hamiltonians obey the time-reversal conditions required at $\phi_x = 0$ or $\phi_x = \pi$. The bold lines indicate paths along which all Hamiltonians are time-reversal invariant, and the disk with horizontal lines indicates, as before, how pairs of points in the second stage are related by time-reversal. Reprinted with permission by *Physical Review*.

effectively closed the effective Brillouin zone to a torus rather than a sphere. One weakness of the above pumping definition, compared to the IQHE, is that obtaining the $\mathbb{Z}_2$ invariant depends on Fermi statistics, so that the above pumping definition cannot be directly applied to the many-body wavefunction as in the IQHE case. We will solve this problem later for the three-dimensional topological insulator by giving a pumping-like definition that can be applied to the many-particle wavefunction.

4.4.3 3D Band Structure Invariants and Topological Insulators

We start by asking to what extent the two-dimensional integer quantum Hall effect can be generalized to three dimensions. A generalization of the previous homotopy argument (Avron et al., 1983) can be used to show that there are three Chern numbers per band in three dimensions, associated with the xy, yz, and xz planes of the Brillouin zone. A more physical way to view this is that a three-dimensional integer quantum Hall system consists of a single Chern number and a reciprocal lattice vector that describes the stacking of integer quantum Hall layers. The edge of this three-dimensional IQHE is quite interesting: it can form a two-dimensional chiral metal, as the chiral modes from each IQHE combine and point in the same direction.

Consider the Brillouin zone of a three-dimensional time-reversal-invariant material. Our approach will be to build on the understanding of the two-dimensional case from the previous subsection: concentrating on a single band pair, there is a $\mathbb{Z}_2$ topological invariant defined in the two-dimensional problem with time-reversal invariance. Readers can find tight-binding band structures for material examples of the three-dimensional invariants in Bernevig and Hughes (2013); we focus here on explaining where they come from. Taking the Brillouin zone to be a torus, there are two inequivalent xy planes that are distinguished from others by the way time-reversal acts: the $k_z = 0$ and $k_z = \pm \pi/a$ planes are taken to themselves by time-reversal (note that $\pm \pi/a$ are equivalent because of the periodic boundary conditions). These special planes are essentially copies of the two-dimensional problem, and we can label them by $\mathbb{Z}_2$ invariants $z_0 = \pm 1$, $z_{\pm 1} = \pm 1$, where $+1$ denotes even Chern parity or ordinary 2D insulator and -1 denotes odd Chern parity or topological 2D insulator. Other xy planes are not constrained by time-reversal and hence do not have to have a $\mathbb{Z}_2$ invariant.

The most interesting 3D topological insulator phase (the strong topological insulator) results when the z_0 and $z_{\pm 1}$ planes are in different 2D classes. This can occur if, moving in the z direction between these two planes, one has a series of 2D problems that interpolate between ordinary and topological insulators by breaking time-reversal. We will concentrate on this type of 3D topological insulator here. Another way to make a 3D topological insulator is to stack 2D topological insulators, but considering the edge of such a system shows that it will not be very stable: since two odd edges combine to make an even edge, which is unstable in the presence of T-invariant backscattering, we call such a stacked system a weak topological insulator.

Above we found two xy planes with two-dimensional $\mathbb{Z}_2$ invariants. By the same logic, we could identify four other such invariants x_0, $x_{\pm 1}$, y_0, $y_{\pm 1}$. However, not all six of these invariants are independent: some geometry[4] shows that there are two relations, reducing the number of independent invariants to four:

$$x_0 x_{\pm 1} = y_0 y_{\pm 1} = z_0 z_{\pm 1}. \qquad (4.33)$$

We can take these four invariants in three dimensions as $(x_0, y_0, z_0, x_0 x_{\pm 1})$, where the first three describe layered weak topological insulators, and the last describes the genuinely three-dimensional strong topological insulator. A more mathematical and rigorous perspective on this homotopy argument is given in Kaufmann et al. (2016). Alternately, the axion electrodynamics field theory in the next

---

[4] Sketch of geometry: to establish the first equality above, consider evaluating the Fu-Kane 2D formula on the four EBZs described by the four invariants x_0, x_{+1}, y_0, y_{+1}. These define a torus, on whose interior the Chern two-form $\mathcal{F}$ is well defined. Arranging the four invariants so that all have the same orientation, the A terms drop out, and the $\mathcal{F}$ integral vanishes as the torus can be shrunk to a loop. In other words, for some gauge choice the difference $x_0 - x_{+1}$ is equal to $y_0 - y_{+1}$. A mathematically rigorous presentation of this argument is given in Kaufmann et al. (2016).

subsection can be viewed as suggesting that there should be only one genuinely three-dimensional $\mathbb{Z}_2$ invariant.

For example, the strong topological insulator cannot be realized in any model with S_z conservation, while, as explained earlier, a useful example of the 2D topological insulator (aka quantum spin Hall effect) can be obtained from combining IQHE phases of up and down electrons. The impossibility of making this kind of strong topological insulator with S_z conservation follows from noting that all planes normal to z have the same Chern number, as Chern number is a topological invariant whether or not the plane is preserved by time-reversal. In particular, the $k_z = 0$ and $k_z = \pm\pi/a$ phases have the same Chern number for up electrons, say, which means that these two planes are either both 2D ordinary or 2D topological insulators.

While the above argument is rigorous, it does not give much insight into what sort of gapless surface states we should expect at the surface of a strong topological insulator. The answer was already mentioned in Chapter 3, and can be obtained by various means (some properties can be found via the field-theory approach given in the next section): the spin-resolved surface Fermi surface encloses an odd number of Dirac points. In the simplest case of a single Dirac point, believed to be realized in Bi_2Se_3, the surface state can be pictured as "one-quarter of graphene." Graphene, a single layer of carbon atoms that form a honeycomb lattice, has two Dirac points and two spin states at each k; spin-orbit coupling is quite weak since carbon is a relatively light element. The surface state of a three-dimensional topological insulator can have a single Dirac point and a single spin state at each k. As in the edge of the 2D topological insulator, time-reversal invariance implies that the spin state at k must be the T conjugate of the spin state at $-k$. An alternative to direct calculation of $\mathbb{Z}_2$ invariants in Brillouin zone planes is to find a path between a given band structure and a reference band structure known not to be topological, and then integrate the four-dimensional second Chern form introduced in the following section.

4.5 Axion Electrodynamics, Non-Abelian Berry Phase, and Magnetoelectric Polarizability

A question that we have dodged so far is, What is quantized in topological insulators, in the manner that the Hall effect is quantized in quantum Hall states? The three-dimensional topological insulator turns out to be connected to a basic electromagnetic property of solids, and this can be used to give an alternate definition of the state. We know that in an insulating solid, Maxwell's equations can be modified because the dielectric constant ϵ and magnetic permeability μ

need not take their vacuum values. Another effect is that solids can generate the electromagnetic term

$$\Delta \mathcal{L}_{EM} = \frac{\theta e^2}{2\pi h} \mathbf{E} \cdot \mathbf{B} = \frac{\theta e^2}{16\pi h} \epsilon^{\alpha\beta\gamma\delta} F_{\alpha\beta} F_{\gamma\delta}. \tag{4.34}$$

This term describes a magnetoelectric polarizability: an applied electrical field generates a magnetic dipole, and vice versa. An essential feature of the above axion electrodynamics (Wilczek, 1987) is that, when the axion field $\theta(\boldsymbol{x}, t)$ is constant, it plays no role in electrodynamics; this follows because θ couples to a total derivative, $\epsilon^{\alpha\beta\gamma\delta} F_{\alpha\beta} F_{\gamma\delta} = 2\epsilon^{\alpha\beta\gamma\delta} \partial_\alpha (A_\beta F_{\gamma\delta})$ (here we used that F is closed, i.e., $dF = 0$), and so does not modify the equations of motion. However, the presence of the axion field can have profound consequences at surfaces and interfaces, where gradients in $\theta(\boldsymbol{x})$ appear.

A bit of work shows that, at a surface where θ changes, there is a surface quantum Hall layer of magnitude

$$\sigma_{xy} = \frac{e^2 (\Delta\theta)}{2\pi h}. \tag{4.35}$$

This can be obtained by moving the derivative from one of the A fields to act on θ, leading to a Chern–Simons term for the EM field at the surface. Such a Chern–Simons term describes the electromagnetic response of a quantum Hall layer. The magnetoelectric polarizability described above can be obtained from these layers: for example, an applied electric field generates circulating surface currents, which in turn generate a magnetic dipole moment (Figure 4.4). In a sense, σ_{xy} is what accumulates at surfaces because of the magnetoelectric polarizability, in the same way as charge is what accumulates at surfaces because of ordinary polarization.

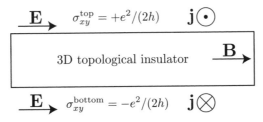

Fig. 4.4 Connection between surface quantum Hall states and bulk magnetoelectric polarizability. The fractional offset to surface integer Hall effect comes from the bulk value $\theta = \pi$ for the $\mathbf{E} \cdot \mathbf{B}$ coupling. The case shown has no net Hall effect for the whole slab, and consequently no IQHE-like gapless edge states are required at the sides. Measurements of low-frequency optical rotation suggesting these half-integer values for a single surface state have been carried out using low-frequency optics by Wu et al. (2016).

The meaning of this for the 3D $\mathbb{Z}_2$ topological insulator is illustrated in the slab geometry in Figure 4.4. To say that a 3D TI has $\theta = \pi$ means that, when its surface states are gapped, they will show a *half-integer* Hall effect, that is, $\sigma_{xy} = (n + 1/2)e^2/h$ at each surface.[5] One possibility is that there is no gapless edge state and σ_{xy} is constant with respect to the outward normal from the region of nonzero θ to the region of zero θ, but there can also be gapless integer Hall states at boundaries where θ defined in this way shifts by a multiple of 2π.

We are jumping ahead a bit in writing θ as an angle: we will see that, like polarization, θ is only well defined as a bulk property modulo 2π (for an alternate picture on why θ is periodic, see Wilczek, 1987). The integer multiple of 2π is only specified once we specify a particular way to make the boundary. How does this connect to the 3D topological insulator? At first glance, $\theta = 0$ in any time-reversal-invariant system, since $\theta \to -\theta$ under time-reversal. However, since θ is periodic, $\theta = \pi$ also works, as $-\theta$ and θ are equivalent because of the periodicity, but are inequivalent to $\theta = 0$.

Here we will not give a microscopic derivation of how θ can be obtained, for a band structure of noninteracting electrons, as an integral of the Chern-Simons form:

$$\theta = -\frac{1}{4\pi} \int_{\text{BZ}} d^3k \, \epsilon_{ijk} \, \text{Tr}[\mathcal{A}_i \partial_j \mathcal{A}_k - i\frac{2}{3}\mathcal{A}_i\mathcal{A}_j\mathcal{A}_k], \qquad (4.36)$$

which can be done by imitating our previous derivation of the polarization formula. This piece of the magnetoelectric effect first appeared in Qi et al. (2008) and its relationship to the theory of polarization in Essin et al. (2009). For derivations, see either Essin et al. (2010), which calculates dP/dB, or Malashevich et al. (2010), which calculates dM/dE. In a general solid, there are additional contributions other than the Chern–Simons form, but these vanish with time-reversal symmetry. This is a difference between the magnetoelectric polarizability and the electric polarization **P**, which are otherwise quite similar as we now show, because the bulk orbital contribution to **P** is given entirely by the Berry phase, independent of symmetry.

Instead we will focus on understanding the physical and mathematical meaning of the Chern–Simons form that constitutes the integrand, chiefly by discussing analogies with our previous treatment of polarization in one dimension and the IQHE in two dimensions. These analogies are summarized in Table 4.1.

Throughout this section,

$$\mathcal{F}_{ij} = \partial_i \mathcal{A}_j - \partial_j \mathcal{A}_i - i[\mathcal{A}_i, \mathcal{A}_j] \qquad (4.37)$$

---

[5] We are assuming implicitly that the excitations of the system continue to have well-defined integer charge. There exist alternative, rather exotic ways to gap the 3D TI surface state for which this is not true and the θ-angle is ill defined.

Table 4.1 *Comparison of Berry phase theories of polarization and magnetoelectric polarizability*

	Polarization	Magnetoelectric polarizability
$d_{\min}$	1	3
Observable	$\mathbf{P} = \partial\langle H\rangle/\partial E$	$M_{ij} = \partial\langle H\rangle/\partial E_i \partial B_j = \delta_{ij}\theta e^2/(2\pi h)$
Quantum	$\Delta\mathbf{P} = e\mathbf{R}/\Omega$	$\Delta M = e^2/h$
Surface	$q = (\mathbf{P}_1 - \mathbf{P}_2)\cdot\hat{\mathbf{n}}$	$\sigma_{xy} = (M_1 - M_2)$
EM coupling	$\mathbf{P}\cdot\mathbf{E}$	$M\mathbf{E}\cdot\mathbf{B}$
CS form	$\mathcal{A}_i$	$\epsilon_{ijk}(\mathcal{A}_i\mathcal{F}_{jk} + i\mathcal{A}_i\mathcal{A}_j\mathcal{A}_k/3)$
Chern form	$\epsilon_{ij}\partial_i\mathcal{A}_j$	$\epsilon_{ijkl}\mathcal{F}_{ij}\mathcal{F}_{kl}$

Note: The rows show the minimum dimensionality for the quantity to be defined; the physically observable response; the ambiguity or quantum; the quantity that accumulates at surfaces; the electromagnetic term generated; the Chern–Simons form of the Bloch electrons that contributes; and the Chern form in one higher dimension that describes changes in the quantity.

is the (generally non-Abelian) Berry curvature tensor ($\mathcal{A}_\lambda = i\langle u|\partial_\lambda|u\rangle$), and the trace and commutator refer to band indices. We will understand the Chern–Simons form $K = \text{Tr}[\mathcal{A}_i\partial_j\mathcal{A}_k - i\frac{2}{3}\mathcal{A}_i\mathcal{A}_j\mathcal{A}_k]$ above starting from the second Chern form $\text{Tr}[\mathcal{F}\wedge\mathcal{F}]$; the relationship between the two is that

$$dK = \text{Tr}[\mathcal{F}\wedge\mathcal{F}], \qquad (4.38)$$

just as $\mathcal{A}$ is related to the first Chern form: $d(\text{Tr}\mathcal{A}) = \text{Tr}\mathcal{F}$. These relationships hold locally (this is known as Poincare's lemma, that given a closed form, it is *locally* an exact form) but not globally, unless the first or second Chern form generates the trivial cohomology class. For example, we saw that the existence of a nonzero first Chern number on the sphere prevented us from finding globally defined wave-functions that would give an $\mathcal{A}$ with $d\mathcal{A} = \mathcal{F}$. We are assuming in even writing the Chern–Simons formula for θ that the ordinary Chern numbers are zero, so that an $\mathcal{A}$ can be defined in the 3D Brillouin zone. We would run into trouble if we assumed that an $\mathcal{A}$ could be defined in the 4D Brillouin zone if the *first or second* Chern number were nonzero. Note that the electromagnetic action above is just the second Chern form of the (Abelian) electromagnetic field.

The second Chern form is closed and hence generates an element of the de Rham cohomology we studied earlier. There are higher Chern forms as well: the key is that symmetric polynomials can be used to construct closed forms, by the antisymmetry properties of the exterior derivative. In physics, we typically keep the manifold fixed (in our Brillouin zone examples, it is usually a torus T^n) and are interested in classifying different fiber bundles on the manifold. In mathematical language, we want to use a properly normalized cohomology form to compute a homotopy invariant (i.e., with respect to changing the connection, not the manifold). This is exactly what we did with the Chern number in the IQHE, which was argued to compute certain integer-valued homotopy π_2 invariants of nondegenerate Hermitian matrices.

More precisely, we saw that the U(1) gauge dependence of polarization was connected to the homotopy group $\pi_1(\mathrm{U}(1)) = \mathbb{Z}$, but that this is connected also to the existence of integer-valued Chern numbers, which we explained in terms of π_2. (These statements are not as inconsistent as they might seem, because our calculation of π_2 came down to π_1 of the diagonal unitary group.) We can understand the second Chern and Chern–Simons form similarly, using the homotopy invariants π_3 (gauge transformation in $d = 3$) and π_4 (quantized state in $d = 4$). The Chern–Simons integral for θ given above, in the non-Abelian case, has a $2\pi n$ ambiguity under gauge transformations, and this ambiguity counts the integer-valued homotopy invariant

$$\pi_3(\mathrm{SU}(N)) = \mathbb{Z}, \quad N \geq 2. \tag{4.39}$$

In other words, there are large (non-null-homotopic) gauge transformations. Note that the Abelian Chern–Simons integral is completely gauge-invariant, consistent with $\pi_3(\mathrm{U}(1)) = 0$.

The quantized state in $d = 4$ was originally discussed in the context of time-reversal-symmetric systems. The set $\mathcal{Q}$ has one integer-valued π_4 invariant for each band pair, with a zero sum rule. These invariants survive even once T is broken, but still require treating a set of bands rather than a single band. In this sense, the four-dimensional quantum Hall effect is a property of how pairs of bands interact with each other, rather than of individual bands. Even if this 4D QHE is not directly measurable, it is mathematically connected to the 3D magnetoelectric polarizability in the same way as 1D polarization and the 2D IQHE are connected.

The above Chern–Simons formula for θ works, in general, only for a noninteracting electron system. This is not true for the first Chern formula for the IQHE, or the polarization formula, so what is different here? The key is to remember that the 3D Chern formula behaves very differently in the Abelian and non-Abelian cases; for example, in the Abelian case, θ is no longer periodic as the integral is fully gauge-invariant. Taking the ground-state many-body wavefunction and inserting it into the Chern–Simons formula is not guaranteed to give the same result as using the multiple one-particle wavefunctions. Practical calculations of the Chern–Simons formula are not easy, unless there is an obvious adiabatic path to a material where θ is clearly 0; for computational methods, see Vanderbilt (2018).

However, we can give a many-body understanding of θ that clarifies the geometric reason for its periodicity even in a many-particle system. Consider evaluating dP/dB by applying the 3D polarization formula

$$P_i = e \int_{BZ} \frac{d^3k}{(2\pi)^3} \operatorname{Tr} \mathcal{A}_i . \tag{4.40}$$

to a rectangular-prism unit cell. The minimum magnetic field normal to one of the faces that can be applied to the cell without destroying the periodicity is one flux quantum per unit cell, or a field strength $h/(e\Omega)$, where Ω is the area of that face. The ambiguity of polarization (4.40) in this direction is one charge per transverse unit cell area, that is, e/Ω. Then the ambiguity in dP/dB is

$$\Delta \frac{P_x}{B_x} = \frac{e/\Omega}{h/(e\Omega)} = \frac{e^2}{h} = 2\pi \frac{e^2}{2\pi h}. \tag{4.41}$$

So the periodicity of 2π in θ is really a consequence of the geometry of polarization, and is independent of the single-electron assumption that leads to the microscopic Chern–Simons formula. Of course, this argument implicitly assumed that the fundamental particle remains an electron, which we will see in the next chapter is not always the case, thanks to the remarkable phenomenon of fractionalization.

5

Hydrogen Atoms for Fractionalization

This chapter presents four settings of fractionalized phases: the Laughlin liquid, which underpins the fractional quantum Hall effect; the resonating valence bond liquid phase, which came to prominence in the context of high temperature superconductivity; the Peierls-distorted and Majumdar–Ghosh chains which exhibit quantum number fractionalization in $d = 1$; and the Coulomb phase of spin ice, the first experimentally known fractionalized magnetic phase in $d = 3$. Not only have the first two played a tremendously important role in the development of the field, they also exemplify the striking phenomena encountered as well as illustrate a number of basic concepts. A more formal, field-theoretic description of fractionalized phases is the subject of the following chapter.

The discovery of these phenomena took place against the background of one of the great successes of twentieth-century physics, namely, the degree to which the behavior of electrons in solids has been understood based on the notion of Fermi liquids and the instabilities of their Fermi surfaces. The formulation of the BCS theory of superconductivity stands out as one of the crowning achievements; but much else was understood in considerable detail, such as corrections to transport properties due to electron interactions or an applied magnetic field, or the Kondo effect, which deals with the role of magnetic impurities in the Fermi liquid. In turn, for the description of ordered phases such as charge-density waves, a satisfyingly complete picture in the form of Landau–Ginzburg theory was available, supplemented by the development of the renormalization group.

The 1980s turned up a couple of surprises which were to extend this worldview in several fundamental respects, and both were in the form of entirely unforeseen experimental results.[1] The first set was provided by semiconductor physics, where it had become possible to confine electrons to two dimensions. In strong magnetic fields, these systems were found to exhibit impeccably quantized values of their

---

[1] It turned out later that even BCS superconductors were already a class apart from conventional ordered states, with aspects of topological order as well (Hansson et al., 2004).

Hall conductivity (Klitzing et al., 1980), $\sigma_{xy} = \nu e^2 / h$, where ν could be an integer or one of a set of simple rational numbers. These would become known as the integer (Chapter 3) and fractional quantum Hall effects, respectively.

Such an unusual and robust type of order was quickly recognized as heralding a new field of physics. A few years later, this intense sense of novelty – Klaus von Klitzing received a Nobel Prize in 1985, comparatively soon after the discovery – was spectacularly surpassed by the immediate impact of a discovery in 1986 of a new class of superconductors. Strikingly raising the maximal critical temperature, which had hovered below 30 K for decades, this set in motion a research effort in materials science and condensed matter physics of an unprecedented intensity. This included the subject of "exotic" (i.e., non-Néel) orders and what we may now call topological magnets.

Much of what this book is about ultimately results from developments rooted in these discoveries. Needless to say, none of these developments took place in a vacuum, and far-sighted precursors were provided by Wegner's gauge theories, Kosterlitz–Thouless physics and the scaling theory of localization, all of which are covered elsewhere in this book. In this spirit, this chapter presents a set of simple and important examples in this topological sector of non–Fermi-liquid physics and an introduction to the new concepts which have flowed from their discoveries, providing concrete physical pictures on which to build a more general mathematical treatment of the field. Several instances, which we term "hydrogen atoms" for their simplicity, historical and conceptual importance, are discussed in turn. The aim is not a mathematically complete or rigorous summary – indeed, we are somewhat permissive about what we include here. In particular, we do not insist on the strict criterion on the low-energy theory having to be a topological field theory. Rather, we have organized the presentation around phenomena, hence the word fractionalization in the chapter title. Under this umbrella, there is the fractional quantum Hall effect as the initial instance, and paragon, of a fractionalized phase; but also fractionalization in $d = 1$, either in the bulk alongside symmetry breaking, or at the edge as symmetry protected topological order (see also Section 11.1); the resonating valence bond liquid as a two-dimensional spin system, and spin ice with its magnetic monopoles in $d = 3$. For this broad class of systems, a few common features and central themes are worth mentioning at the outset.

One is that each of the examples we encounter can be understood based on a simple model Hamiltonian that makes the underpinning microscopic physics relatively transparent. The corresponding experimental systems, or more realistic model Hamiltonians, are then (supposed to be) related by adiabatic continuity. The need for such a procedure arises from the fact that these states are often not in a perturbative sense close to better-known conventional states such as a Fermi liquid.

Another recurrent feature is the existence of an initially huge degeneracy as a central ingredient for such a construction, which may take the form of a Landau level with its degenerate single-particle orbitals, or a frustrated magnet with its apparent zero-point entropy. The ability of fluctuations to act in this huge set of states appears conducive to the existence of the resulting exotic phases. Of particular importance is the constraint these spaces incorporate, as will be discussed more formally in the following chapter.

What Is Fractionalization?

In each of these instances, one finds that the low-energy excitations are fundamentally different from the high-energy degrees of freedom in terms of which the model was originally formulated. This phenomenon is of course not at all uncommon: for instance, phonons or magnons, the low-energy excitations of a (magnetic) solid, look nothing like its constituent atoms, and can indeed be studied without even any knowledge of the atomic nature of matter. The fact that effective low-energy degrees of freedom bear little resemblance to the underlying microscopic ones is an instance of what has become known as emergence. This idea was formulated in an influential philosophy of science article entitled "More Is Different" (Anderson, 1972).

What is crucially special for fractionalized degrees of freedom, is that their quantum numbers are not even representable as *sums* of those of the high energy degrees of freedom. Thus, one finds electrons that break up into several Laughlin quasiparticles; or whose spin and charge propagate separately (spin–charge separation); or magnetic dipoles that fractionalize into magnetic monopoles.

5.1 The Fractional Quantum Hall Effect

The plateaux in the magnetoresistance traces discovered by von Klitzing in the integer quantum Hall effect (Section 3.1) were not to be the end of the story. In 1982, Horst Störmer and Daniel Tsui presented the results of an analogous measurement on a much cleaner two-dimensional electron gas, which had become available thanks to the invention of a new growth technique, modulation doping (see Box 3.2). This showed a plateau not just at integer values of the filling factor ν, but also at the fractional filling $\nu = 1/3$ (Tsui et al., 1982); this, along with much additional structure since discovered, is shown in Figure 5.1. So, whereas charge and flux are still commensurate, the proportionality is no longer an integer but now a rational fraction, whence the name fractional quantum Hall effect: at $\nu = 1/3$, there is a third of an electron for each flux quantum threading the system.

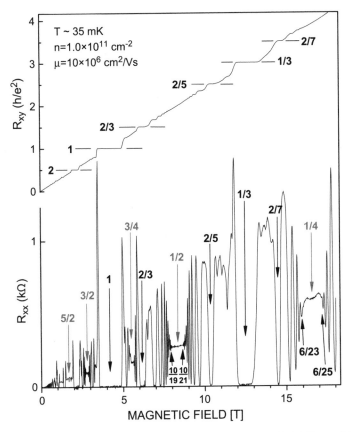

Fig. 5.1 The resistance of a 2DEG as a function of magnetic field (Stern, 2008). In addition to the plateaux at integer filling fraction v, much new structure appears, with a particularly robust feature around $v = 1/3$. Reprinted with permission by Elsevier.

The following account of the fractional quantum Hall effect is centered around a study of Laughlin's wavefunction: its form, its energetic origin, and the various intriguing properties it displays. These include topological order, and excitations with fractional charge and statistics.

5.1.1 The Challenge of Fractional Quantum Hall Physics

The understanding of the IQHE at this stage was sufficiently advanced that it immediately became clear that it alone was insufficient to account for this new phenomenon. While $v = 1$ corresponds to a fully filled Landau level, and hence a jump in the chemical potential as a further electron is added, there was simply no known mechanism which made $v = 1/3$ a similarly gapped state: in a non-interacting picture, $\nu = 1/3$ looks spectacularly unremarkable.[2] The following little detour provides an account of what lies at the origin of both the exceptional interest, and the exceptional difficulty, of the fractional quantum Hall regime.

Quantum Hall physics is mainly concerned with the situation where the cyclotron energy $\hbar\omega_c$ is the largest energy scale, so that other terms in the Hamiltonian – such as interactions or disorder – can be considered as acting only after the kinetic energy has been minimized. This is in spirit not unlike the study of Fermi liquids, where the kinetic energy is initially minimized by the – essentially unique – state obtained by filling up the Fermi sphere up to the Fermi energy.

Crucially, however, in the FQH regime there is generically a *huge* degeneracy of configurations minimizing the kinetic energy. For a partially filled Landau level, there is a degenerate space spanned by $\binom{N}{N\nu}$ Slater determinants obtained by distributing the $N\nu$ electrons over N single particle states.

In the presence of a perturbation, one needs to solve the many-body problem in this fabulously degenerate manifold. This is at the origin of much non–Fermi-liquid physics: after restriction to an effective low-energy space of states, a simple and unique "noninteracting" ground state is no longer available. It is this lack of a tractable starting point which led to the failure of approaches which are ultimately perturbative in nature, such as diagrammatic approaches developed in the context of Fermi liquid physics.

One feature does turn out to be helpful for the theoretical treatment of quantum Hall physics: unlike in the case of the Fermi sphere, where there are always gapless particle-hole excitations on the Fermi surface, there is a gap of size $\hbar\omega_c$ to the other Landau levels, so that the problem can be projected onto the partially filled Landau level in a controlled way. Much progress has been made by analyzing how the nature of, say, an interaction Hamiltonian, is altered by such a projection.

An exception to the remarkable difficulty of quantum Hall physics is the case of integer ν, a filled Landau level, which is not just uniquely defined but also gapped, with a many-body state taking the form of a simple Slater determinant: this is the case of the integer quantum Hall effect discussed in Section 3.1. In first-quantized notation, the many-body wavefunction describing the fully occupied first Landau level can be written in shorthand as an analytic function of the form

$$\Psi_1(\{z_i\}) = \prod_{i<j}(z_i - z_j) , \qquad (5.1)$$

where $z = x + iy$ parameterizes the two-dimensional plane.

---

[2] Indeed, among the more entertaining dinner conversations are the reminiscences of the scientists who were around at the time, retailing why none of the standard techniques of the time manage to crack this problem, expressed in sentences such as "Real theorists did not use first quantization" or "You write down $a^\dagger$, and it's game over." Mature fields of science tend to generate an inertia of their own.

More precisely, this expression encodes the following (unnormalized) wavefunction of a finite number N of electrons. It uses the single-particle basis functions from Section 2.2, where we have already introduced the notation to be used in the following analysis. The full expression for the wavefunction includes an additional exponential factor which depends on the magnetic length ℓ and is often dropped for convenience, as it is the same for all wavefunctions in a given Landau level:

$$\Psi_1(\{z_i\}) = \left[\prod_{1 \leq i < j \leq N} (z_i - z_j) \right] \exp\left[-\sum_{i=1}^{N} \frac{|z_i|^2}{4\ell^2} \right]. \qquad (5.2)$$

The prefactor depending on a product of functions of differences of particle coordinates is said to be of a Jastrow form, which is a popular ansatz for many-body wavefunctions in computational many-body physics. This polynomial can also be obtained by multiplying out a Slater determinant of N electrons in the N orbitals $|0, m\rangle$ (Eq. 2.36), with $m = 0, 1, \ldots, N - 1$.

We mention in passing that one attraction of focusing on the analytic function in Eq. 5.1 has the advantage that this also gives a legitimate wavefunction for higher Landau levels. What changes is the exponential in Eq. 5.2, which gets replaced by higher excited simple harmonic oscillator wavefunctions. The energetics, however, do change as a result, so that the Laughlin states tend not to be observed in high Landau levels. This will also play a role for the Pfaffian state at $\nu = 5/2$ (see Section 9.5.1).

5.1.2 The Laughlin State

The reason the fractional quantum Hall effect was so unexpected in large part traces back to the intractability of the general problem of how interactions lift degeneracies. It was only after its experimental discovery that, in a most ingenious piece of work, Bob Laughlin (1983) presented a wavefunction in which the salient properties of the groundstate are manifest. This wavefunction now bears his name.

The form of Laughlin's wavefunction, again up to an overall normalization, reads

$$\Psi_{1/m}(\{z_i\}) = \left[\prod_{i<j} (z_i - z_j)^m \right] \exp\left[-\sum_i \frac{|z_i|^2}{4\ell^2} \right], \qquad (5.3)$$

where m is an integer (odd for fermions, even for bosons). This wavefunction is at first sight remarkably similar to the filled Landau level (Eq. 5.1). The main distinction is that the electron density is lower – it describes a partially filled Landau level – as wavefunctions with a larger value of m require more space, as will be described below. This wavefunction corresponds to a filling fraction of $\nu = 1/m$ in the lowest Landau level. However, behind this superficial similarity hide profound differences: the Laughlin state exhibits topological order as well as fractionalized quasiparticles, two concepts which are the subject of the remainder of this account.

The Laughlin state is a good approximation for the ground state of a Coulomb interaction potential for an electron density distribution $\rho(\mathbf{r})$:

$$H_{\text{int}} = \int d\mathbf{r}d\mathbf{r}'\rho(\mathbf{r})\mathcal{V}(\mathbf{r}-\mathbf{r}')\rho(\mathbf{r}') = \frac{e^2}{4\pi\epsilon}\int d\mathbf{r}d\mathbf{r}'\frac{\rho(\mathbf{r})\rho(\mathbf{r}')}{|\mathbf{r}-\mathbf{r}'|}. \qquad (5.4)$$

Note that this is a three-dimensional Coulomb interaction of form $\mathcal{V}(\mathbf{r}) \propto 1/r$, rather than a two-dimensional one, $\mathcal{V}(\mathbf{r}) \propto \ln r$: even though the electrons are confined to two dimensions, their electric field lives in three dimensions, with its field lines penetrating space outside the two-dimensional plane. The field strength thus decreases like r^{-d+1} with $d = 3$.

The ground state for (classical) pointlike particles subject to a Coulomb interaction is known to be a Wigner crystal, in which each particle occupies the site of a hexagonal/triangular crystal:[3] there is symmetry breaking and ordering in real space. The Laughlin wavefunction, by contrast, represents an entirely different state, one that does not break any such global symmetry, but which exhibits a new, topological, type of order instead.

5.1.3 *Energetics of the Laughlin State: Haldane Pseudopotentials*

Laughlin guessed the form of his wavefunction based on an analytical study of the two particle problem in combination with a numerical analysis of the three-body problem. The former can be solved by appealing to the solution of the single-particle problem in a magnetic field described in Section 2.2: the interacting problem of two particles in the same Landau level (whose kinetic energy is thus fixed and can be dropped) can be solved by transforming into center-of-mass, Z, and relative, z coordinates

$$z = z_1 - z_2; \quad Z = (z_1 + z_2)/2 \qquad (5.5)$$

so that interaction term of the Hamiltonian only involves the relative coordinate z, not the center-of-mass Z. This can be analyzed with an operator algebra identical to the one used for the single-particle problem, with $b_r = \frac{b_1-b_2}{\sqrt{2}}$, $b_R = \frac{b_1+b_2}{\sqrt{2}}$, and a new set of magnetic lengths $\ell_R = \ell/\sqrt{2}$ and $\ell_r = \ell\sqrt{2}$.

---

[3] The hexagonal crystal derives its name from the symmetry of sixfold rotations about a lattice site. While this nomenclature is standard in three dimensions (cf. hexagonal close packing), in two dimensions, the lattice is more familiar under the name triangular lattice, on account of being composed of up- and down-pointing triangular plaquettes. This is not to be confused with the honeycomb lattice, which consists of hexagonal plaquettes.

The choice of circular gauge, $A = (-y, x, 0)B/2$, is very helpful here, as the rotational invariance about the origin of the relative-motion Hamiltonian remains manifest, which is not the case for the Landau gauge. This straightforwardly permits diagonalization in terms of a relative angular momentum variable:

$$H_{mn} = \langle n|V|m \rangle \equiv V_m \delta_{mn} .$$ (5.6)

Here, we have used the fact that the Hamiltonian projected to a given Landau level only contains the interaction term V. The resulting eigenvalues are known as the Haldane pseudopotentials:

$$V_m = \int \frac{d^2q}{(2\pi)^2} V_q e^{-q^2} L_m(q^2)$$ (5.7)

Here, L_m are the Laguerre polynomials, and V_q is the Fourier transform of a real space potential like $\mathcal{V}(\mathbf{r})$ in Eq. 5.4, as sketched in Figure 5.2. As the pair wavefunction needs to be odd (even) for fermions (bosons), so are the respectively allowed values of m. The eigenfunctions $|m\rangle$ corresponding to eigenvalues V_m thence have the form

$$\psi_{1/m} \propto z^m \exp(-|z|^2/4l_r^2) .$$ (5.8)

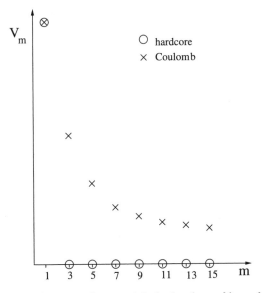

Fig. 5.2 Sketch of Haldane pseudopotentials for hardcore, $\mathcal{V}_1$, and Coulomb interactions. At large relative angular momenta, the Coulomb pseudopotentials decay as $1/\sqrt{m}$, since the two particle separation $\langle r \rangle \sim \sqrt{m}$. Figure from Moessner (1997).

Note the vanishing of the relative wavefunction as $z \to 0$ with power m: particles avoid each other, and hence reduce any repulsive interaction energy, increasingly efficiently so, as m increases.

In particular, for a repulsive contact potential $\mathcal{V}_0(\mathbf{r}) = \delta(r)$, which exacts an energy cost if two particles occupy the same place in space, one finds $V_m \propto \delta_{m,0}$. Since spin-polarized fermions at any rate have m odd, and in particular $m > 0$, on account of the antisymmetry of ψ_m required by the Pauli principle, it is convenient to define a potential $\mathcal{V}_1(\mathbf{r})$ which is even more singular than $\mathcal{V}_0$, namely, one which is formally the first derivative of a δ-function. This yields $V_m \propto \delta_{m,1}$.

In order to return from the few- to the many-body problem, one can now multiply together the first quantized single-particle wavefunctions, obtaining a many-body wavefunction

$$\psi(\{z_i\}) = f(\{z_i\}) \exp\left(-\sum_i \frac{|z_i|^2}{4\ell^2}\right). \tag{5.9}$$

Here $f(\{z_i\})$ needs to be a homogeneous polynomial in the $\{z_i\}$ whose degree is the total angular momentum. Also, quantum statistics requires the wavefunction to be (anti-) symmetric for (fermions) bosons under particle exchange.

The Laughlin state satisfies these requirements, and in addition it is optimized energetically: $\psi_{1/3}$ has vanishing energy for the interaction $\mathcal{V}_1$ because the factor $(z_i - z_j)^3$ ensures that no pair of particles ever occupies a state with relative angular momentum less than $3\hbar$. As $\mathcal{V}_1$ is positive definite, the Laughlin state is therefore a ground state. (Similarly, for repulsively interacting bosons subject to potential $\mathcal{V}_0$, one finds a ground state in form of the Laughlin state $\Psi_{1/2}$.)

The relationship between a filling fraction and its Laughlin state is established by recalling that the guiding centers of a particle pair with relative angular momentum $m\hbar$ are separated by a distance growing as $\sqrt{m}$, as follows from locating one of the pair at the origin and thinking about the rotational-gauge wavefunctions. The higher m, the lower the density that can be accommodated to allow for a large inter particle separation, and one finds that above $\nu = 1/m$, it is no longer possible to avoid relative angular momenta less than m altogether. To see this, note that in Eq. 5.3, each coordinate z_i appears in $N - 1$ terms in the product in brackets, so that its maximal power in the homogeneous polynomial is $m(N - 1)$. The area needed to accommodate this wavefunction is given by $\pi \langle z_i z_i^* \rangle$, the surface of a disk with radius of the root-mean square displacement of the i-th electron, $\mathcal{R}^2 = \langle z_i z_i^* \rangle = \langle x^2 + y^2 \rangle$, which can be evaluated as $\mathcal{R}^2 \sim 2m N \ell^2$. This corresponds to $\nu = N/N_\phi = 1/m$.

While such an argument – comparing extensive quantities $\mathcal{R}$, N and N_ϕ – is obviously too crude to make detailed statements about *single-particle* excitations (which we will discuss separately below) it does motivate very well how an "ordering" in relative angular momentum space can in principle lead to an energy gap at special fillings. Given that the noninteracting problem was hugely degenerate, the next question to be addressed is whether such a ground state at filling $v = 1/m$ is in fact unique, and also, whether the state is gapped. Several types of excitations are possible: charged (adding an extra particle), or neutral (creating particle-hole pair).

In an algebraic tour de force, Girvin, Macdonald, and Platzman (Girvin et al., 1986) showed that the spectrum of charge neutral excitations of the Laughlin state is gapped within the single mode approximation (SMA). The basics of the approach are explained in Box 5.1. The variational excited state wavefunction at momentum q is $|q\rangle = \bar{\rho}_q \Psi_{1/m} \equiv \mathcal{P}_{LLL} \rho_q \mathcal{P}_{LLL} \Psi_{1/m}$. $|q\rangle$ differs from the conventional "phonon" trial state considered in the simplest application of the SMA, $|q\rangle = \rho_q |0\rangle$, in that this state is then projected to lie in the lowest Landau level by $\mathcal{P}_{LLL}$, as occupancy of higher Landau levels involves states at an energy $\hbar \omega_c$ – the largest energy scale of quantum Hall physics – above the ground state.

The projection operation $\mathcal{P}_{LLL}$ really encodes two distinct items. One is that it takes account of the *magnetic* translations (Zak, 1964) in a Landau level, that is, the fact that in the presence of a magnetic field, two nonparallel translations $T_{\mathbf{r}}$ and $T_{\mathbf{s}}$ no longer commute:

$$T_{\mathbf{r}} T_{\mathbf{s}} = T_{\mathbf{s}} T_{\mathbf{r}} \exp\left[i \frac{\hat{\mathbf{z}} \cdot (\mathbf{r} \times \mathbf{s})}{\ell^2}\right], \tag{5.10}$$

where $\hat{\mathbf{z}}$ is the unit vector in the field direction.

This is just a restatement of the fact that an electron executing a closed orbit in a plane perpendicular to a magnetic field picks up a phase proportional to the area A of the orbit (see also Section 2.2). This feature is universally encountered in a magnetic field, and is the same in different Landau levels, and even for the multicomponent relativistic fermions in graphene.

The second item is the form factor of the wavefunctions in a given Landau level, which does depend on the details of the setting. These are reflected in the exponential factors in the wavefunctions in Eq. 5.1, and, for example, differ for graphene, and also change with Landau level as mentioned above, since the cyclotron orbits grow in size as their energy increases.

This feeds through to a W_∞ algebra of the projected density operators,

$$[\bar{\rho}_q, \bar{\rho}_p] = 2i \sin\left[\frac{\hat{\mathbf{z}} \cdot (\mathbf{q} \times \mathbf{p}) \ell^2}{2}\right] \bar{\rho}_{q+p}. \tag{5.11}$$

The variational excitation energy is now given by the expression

$$E(q) - E(0) = \frac{\langle 0 | [\bar{\rho}_{-q}, [\mathcal{H}, \bar{\rho}_q]] | 0 \rangle}{\langle 0 | \bar{\rho}_{-q} \bar{\rho}_q | 0 \rangle} = \frac{\bar{f}(q)}{\bar{s}(q)}. \tag{5.12}$$

Studiously Taylor expanding the above expressions for an isotropic liquid wave-function such as $\Psi_{1/m}$ yields both $\bar{s}(q) \propto q^4$ and $\bar{f}(q) \propto q^4$. This implies a variationally gapped excitation spectrum as their ratio is nonvanishing in the limit of $q \to 0$. While this does not rule out gapless states other than $|q\rangle$, numerical studies have so far confirmed this picture.

We note that the fast vanishing of $\bar{s}(q)$ in this limit is an indication of the incompressibility of the Laughlin liquid: long-wavelength density fluctuations are suppressed in this state compared to "normal" systems with short-range interactions, where $s(q)$ vanishes only as q^2.

Box 5.1 Single-Mode Approximation

Single mode approximation (SMA) is the name of a variational approach that con-structs a wavefunction, $|q\rangle$, orthogonal to a (putative) ground state $|0\rangle$, and provides an *upper* bound to its energy. It does this by only evaluating expectation values in the state $|0\rangle$, providing an example of information about the excitation spectrum being encoded in the ground state. If the SMA yields a gapped excitation branch, the system can nonetheless be gapless – unless backed up by other methods, for example, numer-ical ones, such a result may just indicate that $|q\rangle$ is not a good approximation to the actual lowest-lying excitation.

Orthogonality to the ground state is often achieved by constructing an excited state $|q\rangle$ in a momentum sector different to the ground state's. To be concrete, consider some density operator, $\hat{\rho}(r)$, of say, electrons, dimers, atoms, . . . , and take its Fourier transform

$$\rho_q \equiv \sum_r \rho(r) \exp(iq \cdot r) , \qquad (5.13)$$

in order to define the (unnormalized) state

$$|q\rangle \equiv \rho_q |0\rangle \qquad (5.14)$$

with, for $q \neq 0$, $\langle 0|q\rangle = 0$, and energy $\langle q|\mathcal{H}|q\rangle / \langle q|q\rangle$, provided of course that the denominator, $s(q) \equiv \langle q|q\rangle = \langle 0|\hat{\rho}_{-q}\hat{\rho}_q|0\rangle$, does not vanish. If it does, one has not constructed a variational state at all.

Then, by the variational principle, the excitation energy $E(q)$ is bounded above by $f(q)/s(q)$ with $f(q) = \langle 0|[\rho_{-q}, [\mathcal{H}, \rho_q]]|0\rangle$, so that

$$E(q) \leq \frac{\langle 0|[\rho_{-q}, [\mathcal{H}, \rho_q]]|0\rangle}{\langle 0|\rho_q \rho_{-q}|0\rangle} . \qquad (5.15)$$

Gapless modes can appear in two ways. First, the *oscillator strength* $f(q)$ can vanish as $q \to 0$ as happens, for example, for a conserved quantity, $[\mathcal{H}, \hat{\rho}_0] = 0$. Alternatively, the *structure factor* $s(q)$ may diverge, corresponding to a "soft mode" in the sense of a symmetry-breaking phase transition.

These represent the two routes to gapless behavior in hydrodynamics. In the FQHE, gapfulness occurs for $s(q)$ and $f(q)$ vanishing with the same power of q.

We mention in passing that the SMA energy appears to decrease with q, before reaching a minimum at a nonzero value of $q = q_{MR}$. In analogy to excitations in liquid helium, where the SMA was developed by Feynman, the excitation at this wavevector is known as a magneto-roton, and it presages the appearance of the Wigner crystal instability as the filling factor v is decreased.

For quantum dimer models, treated later on in this chapter, both routes to gaplessness appear (Moessner and Sondhi, 2003b). There, it is the dimer density operators which replace the charge density ρ_q. For the square lattice quantum dimer model, this has two special locations in the Brillouin zone. One corresponds to the "photon," known as resonon in that context (Rokhsar and Kivelson, 1988). As the square lattice QDM has only solid phases, the resonon exists only precisely at the RK point, $v = t$, separating two of these, where it is anomalously soft, with frequency $\omega(q) \sim q^2$. The other resides at wavevector $(\pi, 0)$. This is the "soft mode" of the instability to the valence bond crystal which exists for $v < t$. Due to its location in the Brillouin zone, this has been christened the $\pi 0n$.

Finally, for quantum spin ice (Section 5.4), the single-mode approximation involves the spin (i.e., the emergent flux) density. At the RK point, this again yields an anomalously soft photon, with the conventional (linearly dispersing) photon existing in the adjacent fractionalized Coulomb phase for $v < t$. As there is no local instability to an adjacent ordered phase, there is no analogue to the $\pi 0n$ in this setting.

However, that is not the end of the story. The single-mode approximation is appropriate for local excitations like magnons, phonons or plasmons. The above calculation does not exclude any degeneracies between states which do not differ by local observables such as density modulations generated by ρ_q. Indeed, whereas the Laughlin state does not go along with local symmetry breaking like the Wigner crystal, it does retain a degeneracy of topological origin. It was in this context that Wen and Niu formulated the concept of topological order (Wen and Niu, 1990), which is one of a cluster of interrelated remarkable features of the Laughlin state to which we turn next.

5.1.4 Properties of the Laughlin State

Quantized Hall Conductance

In the case of superconductivity, the question, Why is the current so nearly dissipationless? is not the easiest to answer simply. The analogous question for the quantum Hall effect, why σ_{xy} is so accurately quantized, is similarly tough. One has to have a clear idea of not only what the ideal many-body state is, but also what the mechanism of the leading *correction* to the quantized value is. In principle, the length of the list of candidate mechanisms – electrostatic disorder, Landau level mixing, lattice defects of the host semiconductor, . . . – is only limited by one's imagination. As a salutary note against any misplaced confidence in one's ability to provide an exhaustive enumeration, we recall the fact that a particular first-order phase transition in Helium at low temperature is nucleated by the absorption of a cosmic ray penetrating the cryogenic experimental setup (Leggett, 1984), which most people would not a priori have included as a legitimate actor in ultra-low temperature physics.

The quantized value itself at filling $\nu = 1/m$ in fact is obtained in an elementary fashion if one is prepared to ignore distractions such as the underlying lattice or the presence of disorder entirely. This appears legitimate insofar as neither have been included in the description thus far. On this level of approximation, one can Lorentz-boost *any* many-body state into a reference frame moving at speed v with respect to the lab frame, as explained in Chapter 8 around Eq. 8.9.

As in the above case of the integer effect, the "miracle" of the quantized Hall effect lies in its stability when quasiparticles are created as the density is *tuned away* from commensurability. There, as here, it is believed that the localization of these quasiparticles leads to their failure to make a contribution to the Hall effect. This issue of localization is addressed in Section 8.2.

Fractional Charge

While a simple change of prefactor in the quantized Hall conductance may not seem much of a surprise, it does herald some fundamental differences between the integer and fractional effects: the arguments proposed originally for explaining the quantization of the integer Hall effect (Section 3.1) turn out to yield some surprising results when applied to the fractional case. Inserting a unit flux at, say, the center of a disk transports a charge $Q = \sigma_{xy}\Phi_0$ from the location of the flux to the boundary of the system. The original argument – that this has to be an integer, so that an entire electron is transported – is now manifestly violated; rather, m such fluxes need to be inserted for an entire electron to be removed from the center of the system.

Upon inserting a single flux, therefore, only a fraction of an electron has been transported. The particle carrying this fractional charge is known as a Laughlin quasiparticle, say a Laughlin quasihole which is left behind at the location of the flux insertion, which goes along with the appearance of a widely separated Laughlin quasiparticle at the edge of the system, that is, at the boundary of a disk. (Note that the term Laughlin quasiparticle is somewhat confusingly sometimes used in distinction to the oppositely charged quasihole, and sometimes as a collective term comprising the two.)

This is our first example of a fractionalized particle, an entity that carries a quantum number – charge/number – which is not a simple sum of the quantum numbers of the original constituents, the electrons, in terms of which the "high-energy" theory, that is, the original microscopic Hamiltonian equation 2.30, was formulated. This is extremely counterintuitive: common sense states that no combination of integer-charged objects can possibly yield a fractional outcome. Indeed, in the Nobel Prize citation "for their discovery of a new form of quantum fluid with fractionally charged excitations" for Laughlin, Störmer, and Tsui, this aspect of the FQHE was emphasized. It was the first instance of this phenomenon in $d > 1$, and as such a true milestone in our understanding of the physical world.

The above arguments are compelling but quite indirect. It is then natural to ask if there is a more direct way of "seeing" the fractional charge of Laughlin's quasiparticle. Unfortunately, it is impossible to extract such a particle from the two-dimensional electron gas, as it only exists in the many-body state hosted there. One possibility is, then, to look for the electrical conduction of the quantum Hall sample. This takes place via edge states, and an idea was to consider not the (mean) current, $\bar{I}$, carried by the edge, but the *noise* of the current. This idea may be familiar from everyday life, when considering precipitation in the form of rain, or perhaps even more impressively, hail. While it is hard to normalize observations from different hailstorms against each other, the sound emanating from a tin roof hit by hail gives information not only on the total amount of hail falling, but also the size of the hailstones.

The basic analysis is then simple: if the quasiparticles are assumed to be non-interacting, they will travel along the edge state independently. In particular, they will arrive at a current measuring device stochastically. Modeling their independent arrival by a Poisson process means that the fluctuations in the current scale like the square root of the number of particles which arrive (Schottky, 1918). Crucially, the number of particles, N_q, arriving in a time window τ is inversely proportional to their charge, q: $I = qN_q/\tau$, so that fluctuations scale as $\delta^2_{N_q} \sim I/q$. The relative noise, $\delta^2_{N_q}/N^2 \sim q$, is proportional to q: the smaller the charge, the smaller the noise. Indeed, the relative noise similarly grows for smaller values of the current, and much ingenuity went into creating an experimental set-up with an appropriately small and well-controlled backscattering current I_B between two edges (see Figure 5.3). The reader is encouraged to appreciate the care taken in isolating and identifying the signal, and excluding alternative explanations for its origin (Saminadayar et al., 1997). This is one of the attractive hallmarks of the field of quantum Hall physics. At any rate, the upshot is that this experiment did extract a value of $q \approx e/3$, with small error bars. A formal definition of the operator measuring such fractional charge directly is deferred to Box 5.3.

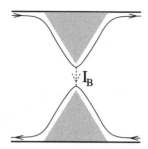

Fig. 5.3 Quantum point contact to generate a tunable backscattering current I_B used for measuring fractional charge. The distance between the counterpropagating edge states (indicated by lines with arrows) can be varied by applying a potential between various regions of the sample. This changes the scattering rate between the two shaded regions, and hence I_B.

Topological order

We now consider how the flux insertion argument introduced earlier is profoundly modified in the FQHE: not only is the Hall conductance modified, but a new property – topological ground-state degeneracy – appears. For a system on a disk, initial and final state after flux insertion differ by the presence of the well-separated Laughlin quasiparticle-hole pair. To recap, the thought experiment of flux insertion through the hole of a torus goes as follows. A flux $\phi(t)$ varying from 0 at $t = 0$ to Φ at $t = T$ gives rise to an electromotive force $\mathcal{E}(t) = -\frac{1}{L}\frac{d\phi(t)}{dt}$, where L is the circumference of the torus. Hence, there is a charge current per unit length $I(t) = \mathcal{E}(t)\sigma_{xy}$. This integrates across the circumference to $Q = L\int_0^T dt' I(t') = \sigma_{xy}\phi$: for $\phi = \Phi_0 = h/e$, the flux quantum, fractional charge e/m has been transported across the circumference of the torus, for $\sigma_{xy} = \frac{1}{m}\frac{e^2}{h}$ of the fractional effect.

As far as the Hamiltonian is concerned, the initial and final Hamiltonian differ only by an unobservable change in vector potential: the Hamiltonian is formulated in terms of electrons only, so that no Aharonov–Bohm effect will occur for an inserted flux of Φ_0, and hence flux insertion can be gauged away. At the same time, if flux insertion is done arbitrarily slowly, by the adiabatic theorem the system has to return to a ground state if no gap closes along the way.[4] Starting off in a unique ground state should thus return the system to itself. This contradicts a fractional charge moving across the cut. Therefore, the state after flux insertion differs from the one before. It must hence be a different but degenerate state: the ground state cannot be unique.

The reader who finds such a verbose argument not entirely convincing has several options for obtaining the same result. One is to squeeze the torus so that one

---

[4] In fact, the sweep has to be fast compared to the splitting of the topologically degenerate multiplet, but slow on the scale of the excitation gap. The former is exponentially small in the linear system size, $L_{x,y}$, while the latter is independent of it, so that this condition does not present a fundamental obstacle.

circumference becomes very small, less than a magnetic length. In this case, the Laughlin state is continuously connected to what is known as the Tao–Thouless state, which looks like a charge-density wave with one in three adjacent Landau orbitals occupied, and the other two empty; there are obviously three such states.

Another is based on the original analysis of Wen and Niu (1990), who constructed an algebra of many-body magnetic translation operators, $\mathcal{T}_{x,y}$, generalizing the single particle version (Eq. 5.10):

$$\mathcal{T}_{\mathbf{r}} = \prod_{j=1}^{N_e} T_{\mathbf{r}}^j , \tag{5.16}$$

where $j = 1 \ldots N_e$ labels the electrons at filling fraction ν, so that there are $N_\phi = N_e/\nu = L_x L_y/(2\pi\ell^2)$ flux quanta piercing the torus, that is, the system with periodic boundary conditions for translations by vectors $\mathbf{L}_{x,y}$ in the x and y-directions, respectively. Crucially, translations of all particles by $\mathbf{l}_{x,y} = \mathbf{L}_{x,y}/N_\phi$ are consistent with the boundary conditions, in that they lead to a displacement effectively involving an area corresponding to an integer number of flux quanta. The concomitant algebra is hence

$$\mathcal{T}_{\mathbf{l}_x}\mathcal{T}_{\mathbf{l}_y} = \mathcal{T}_{\mathbf{l}_y}\mathcal{T}_{\mathbf{l}_x} \exp\left[i\frac{N_e(\mathbf{l}_x \times \mathbf{l}_y)}{N_\phi^2\ell^2}\right] = \mathcal{T}_{\mathbf{l}_y}\mathcal{T}_{\mathbf{l}_x} \exp\left(i2\pi\nu\right) . \tag{5.17}$$

For the case $\nu = 1/q$, given a ground state Ψ^0 which is also an eigenstate of $\mathcal{T}_{\mathbf{l}_y}$, one can thus generate $q - 1$ further degenerate states $\Psi^n = \mathcal{T}_{\mathbf{l}_x}^n\Psi^0$, which have different eigenvalues with respect to the unitary operator $\mathcal{T}_{\mathbf{l}_y}$, and are therefore orthogonal.

The resulting m-fold degeneracy depends on the topology of the surface on which the 2DEG resides–a sphere does not offer any noncontractible loop which could enclose a flux. Indeed as detailed in the following Chapter 6, another derivation of the topological degeneracy proceeds uses Chern–Simons theory (Section 6.6.1) as the effective topological field theory. This allows a direct computation of the topological degeneracy on a manifold of genus g. This turns out to be 3^g, as desired for the torus with $g = 1$; it vanishes for a sphere, $g = 0$, reflecting the fact that a sphere has no noncontractible loop through which a flux could be threaded before being gauged away.

This degeneracy is hence known as a topological degeneracy. We will encounter a more pictorial example of a topological degeneracy in the resonating valence bond (RVB) liquid later in this chapter. We close by noting that this topological degeneracy is a useful fingerprint of topological states in practice, for example, in numerical simulations, so that it can be used both as diagnostic and an element in a systematic classification.

Fractional Statistics

The basic concept of fractionalization is reasonably straightforward as far as a quantity such as particle number is concerned – an electronic charge can "simply" be divided equally between m-fractionalized Laughlin quasiparticles. The next level of complexity concerns the issue of how particles interact with each other. The most fundamental aspect is the issue of relative statistics: the phase change γ_e that occurs in a wavefunction when two identical quasiparticles are exchanged is a fundamental property of quantum mechanics, which distinguished bosons ($\gamma_e = 0$) from fermions ($\gamma_e = \pi$) even when the particles are otherwise entirely noninteracting. The range of possible outcomes in $d = 2$ – in particular the emergence of new types of quantum statistics – is discussed in Box 5.2.

To see how statistics different from bosonic or fermionic come about in principle in $d = 2$ is easy; indeed, Leinaas and Myrheim (1977) identified this possibility a few years before the discovery of the fractional quantum Hall states. Consider a pair of particles at (complex) locations $\pm|z_0|$, which can be exchanged, that is, moved to locations $z_1 = \mp|z_0|$, by adiabatically following a trajectory from $t = 0$ to $t = 1$

$$z(t) = \pm|z_0|e^{i\theta(t)} , \tag{5.18}$$

with $\theta(0) = 0$ and $\theta(1) = 2n\pi + \pi$ yielding $z(0) = \pm|z_0|$ and $z(1) = \mp|z_0|$ as desired.

In $d = 2$, the paths $\theta(t)$ can be classified topologically by a winding number $n \in \mathbb{Z}$, as it is possible to tell how often the particles encircled each other before reaching their final position.

Therefore, in $d = 2$, executing two successive exchanges can be distinguished, so that, unlike in $d = 3$, $\mathcal{P}^2 = 1$ is no longer obeyed by the exchange operator. Hence, its eigenvalues can be not just $+1$ (bosons) or -1 (fermions) but also any "angle" $e^{i\gamma_e}$ with $\gamma_e \in [0, 2\pi)$. The origin of such "anyons" is discussed in more detail in Box 5.2. Non-Abelian fractional statistics and non-Abelian anyons also exist, and we will say more about their physical emergence in Chapter 9.

Box 5.2 Fractional Statistics of Particles in Two Dimensions

It is a well-known consequence of the spin-statistics theorem for relativistic field theories in four spacetime dimensions that there are only two types of quantum statistics: bosonic ($\gamma_e = 0$) or fermionic ($\gamma_e = \pi$); and that these go along with particles of integer or odd half-integer spin, respectively. Condensed matter physics offers two escape routes from this straightjacket of mathematical physics. First, in the absence of Lorentz invariance, the connection between spin and statistics is not operative for quasiparticles. Second, not only are space-time symmetries not sacred, not even the dimensionality of space is: Laughlin quasiparticles in two-dimensional electron liquids happily exist embedded in three-dimensional space.

The latter feature removes the restriction to fermions and bosons, which follows from the fact that the operation P_{ij} of interchanging particles i and j in $d = 3$ squares to an identity operation: $P_{ij}^2 = 1$, so that it has eigenvalues ± 1, and hence corresponds to phases $\gamma_e = 0, \pi$ picked up by the wavefunction as two identical particles are exchanged. The hidden assumption in this argument, which is often accepted without question in quantum mechanics classes, is that the only information about an exchange that can matter in the result is in how particle labels were "permuted" (which initial particles went to which final particles), and fermions and bosons are different kinds of representations of the permutation group. If there are topologically distinct ways of rearranging the particles, then physics could be sensitive to that history.

The example of this we have mentioned above is that, in $d = 2$, there is a distinction between clockwise and counterclockwise particle exchange, which gives rise to a winding number as described in Chapter 2. If we just thought of the exchange as switching the two particles, then it would square to the identity, but exchanging the two particles twice by moving them clockwise around each other might not return the system to the original state. The exchange angle γ_e can thus, as a matter of principle, take on any value $\gamma_e \in [-\pi, \pi)$. The resulting particles are known as anyons, and it is a remarkable fact that the recognition that anyons can exist only appeared four decades after Bose and Fermi statistics were formulated.

From a group-theoretical perspective, it is representations of braid groups which appear in two dimensions, instead of the permutation group. Its basic ingredient is the exchange of two particles described above. This is supplemented by rules for extending the exchange process to N particles. Having chosen an ordering $i = 1 \ldots N$, exchanges are restricted to operate between neighboring particles, $P_{i,i+1}$ or $P_{i,i+1}^{-1}$ only. These respectively correspond to a (counter)clockwise exchange, denoted pictorially by an over(under)pass (see Figure 5.4).

These exchanges are then posited to be independent if they only involve different particles, that is,

$$P_{i,i+1} P_{j,j+1} = P_{j,j+1} P_{i,i+1} \qquad (5.19)$$

unless $|i - j| = 1$; and that the order in which exchanges are done, as long as the same particle pairs are exchanged in the same way, does not matter,

$$P_{i,i+1} P_{i+1,i+2} P_{i,i+1} = P_{i+1,i+2} P_{i,i+1} P_{i+1,i+2} \ . \qquad (5.20)$$

In mathematical terms, the braid group for N particles is the free group on $N - 1$ elements modulo the above braid relations.

The anyons described above correspond to Abelian braid groups, in which each exchange of two particles picks up a phase γ_e, and the phases of successive interchanges simply add up. In addition, there are other, more complex, representation of

Fig. 5.4 Exchange operations and braiding in two dimensions. The vertical direction is to be read as time, with the lines denoting the trajectories of the individual particles. An over(under)pass from left to right encodes a (counter)clockwise exchange. The two leftmost exchange diagrams are therefore distinct in two, but not in three, dimensions: the second diagram corresponds to two counterclockwise exchanges, and hence a phase factor $P_{12}^2 \sim \exp(2i\gamma_e)$ for Abelian anyons; while the first diagram, $P_{12}P_{12}^{-1} \sim 1$ corresponds to no net exchange. (In $d = 3$, both diagrams are topologically identical, hence only bosonic and fermionic statistics are possible.) The middle pair of diagrams correspond to the same exchange processes, and need to yield identical outcomes as a consistency requirement for the braid group representation (Eq. 5.20). The rightmost pair of processes both depict two clockwise exchanges. They are the same for Abelian anyons. However, they can differ for non-Abelian anyons, as they correspond to distinct processes: two overpasses of the initially leftmost particle, rather than two underpasses of the initially rightmost one: $P_{12}P_{23} \neq P_{23}P_{12}$ in general.

the braid group. These are known as non-Abelian and involve higher-dimensional irreducible representations so that they are therefore no longer represented by a simple parameter γ_e but instead by a matrix, as follows.

A set of n non-Abelian anyons is thus associated with a particular braid group. Having fixed their n location vectors r_n, the joint state of the non-Abelian anyons will in general have degeneracies, such as those associated with the Majorana zero modes discussed in Chapter 9 on topological quantum computing. Denoting the dimensionality of this degenerate space by d_0, each braiding operation $P_{i,i+1}$ is represented by a $d_0 \times d_0$ matrix $\rho_{P_{i,i+1}}$. This describes how states of the *multiplet* transform *into each other* in the process of braiding. In other words, the wavefunction does not just pick up a phase γ_e, but considerably more complex consequences, in particular transitions between different states of the multiplet, now result from the braiding operation.

A fuller specification of the braid group is given in a review article on topological quantum computing (Nayak et al., 2008). We also note that the braid group is fundamental to the problem of classifying knots, one of the classic topics in algebraic topology (Simon, 2020).

To determine the exchange angle between a pair of charged excitations, let us consider the phase picked up as one Laughlin quasihole is adiabatically transported around another. In the spirit of the Berry phase (Section 2.1), we identify these two quantities. As a trial wavefunction for a Laughlin quasihole at location ζ, consider modifying (5.3) to

$$\Psi^{\zeta}_{1/m} = \alpha_1 \prod_{j=1}^{N} (z_j - \zeta) \Psi_{1/m}(\{z_i\}) \, , \qquad (5.21)$$

where α_1 is a normalization factor.

This has the desired charge e/m. To see this, consider a quasihole at the origin, $\zeta = 0$: the effect of the new factor is to increase the degree of the homogeneous polynomial by 1, so that there is now one orbital at the center which is never occupied. As each orbital has an average occupancy of $1/m$, the missing charge is just e/m.

A beautiful argument (Arovas et al., 1984) derives both fractional charge and fractional statistics for such quasiholes. As one is only interested in Aharonov-Bohm/Berry phases, it turns out to be sufficient to work with unnormalized wavefunctions in the following.

To compute the Aharonov–Bohm phase, we transport ζ around the origin, such that $\zeta(t) = |\zeta| e^{2\pi i t/T}$, $0 \leq t \leq T$, thereby enclosing flux $\phi = \pi |\zeta|^2 B$. In order to compute the geometrical phase, we need to compute the time dependence of the wavefunction:

$$\frac{d\Psi^{\zeta}_{1/m}}{dt} = \sum_{j=1}^{N} \frac{d}{dt} \ln \left(z_j - \zeta(t) \right) \Psi^{\zeta}_{1/m} \, . \qquad (5.22)$$

This produces a Berry phase (Eq. 2.2) of

$$\frac{d\gamma}{dt} = i \left\langle \Psi^{\zeta}_{1/m} \left| \frac{d}{dt} \sum_j \ln \left(z_j - \zeta(t) \right) \right| \Psi^{\zeta}_{1/m} \right\rangle$$

$$= i \int d^2 r \rho_{\zeta}(z) \frac{d}{dt} \ln(z - \zeta(t)) \, , \qquad (5.23)$$

where $\rho_{\zeta}(z)$ is the electron density at z in the presence of a quasihole at $\zeta(t)$. This approximately equals the uniform density ρ_0 at points more than a magnetic length away from $\zeta(t)$. As we compute the phase proportional to the area of the circle enclosed by the quasihole, we can ignore the subleading correction coming from the edge of width ℓ. Thence,

$$\gamma_{AB} = \int_{t=0}^{T} dt \frac{d\gamma}{dt} = \rho_0 \, i \int d^2 r \ln \left(\frac{z - \zeta(1)}{z - \zeta(0)} \right) = -2\pi \rho_0 A = -\frac{2\pi \phi}{\phi_0} \frac{1}{m} \qquad (5.24)$$

where only the interior of the circle has contributed: outside, the argument of the logarithm is unchanged between times $t = 0$ and $t = T$, whereas it "winds" by 2π inside.

The Aharonov–Bohm phase γ has thus been reduced by a factor $\frac{1}{m}$ compared to that of an electron, confirming the notion that the charge of a Laughlin quasiparticle is $-e/m$.

This computation of the Aharonov–Bohm phase straightforwardly generalizes to that of the anyonic exchange angle γ_e. In analogy to (5.21), the two-hole wavefunction reads[5]

$$\Psi_{\{1/m\}}^{\{\zeta_1,\zeta_2\}} = \alpha_2 \prod_{j=1}^{N} (z_j - \zeta_1)(z_j - \zeta_2)\Psi_{1/m}(\{z_i\}) . \qquad (5.25)$$

Fixing ζ_1 and taking ζ_2 around it yields an equation just like (5.23), where now $\rho_\zeta(z)$ is reduced by $-e/m$ as the quasihole at ζ_1 goes along with a missing charge. The total phase picked up is the sum of the Aharonov–Bohm and the exchange phase, from which it follows that the exchange phase equals

$$\gamma_e = 2\pi/m . \qquad (5.26)$$

This is just the anyonic fractional statistics. This connection between flux density and charge and statistics will become mathematically precise in the context of Chern–Simons theory (Chapter 6), and it also underpins the physics of Skyrmions (Section 3.7.3).

Edge States

Given we have already invoked the existence of edge states carrying fractionally charged quasiparticles, the reader may expect a simple explanation of their origin to be part of this account. Unfortunately, this is not straightforwardly achieved beyond the general statement that the noninteracting Laughlin quasiparticles do not behave all that differently from normal electrons, except for their fractional charge. Such a line of argument does not carry much power beyond rationalizing a previously known result. It is perhaps fair to say that it is at this level that the approach via a Chern–Simons field theoretic treatment in Chapter 6 really becomes compelling. With a number of robust qualitative features as input, the existence of the edge states drops out from general principles, that is, essentially from demanding internal consistency of the theory. In the following, however, we continue to pursue our more qualitative discussion of the basic phenomenon of fractionalization, turning to $d = 1$ next.

---

[5] In case readers are wondering about the existence of a quasi-electron, with opposite charge to the quasi-hole. This entity does exist, and it also has fractional statistics; it is just less straightforward to write down as a model wavefunction.

5.2 Fractionalization, Order, and Topology in $d = 1$

Fractionalization was first discovered in one dimension, in connection with studies of polyacetylene. Although the focus of this book is on the considerably less mature case of dimension $d \geq 2$, there are two reasons for diverting into a discussion of $d = 1$ here. One is historical completeness. The other is pedagogical: whereas the interplay of symmetry breaking (i.e., conventional) and topological forms of order are quite different in $d = 1$, the pictures developed there can nonetheless be transposed to higher dimensions if one takes care to allow the linear structure of the $d = 1$ chains and ladders to fluctuate appropriately (see, e.g., Figure 5.16).

5.2.1 The Peierls Instability

Noninteracting electrons hopping on an elastic chain are described by the Hamiltonian

$$\mathcal{H}_p = -\sum_i t_{x_{i+1}-x_i}\left(c_{i+1}^\dagger c_i + h.c.\right) + \frac{1}{2}k(x_i - x_{i+1})^2. \tag{5.27}$$

Here, x_i denotes the displacement of the ith lattice site from its equilibrium position $a \times i$, where a is the lattice constant, which we henceforth set to unity, $a = 1$; and k is the "spring constant" of the bond joining two adjacent lattice sites.

The hopping matrix element depends on these lattice displacements as $t_y = t_0 + t'y$. Peierls's seminal work (Peierls, 1955) showed that the system can lower its energy by distortion into a dimerized pattern $x_i = \delta(-1)^i$, so that

$$t_{i,i+1} = t_0 + 2(-1)^i t' \delta. \tag{5.28}$$

This is explained by the following energy balance consideration. In the limit of small δ, such a distortion costs elastic energy $2k\delta^2$ per site. The gain in zero-point energy for a half-filled band (one spin-polarized electron per pair of sites) is parametrically larger, $O(\delta^2 \ln \delta)$, so that a distortion will always be favorable. This can be seen from integrating the kinetic energy difference over the filled states. This involves comparing the dispersion relations for the doubled unit cell in presence and absence of the distortion, and expanding the elliptic integral in small $\Delta = 2t'\delta$:

$$E_K = \int\limits_{-\pi}^{\pi} dq \left(\cos\frac{q}{2} - \sqrt{\cos^2\frac{q}{2} + \Delta^2 \sin^2\frac{q}{2}}\right) \sim \delta^2 \log \delta. \tag{5.29}$$

The resulting state has a doubled unit cell, and hence there exist two ground states related by translational symmetry. The extreme cartoon limit, $t_{i,i+1} = [1 + (-1)^i]t_0$, corresponds to a dimerization in which the hopping on alternate bonds is doubled/switched off, respectively. The genesis of fractional charge is

Fig. 5.5 Fractionalization in $d = 1$ goes along with symmetry breaking. The figure shows a dimerized chain (top row) with two domain walls (bottom row). A dimer may represent either an electron preferentially hopping between its two endpoints in the spontaneously Peierls-distorted chain; or a singlet bond in the ground state of the Majumdar–Ghosh chain. In the Peierls case, two domain walls arise upon the removal of an electron, which thus have charge $e/2$ each. In the Majumdar–Ghosh chain, a singlet may be broken into a triplet, in which case the crosses of the domain wall denote spinons; or one electron may be removed from a singlet bond, leaving behind an unpaired spin, i.e., a spinon, and a mobile hole, also known as a holon. The independent motion of the two is known as spin–charge separation.

evident when removing the electron on one dimer, so that one site each with even and odd i is unoccupied (Figure 5.5). These two sites can now be separated without further energy cost by considering them as domain walls between the two degenerate states. Since the two empty sites are symmetry-equivalent, they must share the missing electron and hence each carry fractional charge $e/2$.

Note that Eq. 5.28 has already appeared in the chapter on integer topological phases (see Box 4.1) for the SSH model. There, the presence of edge states was noted. What has happened in the preceding discussion is that the dimerization arises spontaneously. This has promoted the edge states of Box 4.1 to domain walls in the interior of the chain.

Away from this cartoon limit, the missing fractional charge will not be perfectly localized but smeared over a broader region. The operator measuring the missing charge is hence not simply $\mathcal{Q}_i = c_i^\dagger c_i$, but instead a coarse-gained version:

$$\mathcal{Q}_i^{cg} = \lim_{\xi \to \infty} \sum_{j=-\infty}^{\infty} \mathcal{Q}_{i+j} \exp\left(-\frac{|i-j|^2}{\xi^2}\right) \qquad (5.30)$$

(see Box 5.3).

Box 5.3 Fractional Quantum Numbers

The total charge, $\mathcal{Q}_{\text{tot}}$, of an electronic system, in units of the electronic charge e, must be an integer – *always*. This is because we believe electrons to be *indivisible* particles with unit charge. It therefore does not seem to make any sense for there to exist quasiparticles with fractional charge, simply because no configuration appears to be available with the requisite quantum number. However, apparently robust arguments –

based on little more than the adiabatic theorem and elementary electromagnetism – do make the conclusion that fractionalized quasiparticles exist rather hard to escape.

The resolution of this paradox stems from the observation that an expectation value in an ensemble can nonetheless be fractional. Such an ensemble may arise due statistical weights provided by the amplitudes of a quantum wavefunction. For the latter case, let us define a *local* charge density operator, $\mathcal{Q}$, which we posit can have fractional eigenvalues:

$$\mathcal{Q}|f\rangle = q_f|f\rangle . \tag{5.31}$$

The global integer charge condition merely translates into the constraint that the total number of fractional particles add up to an integer. This is not entirely unlike a condition that the universe must be globally charge neutral, which does not preclude the existence of local accumulations of negative and positive charges.

However, to appreciate the subtlety of the situation, it is useful to make the distinction between eigen- and expectation values of such an operator. Having a fractional expectation value q_f is straightforwardly arranged, by, say, acting on an equal-amplitude superposition of two different states, one with charge 0 and the other with charge 1, so that $\langle \mathcal{Q} \rangle = 1/2$. However, this is clearly not an eigenvalue, as it fails the requirement that its variance vanish, $\langle \mathcal{Q}^2 \rangle - \langle \mathcal{Q} \rangle^2 = 0$ as implied by (5.31): $\mathcal{Q}^2|f\rangle = q_f^2|f\rangle$.

From the cartoon of the Peierls chain (Figure 5.5), it is clear that fractional charges do exist – the removal of one electron shared between two sites has generated two unpaired, well-separated domain walls, which by symmetry must be sharing the charge equally: their charge must be $1/2$.

Actually, fractional charge is a cooperative phenomenon, in this case requiring symmetry breaking (and, in higher dimension, some topological form of order). For an emergent cooperative phenomenon, it is rather more natural to consider a long-wavelength observable, rather than a lattice scale density operator like the number operator $\mathcal{Q}_i = c_i^\dagger c_i$.

An candidate operator is therefore a coarse-grained version of $\mathcal{Q}_i$, which samples the charge density around site i with an envelope decaying with a characteristic lengthscale ξ, formally taken to $\xi \rightarrow \infty$:

$$\mathcal{Q}_i^{cg} = \lim_{\xi \rightarrow \infty} \sum_{j=-\infty}^{\infty} \mathcal{Q}_{i+j} \exp\left(-\frac{|i-j|^2}{\xi^2}\right) . \tag{5.32}$$

As perhaps the simplest example, consider the two cases of alternating charges on a line, and the same with a domain wall in the charge order (Figure 5.6). Without the cut-off ξ, the charges in either case can be paired off into neutral pairs so as to leave behind none, or one, unpaired charge at the origin, so that the charge of a domain wall looks as indeterminate as the sum $\sum_{j=0}^{\infty}(-1)^j$. Computing the coarse-grained charge

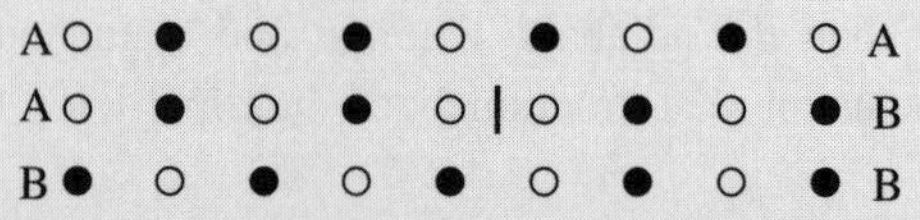

```
A ○  ●  ○  ●  ○  ●  ○  ●  ○ A
A ○  ●  ○  ●  ○ | ○  ●  ○  ● B
B ●  ○  ●  ○  ●  ○  ●  ○  ● B
```

Fig. 5.6 A domain wall between the two symmetry breaking states, A and B, of a charge density wave with period two. Removing one particle creates two identical domain walls, which therefore each have charge $1/2$. The number difference j sites to the right (left) of the wall compared to state A (B) is an alternating, nonconvergent series $(-1)^j$. The coarse-grained charge-density operator $\hat{Q}^{cg}$ (Eq. 5.33), however, yields a well-defined charge of $1/2$.

density operator centered on the origin, however, gives the domain wall charge relative to the ground state as

$$Q^{cg} = \sum_{j=0}^{\infty} (-1)^j \exp\left(-j^2/\xi^2\right) \rightarrow 1/2 \qquad (5.33)$$

with corrections which vanish exponentially rapidly in ξ^2. So, for this configuration of integer charges already, one finds an eigenvalue of Q^{cg} which is fractional.

This consideration is an essentially classical one: the "wavefunction" is a simple collection of immobile charges, so that it is an eigenfunction of each on-site charge density operator individually. This sweeps the added complexity of quantum fluctuations under the carpet. If the state is not an eigenfunction of the local charge-density operators Q_i, the result of the measurement also depends on the timescale of the measurement. This happens because acting with Q_i for a time t will by Heisenberg's uncertainty relation allow excitations to be created with an energy $\sim h/t$. Thus, for a sufficiently gentle measurement, with h/t smaller than the many-body gap Δ, the charge measured would be $1/2$; whereas for a more intrusive fast measurement, charge fluctuations of higher excited states would become visible. (Incidentally, the preceding has assumed the domain walls to be pinned for simplicity, so as to avoid the incidental complications of considering a moving wavepacket with a gapless kinetic energy.)

5.2.2 Irrational Charge

The term fractional charge is mildly misleading as the same picture would hold for particles without electrical charge. The terminology fractional number would therefore be more appropriate.

In fact, it is not even the fractional charge which is fixed at $|q| = \frac{|e|}{2}$. Rather, the charge can drift *continuously*. This can, for example, be achieved by adding a staggered potential $\mathcal{V}_{st} = \sum_i (-1)^i V\, c_i^\dagger c_i$ to the Peierls-distorted Hamiltonian

(Eq. 5.26; see also Eq. 4.15). Now, each dimer also goes along with a dipole moment, **P**, as sites with odd i host more charge than the even ones. Upon coarse-graining, the change in orientation of the dipole moment at the domain wall then contributes to its charge, as encoded by the relation $Q = \nabla \cdot \mathbf{P}$ between charge and dipole moment (densities): the charge of the domain wall can be tuned continuously away from $-\frac{e}{2}$ as **P** is a continuous function of V. For an explicit calculation, see Section 5.4.4.

Therefore, the "fractional" charge is generally *irrational*. (Note that, unlike in the case of an irrational charge arising from screening of the long-ranged Coulomb interaction as encoded by a relative permittivity $\epsilon_r > 1$, the irrational charge here occurs for a purely local Hamiltonian with only short-range interactions.) However, the sum of two domain wall charges on opposite sublattices always adds up to e, as it is one electron which is missing. In a gauge theoretic descriptions of fractionalized phases, the domain walls do carry a new, emergent charge, which simply encodes whether the domain wall is located on an even or odd sublattice. This emergent gauge charge is oblivious of any electric dipole moment.

5.2.3 The Majumdar–Ghosh Chain

This discussion has ignored the spin of the electrons, that is, it has assumed them to be spin polarized, for example, by a strong external Zeeman field. For spinful electrons, it turns out that dimerization occurs even in the absence of coupling to a lattice, and in the presence of a simple magnetic Hamiltonian for local moments (Majumdar and Ghosh, 1969).

Consider the Hamiltonian

$$\mathcal{H}_{MG} = \sum_i \left[(\mathbf{S}_{i-1} + \mathbf{S}_i + \mathbf{S}_{i+1})^2 - \frac{3}{4} \right] , \qquad (5.34)$$

which describes a Heisenberg $S = 1/2$ chain with nearest and next-nearest exchange interactions of ratio $J_{nn}/J_n = 1/2$. Its ground states are again dimerized, with a dimer now denoting singlet bond with two adjacent spins forming a spin singlet: $\mathbf{S}_i + \mathbf{S}_{i+1} = 0$. This follows from the fact that for three spins $S = 1/2$ the total spin $S_{\text{tot}}(S_{\text{tot}} + 1) = (\mathbf{S}_1 + \mathbf{S}_2 + \mathbf{S}_3)^2 = \frac{3}{4}$ or $\frac{15}{4}$, depending on whether they sum to $S_{\text{tot}} = 1/2$ or $S_{\text{tot}} = 3/2$. In the dimerized states, all the terms in the sum represent a sum of $1/2$, and hence we have a ground state.

One can now argue analogously that breaking one singlet bond by turning it into a triplet, amounts to creating a *pair* of domain walls, each of which carries spin $S = 1/2$. Domain wall states are no longer eigenstates of $\mathcal{H}_{MG}$, and the fact that there is a gap above the ground states was shown by Shastry and Sutherland (1981), who estimated that the "kinetic" energy gain of a hopping excitation cannot offset its creation energy cost.

Unpaired spins of this type are referred to as spinons. If a dimer is broken by removing one electron that forms it, what is left behind is one site with a spinon, and another hosting a missing electron, a holon. These two particles can move away from each other, thereby decomposing an electron into two particles, the spinon carrying its magnetic moment and the holon its electrical charge – this is known as spin–charge separation (Figure 5.5).

Like in the case of the Peierls-distorted chain, separating the fractionalized defects corresponds to growing a domain of one dimerization inside the symmetry-equivalent other one. This reflects the fact that in $d = 1$, fractionalization and symmetry are intertwined. We turn to another facet of this intertwinement next.

5.2.4 The AKLT Chain

In the instance of the Peierls/Majumdar–Ghosh chain, the fractionally charged excitations are intimately related to local symmetry breaking: they take the form of a domain wall between two different but symmetry-related ground states, namely, the pair of dimerizations differing by a lattice translation. Such defects are mobile: by shifting a dimer along, they can hop (while remaining on the same sublattice), and topological in that they can only pair-annihilate with an oppositely gauge-charged defect, that is, one on the opposite sublattice. Also, they can not be followed by an identically charged defect without an intervening oppositely charged one.

A one-dimensional example inspired by a search for generalizations of the Majumdar-Ghosh model is afforded by the $S = 1$ Heisenberg chain. This turns out to harbor fractionalized *immobile* defects localized at the edges of the system. These occur despite the absence in this model of both local symmetry breaking (as in the Majumdar-Ghosh chain) and topological order (as in the quantum Hall effect). However, it turns out that one needs protection by unbroken symmetries, as discussed separately in Section 11.1.

This is easiest to analyze for a $S = 1$ Heisenberg chain at its exactly soluble Affleck–Kennedy–Lieb–Tasaki (AKLT) point. As in the Majumdar–Ghosh chain above, the Hamiltonian is a projector of the Klein type (Box 5.4), which amounts to a bilinear-biquadratic combination of nearest-neighbor interaction terms:

$$\mathcal{H}_{\text{AKLT}} = \sum_i P_{i,i+1}^{L=2} = \sum_i \left[(\mathbf{S}_i \cdot \mathbf{S}_{i+1}) + \frac{1}{3} (\mathbf{S}_i \cdot \mathbf{S}_{i+1})^2 \right] . \quad (5.35)$$

AKLT found this exact solution for a gapped phase when the biquadratic term is present soon after Haldane (1983b) had conjectured that the bilinear-only antiferromagnet would be gapped for integer spin, in contrast to the known gapless-ness of the $S = 1/2$ Heisenberg antiferromagnet. Haldane's argument is based on mapping the path integral for the spin chain in the limit of large S to the nonlinear σ-model, which is gapped, for integer spin, but which acquires a nontrivial topo-logical term for half-integer spin, rendering it gapless. This beautiful field theory is explained pedagogically in Auerbach (1994).

Box 5.4 Klein Models

In the spirit of universality, when seeking a Hamiltonian to account for a particular phase (transition), there is considerable freedom of choice. Usually, it is convenient to select a particularly simple model Hamiltonian. For example, the ferromagnetic Ising transition in $d = 2$ is exhibited by a model $\mathcal{H} = -\sum_{\langle ij \rangle} \sigma_i^z \sigma_j^z$ incorporating only nearest-neighbor interactions, but also by one in addition featuring further-neighbor ones. However, the former is exactly solvable, while the latter in general is not. A word of caution is in order here: exactly soluble points do come with baggage of their own, and can in particular behave nongenerically on account of the fine-tuning implicit in them. The RK point with its anomalously soft photons and deconfined spinons in the square lattice quantum dimer model in the next section is a case in point.

Such solvability is rare for quantum spin systems, where it is already optimistic to hope for the availability of an explicit ground-state wavefunction. As for Laughlin's wavefunction, which is an exact ground state of Haldane's pseudopotential Hamilto-nian, such knowledge can be invaluable. A family of Hamiltonians, the Klein models, provide known ground states in this sense for quantum spin systems.

They generalize the Heisenberg Hamiltonian, which can be thought of as a projector which when acting on a spin pair projects out (exacts no energy cost from) a state with the pair in a singlet state:

$$\mathcal{H}_{ij} = J\mathbf{S}_i \cdot \mathbf{S}_j = P_{ij}^0 - \frac{3}{4} \, . \tag{5.36}$$

Thus, a singlet is the ground state of a Heisenberg interaction between two spins. Sim-ilarly, the chain Hamiltonians $\mathcal{H}_{MG}$ and $\mathcal{H}_{AKLT}$ are based on projectors: $\mathcal{H}_{MG}$ is a projector onto the space with $S_{\text{tot}} = 3/2$ formed by three consecutive spins, so that a state with a singlet bond between any two of three successive spins is a ground state. For $\mathcal{H}_{AKLT}$, the projector similarly requires $S_{\text{tot}} = 0, 1$ for a pair of neighboring spins $S = 1$.

The generalized Klein Hamiltonians from which the dimerized ground states fol-low have the form $\mathcal{H}_K = \sum_i P_{\Lambda(i)}^{\frac{z+1}{2}}$, where $P_{\Lambda(i)}^{\frac{z+1}{2}}$ projects the total spin of the cluster consisting of site i and its z neighbors onto the value $S_{\text{tot}} = \frac{z+1}{2}$, that is, it exacts no energy cost if the total spin is less than maximal. Therefore, if two spins in the cluster form a singlet (dimer), there is no energy cost.

We comment that, somewhat confusingly, projector Hamiltonians with a zero-energy ground state are known as frustration free in the quantum information community, while these often correspond to highly frustrated models in condensed matter language.

A cartoon of the ground state of $\mathcal{H}_{AKLT}$ is shown in Figure 5.7, where each spin $S = 1$ is represented by two spins $S = 1/2$; each of these forms a singlet bond with a $S = 1/2$ derived from a neighboring $S = 1$; a symmetrization projecting the pair of $S = 1/2$ spins onto total $S = 1$ is implied by the vertical boxes.

A more formal discussion of the AKLT chain as a symmetry-protected topological states will be given in Section 11.1. The remainder of this account takes a more pictorial approach to discussing the properties of the AKLT chain. Indeed, the cartoon of dimerized $S = 1/2$ spins making up the original $S = 1$ degrees of freedom is useful as it suggests the appearance of a fractional edge state: by terminating or cutting an AKLT chain, the $S = 1$ site at the edge of the chain loses a neighbor, and hence the imagined $S = 1/2$ loses the partner of its singlet bond, so that it appears to become an unpaired *physical* edge degree of freedom.

This spin $S = 1/2$ edge degree of freedom appears to be effectively disconnected from the bulk, as the interior spins $S = 1/2$ are fully dimerized, and the bulk hence gapped. This is where the limits of such a cartoon picture arise: the gap Δ_H is of course finite, as behooves a local bounded Hamiltonian, so that there will be a nonzero correlation length set by the inverse of the gap scale, $\xi = 1/\ln 3$. The spin density of the edge state thus leaks into the bulk, in the form of an oscillatory function with an exponentially decaying envelope, set by the same bulk correlation length. This is illustrated in Figure 5.8 for a $S = 1$ Heisenberg chain.

Fig. 5.7 Cartoon of the AKLT state and its edge states. Each vertical box is to be thought of as two spins $S = 1/2$ symmetrized to give the microscopic $S = 1$ degree of freedom. Each member of the pair is then involved in a singlet bond with one of the spins $S = 1/2$ from a neighboring $S = 1$. The actual wavefunction can efficiently be written down in MPS form (Eq. 5.39). Terminating the chain, or cutting it, thus leaves behind unpaired effective spins $S = 1/2$ at the endpoints. In practice, the $S = 1/2$ degree of freedom is localized near the edge, over a distance set by the inverse gap, $\xi \sim 1/\Delta_H$.

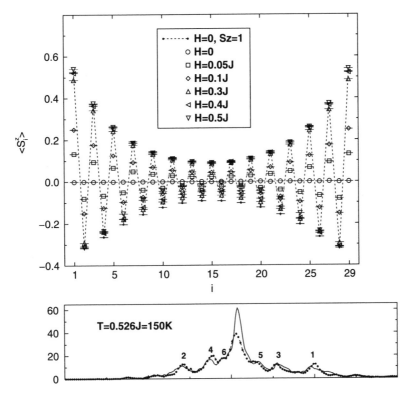

Fig. 5.8 (top) Edge modes of a finite-length $S = 1$ chain in a magnetic field at nonzero temperature (Alet and Sorensen, 2000). The edge modes hybridize in the chain center as a finite-size effect. (bottom) NMR signal of spin density distribution at an edge of the $S = 1$ chain compound $Y_2BaNi_{1-x}Mg_xO_5$ by Tedoldi et al. (1999), compared to QMC results (Alet and Sorensen, 2000). The numbers label the successive local magnetizations at sites moving away from the edge. Reprinted with permission by *Physical Review*.

The total spin of this oscillating cloud of spin density still adds up to exactly $S = 1/2$, well defined due to the SU(2) invariance of $\mathcal{H}_{AKLT}$. This also implies in particular that two edges of a finite chain of length L experience an interaction on a scale set by $\exp(-L/\xi)$. As the system is fully Heisenberg symmetric, this will manifest itself in an exponentially suppressed splitting between a low-lying singlet and triplet formed by the two edge spins.

It is thus only for $L \rightarrow \infty$ that these edge spins act like entirely free spins $S = 1/2$. One way to look for those is to detect their signature in the susceptibility. An edge spin, being effectively free, will thus make a contribution diverging at low temperature T in the usual Curie form

$$\chi_{\text{edge}} = \frac{C_{\text{edge}}}{3T} \, , \qquad\qquad (5.37)$$

with C_{edge} the Curie constant of the effective free edge spins. With the bulk response of the chain suppressed due to the presence of the gap, the low-T signal can thus be dominated by the response of the edge spins.

Strictly speaking, this requires a finite density of edges, which can for example be obtained by inserting internal edges into the chain via chemical substitution of magnetic ions by nonmagnetic ones. This was done for $Y_2BaNi_{1-x}Mg_xO_5$, where the $S = 1$ moments of the Ni^{2+} ions form approximately decoupled chains, which are disrupted by the nonmagnetic Mg^{2+} ions. Of course, an actual compound is not going to realize the AKLT Hamiltonian exactly, but thanks to topological stability, the properties outlined here will be robust for a ratio of biquadratic (J_4) to bilinear (J_2) couplings in the interval $J_4/J_2 \in (-1, 1)$. This interval thus includes not only the AKLT model, $J_4/J_2 = 1/3$, but also the pure Heisenberg point, also known as the Haldane chain, $J_4/J_2 = 0$.

For a spatially nonuniform system, however, a bulk quantity is not the optimal diagnostic. A spatially resolved measurement would allow access to the structure of the magnetization at the edge. This would have the additional advantage that the staggered signal may be considerably larger than the uniform one, as the oscillating nature of the spin density leads to cancellation between adjacent sites.

The technique of nuclear magnetic resonance (NMR) provides such a probe. It measures the size of local magnetic fields at the location of a nucleus of an ion by measuring the transition energies between levels of the nuclear spin split by the fields. The result of such an experiment (Tedoldi et al., 1999), which reflect the oscillating spin density of the edge states, is shown in Figure 5.8. This is compared to the results of a numerical model for the chain by Alet and Sorensen (2000).

As a historical aside, we mention that the AKLT state has played another important role in the intellectual development of condensed matter and many-body physics. This revolves around the possibility of writing its ground-state wavefunction, Ψ_{AKLT}, explicitly in what has become to be known as *matrix-product* form. For periodic boundary conditions, it reads

$$|\Psi_{\text{AKLT}}\rangle = \sum_{\alpha_i \in \{0, \pm 1\}} \left[\text{Tr} \left(\prod_{i=1}^{L} A_{\alpha_i} \right) |\alpha_1 \dots \alpha_N\rangle \right], \qquad (5.38)$$

where $\alpha_i = 0, \pm 1$ denote the z-component of the spin $S = 1$ at site i. Equation 5.38 is the general form of matrix product states (MPS) which underpin the use of the density matrix renormalization group (DMRG) in its modern formulation (Schollwöck, 2011). The basic structure is that the amplitude of each basis state of the d^L-dimensional Hilbert space is given by the product of the relevant A-matrices attached to each site.[6] The size of the matrices, known as bond dimension χ, varies. For a product state, one can choose $\chi = 1$. As the complexity of the state increases, so does the necessary bond dimension, and much of the effort in DMRG is devoted to using a computationally tractable small χ to achieve maximal accuracy.

For the AKLT state, things are still benevolent, and $\chi = 2$ suffices for its exact representation. The three 2×2 matrices A are given by

$$A_{\alpha=-1} = \frac{1}{\sqrt{2}} \begin{pmatrix} 0 & 1 \\ 0 & 0 \end{pmatrix} ; A_{\alpha=0} = \frac{1}{2} \begin{pmatrix} -1 & 0 \\ 0 & 1 \end{pmatrix} ; A_{\alpha=1} = \frac{1}{\sqrt{2}} \begin{pmatrix} 0 & 0 \\ -1 & 0 \end{pmatrix} \tag{5.39}$$

The form of this wavefunction manifestly enforces the constraint that $\alpha_i = \pm 1$ must be followed by $\alpha_j = \mp 1$, with any number of intervening $\alpha_{i<k<j} = 0$; successive identical $\alpha_i = \pm 1$ are forbidden. This implies that a string order parameter can be defined,

$$\mathcal{S}_{i,i+r} = S_i^z \left(\prod_{j=i+1}^{i+r-1} \exp(i\pi S_j^z) \right) S_{i+r}^z , \tag{5.40}$$

with $\lim_{r \to \infty} \mathcal{S}_{i,i+r} = -4/9$ for the AKLT state. The existence of a nonzero expectation value is not at odds with the absence of local symmetry breaking, as this is not a local order parameter.

The Majumdar–Ghosh state, incidentally, is also readily written as a MPS, now with $\chi = 3$ but of course only 2 different matrices, corresponding to spin-up and spin-down:

$$A_{\alpha=-1/2} = \begin{pmatrix} 0 & 1 & 0 \\ 0 & 0 & 1 \\ 0 & 0 & 0 \end{pmatrix} ; A_{\alpha=1/2} = \begin{pmatrix} 0 & 0 & 0 \\ -1 & 0 & 0 \\ 0 & 1 & 0 \end{pmatrix} . \tag{5.41}$$

The AKLT model has yielded tremendous insights into the physics of exotic one-dimensional magnets. As a note of caution, we would like to mention that this approach to generating nonmagnetic states is not guaranteed to work universally. For any lattice with coordination z, one can always generate a spin $S = z/2$ model wavefunction by the AKLT prescription. This thus involves decomposing the large

---

[6] A little confusingly, in quantum information theory, the d denotes not spatial dimensionality but the "size" of the local so-called qudit. The term qudit is a generalization of the conventional $S = 1/2$ qubit, for which $d = 2$.

spin into z spins $S = 1/2$, forming a singlet of two spins $S = 1/2$ at the end-points of each bond, and finally symmetrizing the z spins $S = 1/2$ on each site. This, however, can end up leading not to a nonmagnetic state, but to one with Néel order (Parameswaran et al., 2009), as for instance happens on the cubic lattice for $S = 3$. This is analogous to the observation that spins forming long-range valence bonds can also encode Néel order on the square lattice (Liang et al., 1988).

5.3 The Resonating Valence Bond Liquid

The resonating valence bond (RVB) liquid has been influential in the study of strongly correlated electron systems: following the original RVB proposal for a triangular lattice Heisenberg magnet (Fazekas and Anderson, 1974), it gained tremendous prominence following its resuscitation in the context of the cuprate superconductors (Anderson, 1987). Attempts to find an RVB liquid have generated much beautiful work. Unlike the Laughlin state, however, there is to this day no generally agreed experimental instance of an RVB liquid. Nonetheless, it provides perhaps the simplest setting in $d \geq 2$ dimensions where phenomena such as spin–charge separation and topological degeneracy can be visualized, and the connection to effective field theories be made transparent. It also shows how to generalize the physics of fractionalization from $d = 1$ to higher d; in particular, it emphasizes the different role played by local order and topology in the two cases.

The desire for a spin liquid phase arose when considering the phase diagram of the high-temperature superconductors, where an antiferromagnetically ordered state on the square lattice gives way to a superconducting state as the doping of holes into the ordered magnet is increased. However, a Néel state breaks trans-lational symmetry, which in turn frustrates hole motion, as a moving hole leaves behind a string of frustrated bonds (Figure 5.9). One possible resolution of this mutual frustration of kinetic and magnetic energy is to consider a spin liquid as a parent state of the superconductor, as this does not break any symmetries. To

Fig. 5.9 Frustrated hole kinetic energy in a symmetry-broken state in $d > 1$. A hole shuffling spins as it hops creates a string of domain walls (dashed lines). This happens not only in the Néel state of spins pointing up and down in an alternating fashion as shown, but also, for example, in a valence bond crystal state, where there is spatial order of singlet bonds.

put this vision into action suggests the search for a spin liquid state which is energetically at least closely competitive with an ordered state already in the undoped system, so that the gain of kinetic energy of a hole can tilt the balance in favor of the spin liquid.

5.3.1 How to Create a Spin Liquid

What Is a Spin Liquid?

Spin liquids, like non–Fermi liquids, do not have a generally accepted definition, in part because the definition in practical use has a negative component to it: minimally, any state which has magnetic long-range order does *not* qualify as a spin liquid. In addition, frequently it is demanded that the state in question not exhibit any other form of local order either. Having said that, an important type of spin liquid is the "chiral spin liquid," which does break time-reversal symmetry but no lattice symmetries, yet at the same time exhibits topological order. Another natural demand is for the phase not to be continuously connected to a trivial phase such as a band insulator, or a paramagnet. Again, this excludes certain models, such as the Shastry–Sutherland model which has been important in the study of exotic magnets, and it also disqualifies those experimental systems the corresponding spin liquid phase exists only at (experimentally unreachable) zero temperature.

So in practice, the term spin liquid denotes an exotic nonmagnetic state, and the tradeoff is between excluding cases which should not be included, and including those which should not be excluded, in particular the exotic spin state under consideration at a given moment in time. This trade-off is a matter of personal choice.[7]

In the context of this book, there is a natural such choice (which however is somewhat on the restrictive side): a spin liquid is a state of spins, the low-energy theory of which is a topological field theory. The most problematic aspect of this definition is the exclusion of gapless spin liquids, of which many interesting instances exist. A less restrictive condition is to demand that the magnet be described by a free emergent gauge field. Such a case is described using the example of spin ice at $T = 0$ toward the end of this chapter. This exhibits gapless excitations in the form of emergent photons mathematically identical to those of the familiar (Maxwell) electro-magnetism (see also Chapter 6).

The remainder of this chapter is devoted to exposing the basic phenomenology, as was done in the case of the Laughlin state above. Several of the ideas already

---

[7] Indeed, more fundamentally, it is not even clear what to call a spin, as SU(2) invariance is not a requirement for all topological spin liquids, so that just about any degree of freedom – such as occupancy of a site by a hardcore boson – can provide a pseudospin degree of freedom.

introduced for $d = 1$ turn out to be useful. The main – and highly nontrivial – innovation consists of giving meaning to them when transposed to higher dimension. Crucially, the role of local symmetry breaking in the emergence of fractional excitations will be entirely different. The fractionalized particles are no longer domain walls between symmetry-broken states. Instead, fractionalized magnets in d>1 are liquids.

Spin Liquids from Dimer Models

In order to discard any form of local spin order, a quantum state of a model with Heisenberg symmetry needs to be in a singlet representation of SU(2), as otherwise it is not rotationally invariant.[8] This immediately excludes wavefunctions based on product states over the states of spins on single sites: any individual spin with nonzero length $S > 0$ cannot represent a singlet. The wavefunction must therefore include some degree of entanglement between different sites; the simplest route turns out is to follow the construction of the Majumdar–Ghosh model and to obtain a set of singlet-bond ground states. What is new is to combine these – there are now exponentially many in $d = 2$ – so that the resulting state is a topological liquid which breaks no spatial symmetries.

This proceeds in three steps. First, find a Hamiltonian whose ground states are represented by dimer coverings. For the Majumdar–Ghosh chain, this was

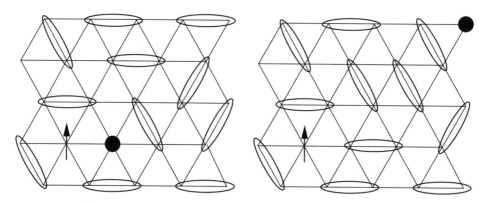

Fig. 5.10 Spin–charge separation and unfrustrated hole motion in a resonating valence bond (RVB) liquid. In the absence of long-range order, motion of a hole does not generate a domain wall so that it can move long distances essentially unimpeded. Removing an electron from a dimer in the bottom left of the figure leaves behind an unpaired spin (spinon) and a charged spinless hole (holon). These can propagate separately, in an instance of spin–charge separation in $d > 1$.

---

[8] To be precise, for finite-size systems, ground states are generically global singlets, $\mathbf{S}_{tot} = 0$. The development of magnetic long-range order like in the Néel state shows up as a degeneracy emerging in the thermodynamic limit between states with different values of $\mathbf{S}_{tot}$ (Anderson, 1952).

depicted in Figure 5.5, where like in Figure 5.10, a singlet bond is denoted by a dimer linking the two spins forming it. Second, add perturbations to this Hamiltonian to lift the degeneracy between dimer states. Third, verify that there exists a set of perturbations for which the ground state is a spin liquid, and study its properties.

Regarding the first step, dimer coverings of the lattice are zero-energy ground states of the Klein Hamiltonian $\mathcal{H}_K = \sum_i P_{\Lambda(i)}^{L=\frac{z+1}{2}}$ (see Box 5.4), where $\Lambda(i)$ denotes the set of spins i along with its z neighbors.

A little additional acrobatics is needed to make sure the dimer states are the *only* ground states of the Klein Hamiltonians. Also, different dimer states are not orthogonal, nor are they necessarily linearly independent. As the following account is interested in the simple physical pictures, we refer the reader to specialized literature on how these issues are addressed in detail (e.g., Moessner and Raman, 2008). We thus skip ahead to a discussion of a popular and historically influential family of resulting Hamiltonians describing the lifting of the ground state degeneracy. These are the Rokhsar–Kivelson quantum dimer models. Their effective dimer Hamiltonian reads,

$$H_{\mathrm{QDM}} = -t \sum_\square \left(|\text{▯}\rangle\langle\text{▭}| + |\text{▭}\rangle\langle\text{▯}| \right) + v \sum_\square \left(|\text{▯}\rangle\langle\text{▯}| + |\text{▭}\rangle\langle\text{▭}| \right). \qquad (5.42)$$

This pictorial representation of a Hamiltonian is to be understood as follows. The Hamiltonian acts on states in a Hilbert space given by all hardcore dimer coverings of the lattice (square and triangular, the cases we address in the following). These imply that each spin S=1/2 forms a singlet bond with one, and only one, of its nearest neighbors.

The sum in Eq. 5.42 runs over all the plaquettes, $\square$, of the lattice. The first term is known as the kinetic, or resonance term: a pair of dimers are reconnected ("resonate") between the four sites they join. (It is by analogy to the picture of the resonance between the two ways of sharing three valence electrons forming dimers between the six sites of the benzene ring that the term resonating valence bond physics originated.)

The second, potential, term counts the number of plaquettes which could participate in such a resonance move, without actually making them resonate. (It is thus diagonal in the dimer basis, just as the potential energy of a particle is diagonal in the position basis.) It is convenient to add this term for a number of reasons. First, it turns out that spin liquid phases are considerably more common for $v \lesssim t$ than for $v = 0$. Second, and most conveniently computationally, for $v = t > 0$, the groundstate(s) of $\mathcal{H}_{QDM}$ are explicitly known: these are the Rokhsar–Kivelson wavefunctions, so named after the inventors of these short-range resonating valence bond models.

Indeed, at the RK point $v = t$, the Hamiltonian turns into (yet another) sum of projectors:

$$H_{\text{QDM}}^{v=t} = v \sum_{\square} \left(|\square\rangle - |\square\rangle \right) \left(\langle\square| - \langle\square| \right). \tag{5.43}$$

For a simple plaquette, this straightforwardly gives the resonating valance bond wavefunction – the equal amplitude superposition of both dimer coverings – as ground state. One can now explicitly verify that

$$|\text{RK}\rangle = \frac{1}{\sqrt{N_c}} \sum_c |c\rangle, \tag{5.44}$$

the equal amplitude sum over all dimer coverings, is a ground state for $\mathcal{H}_{RK}$ on any lattice. This follows from the fact that it has zero energy, the minimum possible for a Hamiltonian consisting of projectors with eigenvalues ≥ 0. Note that expectation values of operators that are diagonal in the basis of dimer coverings: $\langle c|\hat{O}|d\rangle = O_c \delta_{c,d}$, can now be obtained from a classical ensemble of dimer coverings:

$$\langle \hat{O} \rangle = \langle \text{RK}|\hat{O}|\text{RK}\rangle = \sum_{c,d} \langle c|\hat{O}|d\rangle / N_c = \frac{1}{N_c} \sum_c \langle c|\hat{O}|c\rangle = \frac{1}{N_c} \sum_c O_c, \tag{5.45}$$

where the final term denotes a straight classical average over the N_c dimer coverings.

The advantage of such a procedure is that it makes available a concrete wavefunction, the properties of which are accessible to *classical* Monte Carlo simulations, or exact solution (Box 5.5), so that a more mechanical approach than ingeniously guessing an appropriate many-body wavefunction is available. Such an approach can then be complemented by purely numerical studies of other Hamiltonians with closer connection to experiment.

5.3.2 Properties of the RVB Liquid

A number of RVB liquids have been discovered in this way. For reasons which will be discussed in the next chapter, in $d = 2$ RVB liquids do not exist on bipartite lattices, that is, lattices which can be subdivided into two sublattices with all neighbors of a site on one sublattice residing on the opposite sublattice. As the square lattice is bipartite, the high-T_c based search for RVB liquids in this setting therefore faltered and was largely abandoned for a few years. It was on the triangular lattice that an RVB liquid phase was eventually discovered (Moessner and Sondhi, 2001b), the simplest, nonbipartite Bravais lattice with coordination six.

The great advantage of the route via a quantum dimer model is that a very physical picture of the topological spin liquid emerges. The next few paragraphs describe its unusual ground-state properties and the low-lying excitations. Here, we appeal to the RK point wavefunction, where many computations can be carried out explicitly. The demonstration that this point extends into a phase rather than just describes an isolated outpost is more delicate, but much numerical evidence is now available to demonstrate that this is the case. We also contrast the behavior of the triangular RK-QDM to that on the square lattice, which exhibits no liquid phase.

The Hamiltonian of the QDM on the triangular lattice is entirely analogous to Eq. 5.42. The fact that a triangular plaquette has an odd number of sites means that resonance moves have to involve a rhombohedral plaquette pair hosting two dimers on its four sites:

$$H_{QDM} = -t \sum \left(|\diamondsuit\rangle\langle\diamondsuit| + |\diamondsuit\rangle\langle\diamondsuit| \right) + v \sum \left(|\diamondsuit\rangle\langle\diamondsuit| + |\diamondsuit\rangle\langle\diamondsuit| \right). \tag{5.46}$$

This Hamiltonian is then to be understood as a sum over the three different orientations of such plaquettes. The projector form of the Hamiltonian at the RK point is again analogous to the one on the square lattice.

Classical Dimer Partition Functions and Correlations

The correlations of the dimer model can be computed using fermionic path integrals, as explained in Box 5.5. The nature of the dimer correlations is encoded in the Kasteleyn matrix, in particular whether its spectrum is gapped or not. The spectrum of the Kasteleyn matrix for the triangular lattice, Figure 5.11, is readily obtained by diagonalizing it via a Fourier transform. Whether or not there is a gap is determined by the existence of zeroes of the quantity

$$\Gamma(k_x, k_y) = \sin^2(k_x) + \sin^2(k_y) + t^2 \cos^2(k_x + k_y). \tag{5.47}$$

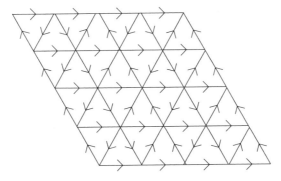

Fig. 5.11 The Kasteleyn sign convention for the triangular lattice. Arrows along the bond from site i to j denote the sign of the matrix entry K_{ij}. A square lattice convention is obtained by omitting one of the three bond directions. Reprinted with permission by *Physical Review*.

For the isotropic triangular lattice, $t = 1$, so that this expression is gapped throughout the Brillouin zone.

From, this fact, it follows that the inverse, K^{-1}, of the Kasteleyn matrix, which encodes the dimer correlation functions (Box 5.5), has matrix elements decaying exponentially in real space. Therefore, so do the dimer correlators. In fact, this exponential decay, by Wick's theorem, holds for *any* multidimer correlator. This proves that the triangular quantum dimer model at RK point represents a genuine *short-range* resonating valence bond liquid. This construction has therefore gifted us – analogously to Haldane's pseudopotentials and the Laughlin state – an exactly soluble instance of an interacting topological phase of matter.

The same construction for the square lattice RK-QDM yields an entirely different result. The spectrum is now given by the same expression (5.47) with $t = 0$, so that there is a vanishing gap at $k_x = k_y = 0$. The dimer correlations are therefore algebraically decaying: here, the RK point is a critical point, sitting at a transition between valence bond crystals. This is evidenced by the appearance of a soft mode, the $\pi 0 n$, related to the incipient ordering at wavevector $(\pi, 0)$. This can be captured by the single mode approximation, Box 5.1, where the variational state $|q\rangle$ is constructed by acting on the ground state at the Rokhsar–Kivelson point with the Fourier transform of the dimer density operator near $(\pi, 0)$.

We note in passing that this quantum critical point is remarkable in that it hosts deconfined fractionalized excitations despite the fact that its neighboring phases are both conventional and confining (Moessner and Sondhi, 2001b). However, this is an effect of fine-tuning implicit in the RK point. A more robust form of deconfined quantum criticality is discussed in Senthil et al. (2004).

Box 5.5 Classical Dimer Models and Their Correlations

There is an elegant way of computing properties of classical dimer models using fermionic path integrals. This then permits the computation of correlators of ground-state wavefunctions of the Rokhsar–Kivelson type, $|RK\rangle$.

The central idea is that the hardcore condition – each site has to be part of one and only one dimer – is like the basic property of Grassmann integrals that the only monomial which does not integrate to zero is of first order in a Grassmann variable η:

$$\int d\eta \, \eta^k = \delta_{k,1}. \tag{5.48}$$

The partition function of a dimer model is then given by (i) defining Grassmann variables η_i on each site of the lattice, and (ii) an antisymmetric matrix K_{ij} known as the Kasteleyn matrix, the entries of which are the weights (fugacities) of the dimers linking sites i and j (for an equal amplitude superposition, $|K_{ij}| = 1$ for bonds of the lattice and $K_{ij} = 0$ otherwise):

$$Z = \int d[\eta] \exp\left(\frac{1}{2}\eta_i K_{ij}\eta_j\right) . \tag{5.49}$$

The matrix K needs to be chosen such that all dimer coverings are counted with a positive weight. This is not trivial, as the order of the η_i in the integrals matters for Grassmann variables: $\int d\eta_2 d\eta_1 \eta_1 \eta_2 = -\int d\eta_2 d\eta_1 \eta_2 \eta_1 = +1$. A theorem proved by Kasteleyn (1961) states that such a matrix exists for planar graphs, that is, in particular for lattices without crossing bonds such as the square, honeycomb or triangular, but not the checkerboard lattice (Figure 5.13).

The choice of sign is given by a simple arrow convention, where an arrow pointing along a bond from site i to j implies $K_{ij} = -K_{ji} > 0$. Kasteleyn's theorem states that Eq. 5.49 holds if there is an odd number of arrows pointing in the clockwise direction for each and every elementary plaquette. To see how this works, consider a single square plaquette, with two dimer coverings $Z = 2$, for which $K_{12} = K_{23} = K_{34} = K_{14} = -K_{21} = -K_{32} = -K_{43} = -K_{41} = 1$. Writing $\frac{1}{2}\sum_{ij} \eta_i K_{ij}\eta_j = \sum_{i<j} \eta_i K_{ij}\eta_j$, one obtains by Taylor expanding the exponential:

$$Z = \int d\eta_4 d\eta_3 d\eta_2 d\eta_1$$

$$\times \left[1 + \sum_{i<j} \eta_i K_{ij}\eta_j + \frac{1}{2}\sum_{i<j}\sum_{k<\ell} \eta_i K_{ij}\eta_j \eta_k K_{k\ell}\eta_\ell + \ldots \right], \tag{5.50}$$

where the first two terms in brackets, as well as the omitted higher order terms vanish by Eq. 5.48. The remaining term in brackets is $2 \cdot \frac{1}{2} [\eta_1 K_{12}\eta_2\eta_3 K_{34}\eta_4 + \eta_1 K_{14}\eta_4\eta_2 K_{23}\eta_3] = [\eta_1\eta_2\eta_3\eta_4 + \eta_1\eta_4\eta_2\eta_3] = 2\eta_1\eta_2\eta_3\eta_4$, so that $Z = 2\int d\eta_4 d\eta_3 d\eta_2 d\eta_1 \, \eta_1\eta_2\eta_3\eta_4 = 2$, as desired.

Unfortunately, lattices with periodic boundary conditions are not planar but it turns out that in $2d$, a sum of 4 terms like the ones above correctly evaluates the partition function. For $3d$, no such exact result is available, unfortunately. In passing, we note that Ising models on planar graphs can be mapped onto dimer models, and thus be solved efficiently. In particular, even the disordered "Ising spin glass" $\mathcal{H} = \frac{1}{2}\sum_{ij} J_{ij}\sigma_i\sigma_j$ on the square lattice can be solved exactly by this method.

For the purpose of computing correlators, Eq. 5.50 is further useful as any dimer correlation can also be computed, as follows. In the simplest case, dimer occupancies, one can evaluate the relative weight of all dimer configurations with dimer (ab) present as

$$W_{(ab)} = \frac{1}{Z}\int d[\eta]\eta_a K_{ab}\eta_b \exp\left(\frac{1}{2}\sum_{ij} \eta_i K_{ij}\eta_j\right) = \left(K^{-1}\right)_{(ab)} K_{ab}, \tag{5.51}$$

which again follows from Taylor expanding the integrand. The last equality follows from the standard properties of Grassmann integrals, where K^{-1} denotes the matrix inverse of K.

This being a quadratic action, multidimer correlations simplify via Wick's theorem. For instance, the correlation between dimers (ab) and (cd) is given by

$$C_{(ab),(cd)} = \frac{1}{Z} \int d[\eta] \eta_a K_{ab} \eta_b \eta_c K_{cd} \eta_d \exp\left(\frac{1}{2} \sum_{ij} \eta_i K_{ij} \eta_j \right)$$

$$= \left[K^{-1}_{(ab)} K^{-1}_{(cd)} + K^{-1}_{(ad)} K^{-1}_{(bc)} - K^{-1}_{(ac)} K^{-1}_{(ba)} \right] K_{ab} K_{cd} \, . \tag{5.52}$$

The first term in brackets accounts for the "disconnected" independent probabilities of having the two dimers present, while the other two terms account for correlations between the dimers. Whether or not the long-distance correlations decay algebraically or exponentially is then a simple property of the structure of the spectrum of K^{-1}, and hence K: gaplessness implies algebraic decay of the correlators with distance.

Topological Order: Winding Parity and Winding Number

A topological property which is particularly easy to visualize in the RVB liquid is that of topological order. In Figure 5.12, a winding parity is defined. This is given by the number of dimers crossing an (imaginary) line wrapping around the torus.

This parity is invariant under any local dimer dynamics. This can be seen for the simple dimer resonance of the RK model. This involves displacing dimers so that the old and new configuration superimposed on each other form a loop which, crucially, is closed. Thus, the winding parity can never change, unless the closed loop itself winds round the system, so that the cut crosses it only once. However, in the RK-QDM, no such nonlocal resonance moves are included. In a more general model, their amplitude would be exponentially small.

As the winding parity is a nonlocal observable which can only be obtained by considering dimer correlations across the system size L, for a liquid with correlation length ξ, any energy splitting between sectors with differing winding parity is itself going to be exponentially small in ξ/L. This is the splitting of topologically quasi-degenerate multiplets captured by the theory of Wen and Niu (1990).

Deconfinement and Spin-Charge Separation

Imagine removing one electron from a dimer, so that a hole is left behind, along with an electron that is now unpaired (Figure 5.10). As in the case of the Majumdar-Ghoash model, the former is known as a holon, the latter a spinon, although for the present purposes we represent both simply as monomers. The

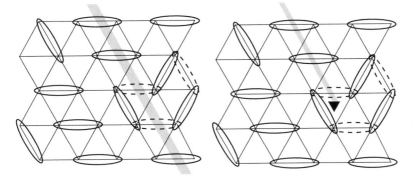

Fig. 5.12 (left) The parity of the number of dimers crossing a cut across the system (say, winding around a cylinder for periodic boundary conditions) is unchanged by local dimer resonance dynamics. For the square lattice, it is a winding number that is conserved because the dimers can be oriented to point from one sublattice to the other. (right) A vison wavefunction (Eq. 5.53) includes the winding parity with respect to a string that emanates from the plaquette hosting the vison (inverted black triangle). It does change sign under local dynamics of the type indicated by replacing the solid dimers by the ones indicated by dashed lines. This can occur by two sequential dimer resonance moves, or via a dimer breaking into a pair of monomers and encircling the vison before recombining. The resulting π phase from changing the parity of dimers crossing the cut amounts to a -1 relative Aharonov-Bohm phase of vison and monomer.

nomenclature conveys the message that these two entities behave like independent emergent quasiparticles. As depicted in this cartoon of spin–charge separation in $d = 2$, electric and magnetic properties of the electron are carried by two distinct quasiparticles.

Exactly at the RK point, the energy of the pair does not depend on its separation as $\mathcal{H}|\text{RK}\rangle = \mathcal{H}|\text{RK}; (i, j)\rangle \equiv 0$ for all locations (i, j) of holon and spinon. This follows from the fact that $\mathcal{H}_{RK}$ (5.43) annihilates the equal amplitude superposition of all dimer coverings consistent with monomers at sites (i, j) regardless of the location of the monomers.

Away from the RK point in the RVB phase, this still holds as a consequence of the finite correlations length in the dimer liquid – for $|i - j| \gg \xi$, the energy of the state $|\text{RVB}; (i, j)\rangle$ depends on $|i - j|$ only exponentially weakly. In particular, it remains finite. This connects to the notion of deconfinement familiar from high-energy physics (see Section 6.2.2).

Visons

Besides the spinons/holons, there exists a second gapped excitation, which is a defect not in the "diagonal" dimer terms, but in the conjugate, "off-diagonal"

kinetic energy. In the gauge theory language of Chapter 6, the former are electric, while the latter are magnetic excitations. These are known as Ising vortices, or
in short, visons, a nomenclature common in condensed matter but not high energy
theory.

Visons reside on plaquettes of the lattice (Eq. 5.54), for which the kinetic energy
is frustrated in that the resonance of dimers around the plaquette costs, rather than
gains, energy. In detail, the wavefunction of a vison is related to that of the ground-
state RK wavefunction by

$$|\{\alpha\}\rangle = \frac{1}{\sqrt{N_c}} \sum_c (-1)^{w_c} |c\rangle \ , \tag{5.53}$$

where w_c is the parity of dimers cut by the string connecting the vison in plaquette
α to the boundary of the sample. (For periodic boundary conditions, both ends of a
string must terminate on a vison, so that the total number of visons must be even.)

Visons play an important role for a number of reasons. First, along with the
monomers, they are the natural excitations of the topological theory describing
the topological RVB liquid. Second, they can be the lowest-energy excitation and
in particular go soft and condense, thereby terminating the deconfined topological phase in favor of, for example, a more conventional valence-bond crystal, the
translational symmetry breaking of which confines the fractionalized excitations.
Third, they can bind to monomers to form composite objects with altered quantum
statistical properties, as discussed next.

Statistical Transmutation

What are the relative statistics of these gapped excitations of the RVB liquid? Let
us first remark is that quantum statistics are not as immutable in condensed matter
physics as in high-energy physics, where the Poincare group dictates them via a
spin-statistics relation. So far, we have not specified the statistics of the monomers
at all – we have indeed not even given them any dynamics which would allow us
to braid them, so far having just kept them static.

As a first step, let us consider the phase picked up by a monomer moving around
a plaquette, depending on whether a vison sits in the plaquette or not. There will
be a contribution to the phase of the hopping process from the hopping matrix
elements, which however does not depend on the presence or absence of the vison.
By contrast, once the monomer has returned to its original location, the dimer parity
of the string of the vison has changed, so that the wavefunction picks up a factor
of -1.

The vison thus provides the same Aharonov–Bohm phase as a π-flux – hence
its name "Ising vortex." It is now possible, by simple energetic means, to bind a
vison to a monomer, thereby altering the relative statistics of the monomers via

this flux attachment. For instance, one can add a resonance term for a length six dimer resonance loop,

$$t_6 | \diagdown \diagup \rangle \langle \diagdown \diagup | + \text{h.c.} \tag{5.54}$$

It then becomes favorable to bind the vison as the resulting wavefunction gains, rather than pays, a (sufficiently large) energy t_6 when the triplet of dimers resonates. However, as the bound object may have a hard time hopping together, the energy gain must also compensate for this possible loss of kinetic energy. If that happens, transporting one monomer around another thus involves an additional phase due to the bound vortices.

The issue of binding a monomer to a vortex is therefore a low-energy property which depends on details. From a topological perspective, such details cannot matter – they can be tuned continuously, and certainly do not depend on global properties such as the genus of the surface on which we have defined the model. Rather, topologically, in the sense of "as seen from a large distance," we really have four distinct possibilities: there are either no excitations; or there is one of three excitations: a vortex, a monomer, or both together. It is the relative energetics of these which appears as a matter of detail from this perspective, although it is clearly of great potential importance for the observable low-energy physics. We will return to the different types of quasiparticles from a topological perspective in the context of superselection sectors in the toric code (Section 6.3).

Gapfulness of Collective Modes

Having taken care of gapped monomers and visions, the electric and magnetic charges, the next question is whether there are any other candidates for gapless modes, involving just the manifold of dimers, such as an electromagnetic photon. Here, no exact results are available, but it is by now well established that all such excitations are gapped in the RVB liquid (Ralko et al., 2006). In particular, the two gapless modes from the square lattice quantum dimer model, the resonon and the $\pi 0n$, are predicted to be gapped, at least within the single-mode approximation (Box 5.1).

5.4 Spin Ice

While fractionalization is not so hard to achieve in $d = 1$, there is a dearth of experimentally accessible topological spin states in two dimensions, $d = 2$. It therefore looks at first rather unpromising instead to look for them in even higher dimension $d \geq 3$. After all, it is also well known that it becomes increasingly hard to destabilize a conventional ordered state in higher dimensions, with the Néel state, the ultimate nonexotic state, even turning into an exact ground state on a hypercubic lattice in the limit $d \rightarrow \infty$. While this is true, destabilizing conventional states is only one part of the story. The other is to find competing unconventional phases which are themselves stable. It is here that $d = 3$ fares better, with even bipartite lattices supporting a Coulomb spin liquid phase. It is named thus due to its effective description is that of a free U(1) gauge field familiar from Maxwell electromagnetism. This even includes emergent photons – linearly dispersing transverse excitations – as collective spin excitations.

While these photons have not yet been observed experimentally, a classical version of a Coulomb spin liquid has been realized in a family of materials known as spin ice (Bramwell and Gingras, 2001), which most prominently includes the rare earth titanates $Dy_2Ti_2O_7$ and $Ho_2Ti_2O_7$, where the original experimental discovery occurred (Harris et al., 1997). Since in these, the physics of the emergent gauge field is most transparent, the remainder of this chapter covers an exposition of theoretical and experimental studies on spin ice. It first discusses how conventional magnetic order is absent on account of geometric frustration of the magnetic interactions in spin ice, then explains how the resulting behavior is well described by an emergent gauge field, before giving an account of the peculiar consequences of this structure, most saliently the emergence of magnetic monopoles and artificial photons as quasiparticles. Spin ice thus provides a natural stepping stone into the following chapter on the gauge- and field-theoretic description of these fractionalized phases. Reviews covering the physics of the emergent gauge field in spin ice are also available (Henley, 2010; Castelnovo et al., 2012).

5.4.1 Frustration Leads to Degeneracy

The microscopic degrees of freedom in spin ice are as simple as they come: Ising spins which can only take on values ± 1. These reside on the three-dimensional pyrochlore lattice, which consists of corner-sharing tetrahedra; Figure 5.13). Their Ising nature is a consequence of the crystal field environment of the rare earth RE^{3+} ions, which forces the electronic magnetic moments to be oriented along the local crystallographic (111) symmetry axes (see Section 8.8.1).

The noncollinearity of these axes, together with the long-range nature and complicated angular dependence of the dominant dipolar interactions between the spins makes for a rather involved spin Hamiltonian:

$$H = \frac{\mu_0 \mu^2}{4\pi} \sum_{i<j} \left[\frac{\mathbf{S}_i \cdot \mathbf{S}_j}{r_{ij}^3} - \frac{3 \left(\mathbf{S}_i \cdot \mathbf{r}_{ij} \right) \left(\mathbf{S}_j \cdot \mathbf{r}_{ij} \right)}{r_{ij}^5} \right] + J \sum_{\langle ij \rangle} \mathbf{S}_i \cdot \mathbf{S}_j . \quad (5.55)$$

Here, $\mathbf{S}_i = \sigma_i \hat{d}_{\kappa(i)}$, with $\hat{d}_{\kappa(i)}$ denoting the local easy axis for the $\kappa = 1 \ldots 4$ sublattices of the pyrochlore lattice (Figure 5.13). The direction of the spin along

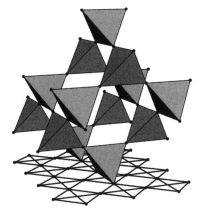 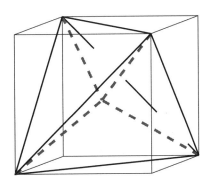

Fig. 5.13 (left) the pyrochlore lattice, a network of corner-sharing tetrahedra, which hosts the spins of spin ice. A projection of four successive layers of the pyrochlore lattice to two dimensions yields the checkerboard lattice, also known as square lattice with crossings, shown below the pyrochlore lattice. From Moessner et al. (2004). (right) The $i = 1 \dots 4$ distinct but symmetry equivalent Ising axes, $\hat{d}_i$ (dashed lines) correspond to body diagonals of the cube, i.e., the $\langle 111 \rangle$ directions. In the ground state, two spins have to point into, and two out of, each tetrahedron. From Moessner and Sondhi (2003a). Reprinted with permission by *Physical Review*.

the easy axis is thus encoded by the Ising variables σ_i. The size of the magnetic moments of the spins is denoted by μ. The dipolar interactions turn out to dominate because the superexchange between f-electrons is quite weak, while the magnetic moments of the Dy^{3+} and Ho^{3+} ions are huge, of order of $\mu = 10\mu_B$, so that a typical nearest-neighbor dipolar energy is in excess of 1 Kelvin.

It therefore came as quite a surprise that the ground state of this model Hamiltonian were rather simply characterized: for each tetrahedron, two spins have to point in, and two spins out. These are called ice rules, as they are isomorphic to the description of the proton positions in ice by Bernal and Fowler. These rules in fact encode the ground states of a much simpler model, namely, the Ising antiferromagnet on the pyrochlore lattice with only nearest-neighbor interactions, obtained when no dipolar interactions are present ($\mu = 0$):

$$\mathcal{H}_{nn} = \frac{J}{3} \sum_{\langle ij \rangle} \sigma_i \sigma_j = \frac{J}{6} \sum_\alpha \left(\sum_{i \in \alpha} \sigma_\alpha \right)^2 . \tag{5.56}$$

The ice rules thus imply $\sum_{i \in \alpha} \sigma_i = 0$: all spins in tetrahedron α have to sum to zero, that is, "two-up, two-down." There are $\binom{4}{2} = 6$ ways of accomplishing this, as there is a choice of which pair of opposite bonds of the tetrahedron to frustrate. Since there are three symmetry-equivalent pairs of bonds, and two Ising directions, six ground states result: the local degeneracy arises from a combination of geometric frustration – not all bonds can simultaneously be satisfied – and local symmetry – the bond pairs are equivalent. For this reason, the highly symmetric – all sites and bonds are symmetry-equivalent – yet strongly frustrated pyrochlore lattice is a particularly prominent place to look for new physics (Anderson, 1956).

For the thermodynamic system, the ground-state degeneracy results in an extensive ground-state entropy, first estimated by Pauling as follows. He decimated the number of microstates – 2^N for a system of N Ising spins – with the constraints imposed by the ice rules: each of the $\frac{N}{2}$ tetrahedra reduces the state space by permitting only 6 out of 16 configurations as ground states. Hence,

$$S_p = \frac{1}{N} \ln \left[2^N \left(\frac{6}{16} \right)^N \right] = \frac{1}{2} \ln \frac{3}{2} \approx 0.2027 \qquad (5.57)$$

is Pauling's estimate of the ground-state entropy per spin, which is about 30% of the entropy, $\ln 2$, of a free spin: the ice rules severely underconstrain the ground state. This number actually provides a rather accurate estimate. For a two-dimensional version of ice, square ice (the six-vertex model on the square lattice), to which this estimate also applies, the exact value is $S_L = \frac{3}{4} \ln \frac{4}{3} = 0.2157\ldots$, while a numerical estimate for spin ice is $S_{SI} = 0.2050$, all close to the experimental value (Ramirez et al., 1999).

Like in the case of the Landau level in the quantum Hall problem, and of the Hilbert space spanned by the dimer coverings, we find a low-energy space which is huge and degenerate. These were just the ingredients identified above for finding novel phases.

Like in the case of the RK wavefunction, the first question is what (unusual) correlations might be encoded in this ensemble of states. Next, as in the FQHE, one wonders how lifting of the degeneracy by perturbations can yield further correlated states. Since the answers to these questions remarkably often turn out to be interesting, frustrated magnets have for a long time played a prominent role in the search for new and exotic states of matter.

5.4.2 *Emergent Gauge Field and Magnetostatics*

In each of the above three cases, the low-energy Hilbert space contains a hard constraint: projection onto a Landau level; the hardcore dimer constraint; or the ice rule. Extracting universal long-wavelength behavior in the presence of such (short-range) constraints is among the central challenges of correlated electron physics. Normally, in a renormalization-style procedure, one coarse-grains degrees of freedom to access longer lengthscales, such as in a block-spin transformation.

However, a "two-in, two-out" constraint does not coarse-grain easily, as it would eventually just lead to a balanced "in/out" spin arrangement, that is, a micro-canonical zero-magnetization ensemble with little structure at long-wavelengths. Sometimes, in related problems, indeed the ensemble of degenerate states is in the same universality class as a conventional paramagnet, for example, in the case of the ground states of the Ising antiferromagnet on the kagome lattice.

In the case of spin ice, however, the ice rule does give rise to considerable additional structure at long wavelengths. This happens because it is equivalent to a conservation law. This conservation law is emergent, that is, enforced by the energetics of $\mathcal{H}_{SI}$, and not a symmetry property of it present at all temperatures. But, regardless of its origin, this conservation law is responsible for correlations in addition to those expected for a paramagnet, as demanded by the general hydro-dynamical viewpoint that it is the ensemble of broken symmetries and conserved quantities which determine the long-wavelength behavior of a physical system, see Chaikin and Lubensky (1995). In the case of spin ice, the correlations can be derived explicitly, as described next. Note that here, quite typically, a key effect of enforcing hard constraints is to generate effective degrees of freedom quite dis-tinct from the original, unconstrained ones: here, we find a divergence-free flux replacing the original Ising spins.

Let the microscopic magnetic moment of each spin $\mathbf{M}$, define a unit of lattice flux along the bond of the lattice formed by the centers of the tetrahedra on which it resides. (This lattice is the familiar diamond lattice, the midpoints of the bonds of which form the sites of the pyrochlore lattice.) The direction of the flux is then simply given by the orientation of the spin along this bond.

The ice rule, $\sum_{i \in \alpha} \sigma_i^\alpha = 0$, amounts to the flux being divergence-free, that is, representing a conserved quantity: at each vertex of the lattice, two units of flux come in, and two go out. This is like Kirchhoff's laws for electrical circuits, where the conservation of charge implies that at each node, the incoming and outgoing electrical currents must be equal.

Such a conservation law, $\nabla \cdot \mathbf{M} = 0$, needs to be maintained when $\mathbf{M}$ is coarse-grained to a long-wavelength $\mathbf{m}$. In other words, the partition function which is initially just a straight sum over all "ice-configurations," becomes

$$Z = \mathrm{Tr}_{\{\sigma_i\}} \left(\prod_\alpha \delta_{\sum_{i \in \alpha} \sigma_i, 0} \right) \rightarrow \int d[\mathbf{m}] \delta \left(\nabla \cdot \mathbf{m} = 0 \right) \exp\left(-S[\mathbf{m}] \right) \ . \qquad (5.58)$$

Here the action

$$S[\mathbf{m}] = \frac{K}{2} \int d^3\mathbf{r} \, |\mathbf{m}(r)|^2 \qquad (5.59)$$

has been assumed to be the simplest analytic function of **m** compatible with all symmetries. More microscopically, if one encounters a configuration with a closed loop of moments arranged head to tail, this loop has zero net moment, as well as providing an obvious second configuration, obtained by inverting all the moments. For a loop around a hexagon of the diamond lattice, this is pictorially represented by $\circlearrowright \longleftrightarrow \circlearrowleft$. A low moment, and more than one microstate, thus seem to go together. An action penalizing a nonzero coarse-grained moment therefore appears reasonable. At any rate, the correctness of this choice of action needs to be checked against numerics or, even better, experiment.

The quadratic action (Eq. 5.59) permits a straightforward computation of the correlations; one needs to invert the coupling matrix $A_{ij} = \delta_{ij}$ implied by the form $m_i A_{ij} m_j$, subject to the constraint $\nabla \cdot \mathbf{m} = 0$. This yields for the correlators of the Fourier transform of **m**

$$\left\langle \tilde{m}_{\mathbf{k}}^i \tilde{m}_{\mathbf{k}}^j \right\rangle = \frac{1}{K}\left(\delta_{ij} - \frac{k_i k_j}{k^2}\right). \qquad (5.60)$$

Back in real space, this corresponds to the form familiar from the dipolar interaction Hamiltonian, which reads for the z-components:

$$m^z(\mathbf{r})m^z(0) \sim \frac{1}{K}\frac{3\cos^2\theta - 1}{r^3}, \qquad (5.61)$$

where (r, θ) is the polar representation of **r** with $\hat{z}$ as the polar axis.

The spins in spin ice can be readily probed via neutron scattering, which measures spin-spin correlations. This experiment proceeds in a way analogous to how X-rays can be used to probe the charge distribution of a crystal structure via Bragg scattering. The neutrons carry no electric charge, but thanks to their magnetic moment, they can couple of the magnetic moment distribution inside the sample. Therefore, in an ordered magnet, one obtains divergent Bragg peaks in the magnetic scattering. Their location in reciprocal space encodes the wavevector of the ordering pattern [$\mathbf{k} = (0, 0, 0)$ for a ferromagnet, $\mathbf{k} = (\pi, \pi, \pi)$ for an antiferromagnet on the cubic lattice], in analogy to the location of the Bragg peaks in X-ray scattering reflecting the properties of the crystal lattice.

The most direct check of the above theory is thence to relate the correlations of **m** back to those of the original spins which defined the microscopic **M**, and compare these to neutron scattering experiment. The necessary calculation involves a number of geometric factors: basically, the correlations between spins on sublattices i, j involve a sum over terms in Eq. 5.60 weighted by the components of the unit vectors, $\hat{d}_i$, corresponding to these sublattices (Hermele et al., 2004). Their asymptotic form can alternatively be computed rather directly by mechanically evaluating them in a self-consistent Gaussian approximation (Isakov et al., 2004).

Crucially, the resulting combinations retain the central feature of Eq. 5.60, namely, the *discontinuity* at $k = 0$. This discontinuity is manifest in, say, the z-component of the correlators, by writing $\mathbf{k}$ in polar coordinates, so that $k^z = k \cos \theta$, and $\left\langle \tilde{m}^i_{\mathbf{k}} \tilde{m}^j_{\mathbf{k}} \right\rangle = \frac{1}{K} \sin^2 \theta$.

This function is independent of the modulus $k = |\mathbf{k}|$ (not to be confused with the stiffness K). Hence, it is discontinuous as it oscillates between 0 and 1 in any arbitrarily small region containing $k = 0$. Put differently, its value depends on the direction from which one approaches this point. In a false-color plot, this discontinuity takes the shape of a butterfly, pinchpoint, or bow-tie motif (see Figure 5.14). On one hand, this discontinuity is in distinction to the much stronger nonanalyticity, the divergence of the Bragg peaks, in an ordered magnet. On the other hand, an ordinary disordered paramagnet has an entirely smooth (in particular, continuous) form of the spin correlations.

Due to the geometrical factors mentioned above, this discontinuity is visible not at $k = 0$ but at the centers of higher-order Brillouin zones. The experimental results for these spin correlations (Fennell et al., 2009), known as the spin structure factor, are plotted in our introductory (Figure 1.3). The agreement with the theoretical form turns out to be very satisfactory, thus validating the approximations made *en route*.

Spin ice thus represents a genuinely new magnetic phenomenon: its long-wavelength properties are captured not by a bosonic (e.g., phonons) or fermionic (e.g., the Fermi gas) field, but by an emergent gauge field.

Fig. 5.14 Pinch-point motif in spin correlations in reciprocal space (Eq. 5.60), signaling emergence of U(1) gauge field. An ordered magnet would exhibit a divergence in the form of a Bragg peak, while a conventionally disordered one would show neither a divergence nor a discontinuity.

5.4.3 Fractionalization

These correlations are just those of a free magnetostatic field, as one might have expected from the quadratic action and the divergence-free constraint familiar from a usual magnetic field.

What we have therefore uncovered is an emergent magnetostatics, with the gauge field emerging because of the enforcement of the ice rules at low temperature. Note that this emergent magnetostatics is *not* a consequence of the magnetostatic nature of the spin magnetic moments in spin ice – any other degree of freedom obeying the ice rule would have given rise to the same emergent gauge field. However, the excitations do remember the nature of the constituent spins, but before addressing this, let us consider their properties without reference to this specific feature.

Given the ice rules lead to an emergent gauge field, one might expect violations of it to correspond to defects in the gauge field (see also Section 6.4.4). Indeed, a tetrahedron for which $\sum_{i\in\alpha} \sigma_i \neq 0$ acts as a source or sink of $\mathbf{M}$, and hence $\mathbf{m}$. It therefore carries an emergent gauge charge since it is charges which are sources and sinks of flux. Flipping a single spin creates equal and opposite gauge charges in the tetrahedra it belongs to (Figure 5.15). Flipping further spins can separate the defect-tetrahedra instead of creating new ones.

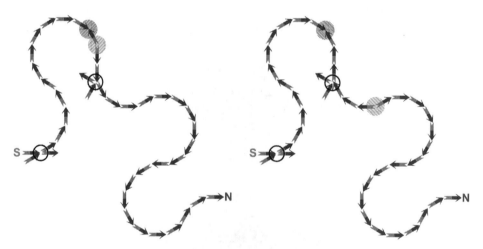

Fig. 5.15 Generation and deconfinement of emergent magnetic monopoles in $d = 3$. Each microscopic magnetic moment is represented by a little bar magnet. One string of spins arranged head-to-tail is shown, with two such strings crossing at each tetrahedron (black circle). (Left) Violating the ice rule by flipping a spin generates two tetrahedra with nonzero gauge charge. (Right) These can be separated by flipping further spins along the string at finite energy cost: the resulting magnetic monopoles are deconfined. Figure courtesy of Claudio Castelnovo.

This is superficially like in the case of the domain walls $d = 1$ (Figure 5.5). However, in $d = 1$, the fractionalized particles act as domain walls between two spatially ordered domains. In stark contrast, and like in the $d = 2$ case of the triangular RVB liquid, in spin ice it is an underlying phase without long-range order that supports fractionalization. Loosely speaking, the confluence of the physics of fractionalization between $d = 1$ and $d > 1$ resides in the nature of the strings which support the fractionalized particles as domain walls. In $d > 1$, the strings form a disordered soup: despite the fact that along a string, one encounters spins arranged head-to-tail, as in the ordered case in $d = 1$, the chaotic arrangement of the strings in $d > 1$ precludes the establishment of long-range order. The necessary ingredient is provided by the degeneracy of the ground states of spin ice. One can construct meandering strings by successive random choices of one of the two out-pointing spins, effectively generating a random walk.

5.4.4 Magnetic Monopoles and "Dirac Strings"

Let us now return to the implications of the magnetic dipole moment of the spins, that is, to the intrinsic gauge structure of Maxwell electromagnetism which is present in $\mathcal{H}_{SI}$ even before any ice rules are introduced.

The results on irrational charge (Section 5.2.2) are readily adapted to the present setting. Upon replacing the electric dipole moment $\mathbf{P}$ by a magnetic one, given by the size of the spin's magnetic moment $\mu \approx 10\mu_B$, we find that the defects carry a charge $Q_m = \nabla \cdot \mathbf{M}$, with $|Q_m| = 2\mu/a_d$, where a_d is the lattice constant of the diamond lattice defined by the midpoints of the tetrahedra.

This result follows in an elementary fashion by considering the process of flipping a string of spins along some *arbitrary* path Λ, which in a continuum approximation sets up a potential distribution of

$$V(\mathbf{r}) = \frac{\mu_0 Q_m}{4\pi} \int_\Lambda d\lambda \cdot \nabla \left(\frac{1}{|\mathbf{r} - \lambda|} \right) \tag{5.62}$$

where $Q_m = 2\mu/a_d$ is the dipole moment density corresponding to flipping a moment μ for every link of length a_d along the path. This uses the fact that the potential set up at location $\mathbf{r}$ of a dipole with moment $\mathbf{p}_m$ located at $\mathbf{r}_p$ is $\mathbf{p}_m \cdot \nabla \frac{1}{|\mathbf{r}-\mathbf{r}_p|}$. Since the line integral of a gradient of a function is the function evaluated at endpoints of the line, this implies

$$V(\mathbf{r}) = \frac{\mu_0 Q_m}{4\pi} \left[\frac{1}{|\mathbf{r} - \mathbf{r}_+|} - \frac{1}{|\mathbf{r} - \mathbf{r}_-|} \right] . \tag{5.63}$$

This is just the potential set up by a pair of equal and opposite *magnetic* charges Q_m located at $\mathbf{r}_\pm$, respectively.

The defects hence experience a *magnetic* Coulomb interaction. As this interaction does not grow without bound – unlike the logarithmic Coulomb interaction in $d = 2$ – they act as magnetically charged *deconfined* quasiparticles, hence their appellation as emergent magnetic monopoles (Castelnovo et al., 2008).

Their magnetic charge is indeed irrational, and can even be tuned, for example, by applying hydrostatic pressure, which leads to a continuous drift of a_d and μ. This seems to be at odds with Dirac's argument that the monopole charge should be quantized, and inverse in size to that of the electric charge: $e \cdot q_D = nh$. Indeed, Q_m is almost four orders of magnitude smaller than q_D, implying a quantization of electric charge in units of $\approx 10^4 e$, clearly at odds with the Millikan experiment.

To resolve this apparent paradox, it is worth recalling Dirac's original argument. He wanted to maintain $\nabla \cdot \mathbf{B} = 0$ in the presence of magnetic charges by supplying the magnetic flux in the form of an *invisible* Dirac string. To make the Dirac string invisible, it had to be infinitesimally thin so that no microscope – regardless of its magnification – could detect it. And it was forbidden from giving rise to an Aharonov-Bohm effect, in which electron paths enclosing the string would pick up a relative phase

$$2\pi\phi_D/\phi = 2n\pi, \tag{5.64}$$

where ϕ_D is the flux carried by the string, and $\phi = h/e$ is the magnetic flux quantum. Hence, the demand $\phi_D = n\phi$, or equivalently a proportionality of q_D and $1/e$.

Since, however, the "Dirac string" in spin ice is observable – the spins flipped along the path Λ are its marker – the invisibility condition does not apply. Another consequence of the observability of the strings as collections of oriented magnetic moments is that one can look for them in neutron scattering, and even orient

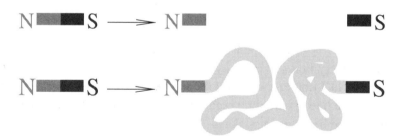

Fig. 5.16 Deconfinement of magnetic monopoles does not involve a high-energy process, like the futile cutting of chemical bonds to separate magnetic monopoles at the two poles of a bar magnet (top), but a cooperative phenomenon, the generation of tensionless strings in a magnetic material, allowing its endpoints to be cheaply separated (deconfined).

them using a magnetic field. The soup-like nature of the assembly of strings can be effectively modeled by an effectively random walk, in good agreement with experiment (Morris et al., 2009).

New Particles in High- and Low-Energy Physics

The relative smallness of Q_m with respect to q_D can be understood in a simple fashion, as a manifestation of the condensed matter origin of the emergent monopole. The link between the two is made via the electron's dipole moment, μ_B, which is of the same order for a free electron, and the spins of the rare earth ions combining several electrons to give $\mu \approx 10\mu_B$.

Viewing the electron as an elementary particle, its dipole moment arises from a combination of a magnetic charge – q_D – and a lengthscale. The only scale available for an isolated electron is its Compton wavelength, the wavelength of a photon corresponding to the rest energy of an electron, $\lambda_c = \frac{h}{mc}$. By contrast $Q_m 2\mu/a_d \approx 10\mu_B/a_d$, so that

$$\frac{q_D}{Q_m} = \frac{\mu_b}{2\mu}\frac{2\pi a_d}{\alpha\lambda_C} \qquad (5.65)$$
$$\approx 8000 \, .$$

In this expression, the lattice constant takes the place of the (reduced) Compton wavelength multiplied by the fine structure constant, $\alpha\lambda_C/2\pi$. The Compton wavelength normally plays no role at the scales of interest in condensed matter physics. As a typical lattice constant is much larger, $a_d \gg \lambda_C$, the disparity between q_D and Q_m arises.

There is a broader message here about the fundamental difference between emergence of quasiparticles, and the elementary particle discovery in cosmic rays or particle accelerators. The former proceeds by cooling (to a temperature below a typical interaction energy scale), so that a new state of matter arises as a collective phenomenon, as illustrated in Figure 5.15. The challenge is to find new states of matter, such as the topological ones with an emergent gauge field. By contrast, the high-energy route aims to probe the universe on ever smaller lengthscales, and thus higher energies, e.g. by increasing the kinetic energy of colliding particles in an accelerator. The fact that particles can be charged under both emergent and "elementary" gauge fields is a special feature linking the two.

The emergence of a fictitious gauge field that is coupled to but distinct from physical electromagnetism is actually a feature of other topological phases. In the following chapter, we will take up this thread and develop a broader view of emergent fields in topological phases. One item will be the quantum Hall effect and its description in terms of a topological field theory of an internal (i.e., emergent) gauge field a_μ that appears alongside the vector potential A_μ of ordinary electromagnetism. The success of Landau–Ginzburg in capturing the properties of symmetry-breaking phases and phase transitions provides ample motivation to pursue such descriptions of topological phases in the hope that they will be similarly simplified and ultimately "correct," in the sense of capturing universal properties of greatest interest.

6

Gauge and Topological Field Theories

The topological phenomena introduced in the previous chapter are placed on a more abstract field theoretical footing. This has the advantage of a higher level of generality and precision than the principally heuristic pictures adduced so far. The analysis of topological quantum field theories (TQFTs) have also managed to suggest new and unexpected phenomena, which can in turn be subjected to tests in numerical or real experiment.

The special feature of these TQFTs is that they are independent of the metric of spacetime and hence of any information that depends on distances. In particular, correlation lengths are zero in the topological-field-theory limit, and when we introduce lattice models in this chapter to make more intuitive some concepts of field theory, these lattice models are often chosen so that correlation lengths are exactly zero. The mathematical beauty of such treatments is quite seductive, and it is worth bearing in mind that any long-wavelength theory is only as good as the continuum limit underpinning its derivation, and the phases it supports only as stable as generic perturbations allow. For example, we expect the field-theory descriptions to be effective on lengths larger than microscopic correlation lengths and at low energy scales. The deeper attraction of the field, as so often in condensed matter physics, manifests itself in the union of microscopic and effective descriptions underpinned by experiment.

In view of the success of Landau–Ginzburg–Wilson theory, it was a bold step of Wegner's in the early 1970s to pose the question whether in fact there did exist continuous phase transitions *not* linked to any spontaneous breaking of a local symmetry. What he did find was ground-breaking in several respects. First, by demonstrating the existence of such transitions, he established the incompleteness of the LGW framework for classifying the states of matter. Second, lattice gauge theories have become a topic of central importance in high energy physics, with considerable cross fertilization between the fields (Kogut 1979). Third, as we now know, gauge theories underpin the description of a number of interesting topological states of matter which are the subject of this book. And finally, as always when a complete classification of anything seems near, it is a reminder that one may simply be missing a broader concept encompassing the known but reaching beyond it.

In this chapter we first introduce an important lattice model, the Ising gauge theory, along with Elitzur's theorem establishing that it cannot exhibit local order. This is followed by the demonstration that it does exhibit a standard phase transition by mapping it onto an Ising model with its ferromagnetic local symmetry breaking. This mapping is used to connect gauge theories to other models of interest, in particular to the lattice models which have appeared in the previous chapter. It provides a clearer picture of their phase structure and in particular of their excitations. We will also encounter a highly influential model Hamiltonian, Kitaev's toric code, which turns out to be related to a simple limit of an Ising gauge theory with matter.

Besides electric and magnetic charges, we will also find emergent photons, that is, gapless excitations of a source-free free $U(1)$ gauge field in the Coulomb phase. This will be supplemented by a presentation of the recently formulated fractonic gauge theories.

It is natural to ask whether these lattice gauge theories, or topological phases in general, have continuum limits, and the second part of the chapter is devoted to such field theories. We begin our study of the field theories that emerge in the continuum limit by reviewing how the Maxwell gauge theory appears as a long-wavelength description of spin ice. Maxwell theory leads naturally to the tensor gauge theories that describe fracton models.

We then motivate the Chern–Simons topological quantum field theory as an effective description of quantum Hall states. As a prelude, we show the existence of bound states in the one-dimensional Dirac equation with spatially variable mass, a form of the model introduced by Jackiw and Rebbi (1976). While this section may be challenging for readers without some prior exposure to field theory, Chern–Simons theory is a remarkably simple and universal framework to capture the key features of the integer and fractional quantum Hall states discussed in Chapters 3 and 5.

In this sense Chern–Simons serves as the minimal description of universal features, analogous to Landau–Ginzburg theory for symmetry-broken phases. It also makes an important prediction for experiments about the nature of fractional quantum Hall edge states. We close with a brief discussion of the generalization of Chern–Simons theory to multicomponent Abelian quantum Hall states, which make an appearance in Chapter 8, and another closely related TQFT known as BF theory.

6.1 Pure Ising Gauge Theory and Absence of Local Order

The basic degrees of freedom of the classical Ising gauge theory are gauge fields taking on values $\sigma_{ij} = \pm 1$. They are defined on the *links* of a lattice Λ with sites $i \in \Lambda$. These links are labeled by the pair of sites, ij, they connect.

For simplicity, in the following we consider a (hyper-)cubic lattice in d dimensions, concentrating on the case $d = 3$, although we will phrase the discussion in terms which should make the generalization to other lattices transparent.

The Hamiltonian consists of a sum of terms defined on *plaquettes,* $\square$, of the lattice. For an elementary square plaquette containing sites i, j, k, l, the term is $W_\square = \sigma_{ij}\sigma_{jk}\sigma_{kl}\sigma_{li}$, giving a Hamiltonian

$$\frac{H_{\mathrm{IGT}}}{T} = -K \sum_\square W_\square, \tag{6.1}$$

where K is a dimensionless coupling constant.

This can be seen as a systematic generalization of more familiar Hamiltonians. Consider a sequence of Hamiltonians on a hypercubic lattice, the terms of which involve combinations of Ising degrees of freedom on increasingly high-dimensional building blocks of the lattice, namely, H_h on a zero-dimensional lattice site .; H_{IM} on a one-dimensional lattice bond, $-$; and H_{IGT} on a two-dimensional plaquette $\square$. These respectively correspond to a spin in a field, $H_h = -|h| \sum \sigma_i$, the conventional Ising ferromagnet with its pairwise interactions, $H_{\mathrm{IM}} = -|J| \sum_- \sigma_i\sigma_j$, and the Ising gauge theory.

The salient point of this construction is the progression of symmetries of this sequence. H_h exhibits no symmetries, and has a unique ground state $\sigma_i = 1$. By contrast, H_{IM} is invariant under global spin inversion $\sigma \to -\sigma$, and thus exhibits two ferromagnetic ground states $\sigma_i = +1$ or $\sigma_i = -1$. It is the simplest case of a model exhibiting spontaneous symmetry breaking, hence its large conceptual (and historical) importance.

H_{IGT} exhibits a much larger set of symmetries. It is invariant under the *local gauge symmetry* of flipping the sign of all σs emanating from a site $i \in \Lambda$. This is generated by $G_i \colon \sigma_{ji} \to -\sigma_{ji}$ for all j neighboring i. Therefore, while in the limit of low temperature, $\langle \sigma_i \rangle_{H_h} \neq 0$, $\langle \sigma_i\sigma_j \rangle_{H_{\mathrm{IM}}} \neq 0$ one finds that $\langle \sigma_{ij}\sigma_{kl} \rangle_{H_{\mathrm{IGT}}} \equiv 0$ always. This follows simply from the fact that under the action of the local gauge transformation G_i, $W_\square \to W_\square$, as the number of bonds flipped on any plaquette is 0 or 2 so that $H_{\mathrm{IGT}} \to H_{\mathrm{IGT}}$. Therefore, for $\{i, j\} \cap \{k, l\} = \varnothing$, $\langle \sigma_{ij}\sigma_{kl} \rangle_{H_{\mathrm{IGT}}} = \langle G_i\sigma_{ij}G_i\sigma_{kl} \rangle_{H_{\mathrm{IGT}}} = -\langle \sigma_{ij}\sigma_{kl} \rangle_{H_{\mathrm{IGT}}} \equiv 0$.

This is a special case of Elitzur's theorem which states that local gauge symmetries cannot be broken, the general proof of which proceeds along very much the same lines. (It is of course still possible to break global symmetries in a theory, if these exist in addition – for the Ising lattice gauge theory, this is not the case.) Also note that one should distinguish between a symmetry relating different physical states of a system, and a gauge invariance, which is a mathematical ambiguity of a system where the physical states are indistinguishable.

6.1.1 Duality and the Existence of a Phase Transition

In the absence of local symmetry breaking, LGW theory would state that there need not be any phase transitions. However, drawing this conclusion would be quite wrong. To see this, we take a small detour of mapping H_{IGT} onto a quantum version, which we then show is equivalent to an Ising model which *does* have a phase transition according to LGW. The quantum form of (6.1) reads (see Box 6.1)

$$H_{\mathrm{IGT}}^{Q} = -\kappa \sum_{\square} W_{\square}^{Q} - \gamma \sum_{-} \sigma_{ij}^{x} \; , \tag{6.2}$$

where $W_{p}^{Q} = W_p$, with the classical σs replaced by the Pauli matrices σ^z representing spin-$1/2$ operators. The first term is thus the regular plaquette as before, while the second term gives a "tension" to the electric field lines: it prefers the vacuum to have $\sigma^x = 1$, at least to the extent that this is compatible with other constraints, such as the Gauss's law (Eq. 6.13).

Box 6.1 Quantum IGT in d Dimensions and Classical IGT in $d + 1$

Here, we present the mapping between the classical IGT, Eq. 6.1, and its quantum version, Eq. 6.2. This is an instance of the standard mapping between a d-dimensional quantum system to a $d + 1$-dimensional, stacked, classical one. This proceeds by identifying matrices which appear in the respective partition functions upon using a Trotter–Suzuki decomposition. This starts with the quantum partition function expressed as

$$Z_{Q} = \mathrm{Tr} e^{-\beta H_{\mathrm{IGT}}^{Q}} = \mathrm{Tr} \exp \left\{ -\beta \left(\upsilon V(\{\sigma_i^z\}) - \gamma \sum_{-} \sigma_{ij}^{x} \right) \right\} \tag{6.3}$$

where $\beta \equiv 1/k_B T$ is the inverse physical temperature of the quantum system.

Let us consider the two terms separately. For generality, we have grouped together all "classical" terms, that is, those depending only on σ_{ij}^z, into $V(\{\sigma_i^z\})$. These are simple in that they take the same form in both formulations, with only their coupling strengths, explicited by υ in the above equation, being altered.

This is turned into a (discretized) path integral via the insertion of complete set of states at each imaginary time slice, which effectively introduces the additional discretized dimension of size β:

$$Z_Q = \sum_{\{\sigma_{ij}=\pm 1\}} \langle \{\sigma_{ij}\} | (\exp(-a_\tau H_{\mathrm{IGT}}^Q))^N | \{\sigma_{ij}\} \rangle$$

$$= \prod_{n=1}^{\beta/a_\tau} \sum_{\{\sigma_{ij,n}=\pm 1\}} \langle \{\sigma_{ij,n}\} | \exp(-a_\tau H_{\mathrm{IGT}}^Q) | \{\sigma_{ij,n+1}\} \rangle \qquad (6.4)$$

$$= \prod_{n=1}^{\beta/a_\tau} \sum_{\{\sigma_{ij,n}=\pm 1\}} \exp\left(-a_\tau \upsilon V(\{\sigma_{ij,n}\})\right) \times \qquad (6.5)$$

$$\times \left(\delta^{(0)}_{\{\sigma_{ij,n}\},\{\sigma_{ij,n+1}\}} + \frac{1}{2} a_\tau \gamma \delta^{(1)}_{\{\sigma_{ij,n}\},\{\sigma_{ij,n+1}\}} + O((a_\tau \gamma)^2) \right)$$

Here, we have introduced the imaginary time step $a_\tau = \beta/N$ and n labels the coordinate of the extra imaginary time dimension, which has periodic boundary conditions. The function $\delta^{(k)}$ is defined to be one if its arguments, the two sets of spin configurations, differ by k entries, and zero otherwise.

The equivalence of this expression of the partition function to that of the classical Hamiltonian

$$H_{d+1} = \sum_{\langle ij \rangle, n} K_V^s V(\{\sigma_{ij,n}\}) + \sum_{i,n} K^\tau \sigma_{ij,n} \sigma_{ij,n+1} \qquad (6.6)$$

is established by expressing the partition sum for the latter,

$$Z_{\mathrm{cl}} = \mathrm{Tr}\, T_Z^N \qquad (6.7)$$

in terms of a transfer matrix, T_Z:

$$T_Z = \exp\left(K_V^s V(\{\sigma_{ij,n}\})\right) \times$$

$$\times \left(\delta^{(0)}_{\{\sigma_{ij,n}\},\{\sigma_{ij,n+1}\}} + \exp\left(-K^\tau/2\right) \delta^{(1)}_{\{\sigma_{ij,n}\},\{\sigma_{ij,n+1}\}} + \right.$$

$$\left. + O\left(\exp\left(-K^\tau\right)\right) \right). \qquad (6.8)$$

The first term on the right hand side of this equation is to be understood as an exponentiated diagonal matrix.

The two expressions for the partition functions have the same form, and are hence equivalent, $Z_Q = Z_{\mathrm{cl}}$, if one chooses $a_\tau \gamma/2 = \exp(-K^\tau/2)$ and $K_V^s = a_\tau \upsilon$. Continuous quantum evolution therefore corresponds to the scaling limit $K_V^s \propto a_\tau \to 0$, $K^\tau \to \infty$, while maintaining

$$2e^{-K^\tau/2}/a_\tau = \gamma, \quad K_V^s \exp(K^\tau/2)/2 = \upsilon/\gamma . \qquad (6.9)$$

As advertised, the terms involving only σ_{ij}^z in the same layer have only changed prefactor, but there is also a ferromagnetic coupling between the layers in the additional imaginary time dimension. This "transverse field term," $-\gamma \sigma_{ij}^x$, has become another Ising exchange between the σ_{ij}^z in adjacent layers – the noncommuting x-components have vanished from the description.

The path back to the plaquette term, $W_\square$, involves one further step: one uses the local gauge invariance to set all the σ_{ij}^z along bonds in the imaginary time direction equal to $+1$. This can (almost) be done: there is one vertical link per site of the lattice, which can be fixed by the one gauge degree of freedom available per site. However, the path integral requires periodic boundary conditions in the imaginary time direction. As the gauge transformation always flips two vertical bonds together, the parity of the product of a closed loop of σ_{ij}^z wrapping the imaginary time direction cannot be changed. This subtlety can be important, giving rise to an additional factor $(-1)^{S_B}$ in the classical path integral, where S_B is the global parity of a given configuration. We do not dwell on this further, besides noting that the presence of this factor distinguishes the "even" from the "odd" IGT (Moessner et al., 2002), which we will encounter further down.

With all of this in hand, we can finally read off the dimensionless inverse of the quantum temperature, $\beta\gamma$, in terms of the classical expression. This is given by the extent L^τ of the system in the time direction: $\beta\gamma = a_\tau\gamma L^\tau = 2L^\tau / \exp(+K^\tau/2)$. The final expression has a physical interpretation: the quantum coupling strength $\beta\gamma$ is given by the size of the system, measured in units of the correlation length of a ferromagnetic Ising chain with the classical coupling K^τ: loosely speaking, the "off-diagonal" quantum energy has become a classical entropy of positioning domain walls.

This also resolves the question why there appear to be more coupling constants in the quantum theory compared to the classical one, which has less terms in the Hamiltonian: the correspondence singles out particular trajectories in the space of coupling constants. This is not the only thing which complicates transfer of intuition between the two settings: the quantum temperature corresponds to the size in the imaginary time direction, and energy terms in the quantum description have an entropic flavor to them classically.

The duality mapping can now most easily be carried out in two steps. First, we make a change of variables and identify each configuration of the old variables with one of the new variables. Then, we evaluate the matrix elements of the H_{IGT}^Q in the new basis. The details of the duality depend on spatial dimensionality, and here we derive the duality of the quantum Ising lattice gauge theory in $d = 2+1$ with the (transverse field) Ising model in $d = 2+1$. This is illustrated in Figure 6.1.

The first step is achieved by introducing spin variables S_p residing on the centers of the plaquettes p of the square lattice, and we define

$$\gamma S_p^z S_q^z = \gamma \sigma_{jk}^x . \tag{6.10}$$

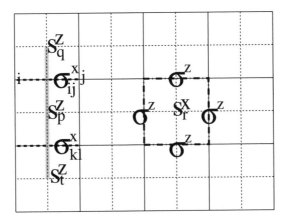

Fig. 6.1 Duality between Ising gauge theory and transverse field Ising model in $d = 2 + 1$. The spins of the Ising model reside on the square lattice (dashed lines) dual to that on the links of which the gauge field resides (solid lines). The eigenvalues of $\sigma_{ij}^x = \pm 1$ denote whether the bond it crosses is satisfied or not, $S_p^z S_q^z = \pm 1$. Flipping a spin using the operator S_r^x toggles the four bonds which emanate from it between being satisfied and frustrated; this implies a change of sign of the four σ^x making up the dot-dashed plaquette surrounding site r. This is achieved by $W_\square = \prod_\square \sigma^z$. The value S_t^z of a spin located at t is given with respect to a reference spin S_q^z by multiplying the values of σ^x of the bonds the string connecting the two crosses, $S_t^z = S_q^z \sigma_{ij}^x \sigma_{kl}^x$.

The second step then amounts to noting that this identification already accounts for the second term in the Hamiltonian of the IGT, which is becomes a simple Ising exchange interaction for the S^z operators; while the first, the plaquette, term of the IGT flips all the σ^x surrounding the spin on plaquette $\square$, and hence can be written as

$$\kappa \, W_p^Q = \kappa \, S_p^x \,. \qquad (6.11)$$

We have thus derived the duality to a transverse field Ising model, Eq. 2.107, with exchange strength γ and transverse field strength κ. The interpretation of the σ_{jk}^x operators is that they keep track of domain walls of the Ising model: whenever the spins at the endpoints of a bond are ferromagnetically aligned, $\sigma^x = 1$. In order to ensure that the configurations map onto each other faithfully, we need to impose a form of Gauss's law: domain walls must never end; rather, at each vertex of the lattice, an even number of domain walls (0, 2 or 4 for the square lattice) must be incident:

$$G_i^p = \prod_+ \sigma_{ij}^x = \pm 1 \,. \qquad (6.12)$$

Here, the product runs over all links emanating from site i, the superscript stands for this being the Gauss's law of the pure theory (without matter), and we will comment on the possible choice -1 on the right-hand side below. Starting from any configuration which satisfies this constraint, the Hamiltonian will only produce other configurations which do so as well.

The crux of the duality mapping lies in the simple observation that there are two phases to the Ising model. At small γ/κ, the spins are polarized by an external field and no spontaneous symmetry breaking is present: this is a paramagnet. By contrast, for large γ/κ, the magnet is in a ferromagnetic state with a spontaneously broken Ising symmetry denoting the preference for $\langle S^z \rangle$ to be spontaneously either positive or negative.

By such arguments, Franz Wegner (1971) showed that (i) the IGT has no order in the LGW sense, and yet, (ii) it does exhibit a phase transition thermodynamically equivalent to that of the standard ferromagnetic transition of the ferromagnetic Ising model.

What looks at first sight as a mere curiosity – some abstract mapping between two Hilbert spaces – turns out to be the tip of an iceberg: the IGT generalizes in many ways, to many settings, which no longer necessarily permit a straightforward identification with conventional LGW phase diagrams.

In fact, much of what we *do* understand about topological physics is related to the phase of gauge theories which corresponds to the paramagnet. It is perhaps surprising that something as featureless as a paramagnet could encompass a new type of order. For one thing, how can the topological degeneracy of a topological phase be consistent with the nondegeneracy of the paramagnetic state? This turns out to be due to the nonlocality of the duality mapping: different sectors of topological phase correspond to paramagnets with different boundary conditions. This is because a configuration in the two sectors of the Ising gauge theory maps onto a magnetic configuration with an even/odd number of domain walls as one goes round the system, which in turn amounts to considering a paramagnet with periodic or antiperiodic boundary conditions.

6.1.2 Odd Ising Gauge Theory and Quantum Dimer Models

In Gauss's law, Eq. 6.12, we chose $G_i^p = +1$, so that an even number of σ_{ij}^x need to take on the value -1. This in particular includes $\sigma_{ij}^x = 1 \ \forall ij$, the perfect ferromagnet. By contrast, the odd choice, $G_i^p = -1$, always requires an odd number, that is, at least one bond, with $\sigma_{ij}^x = -1$. It is a property of frustrated magnets that not all bonds can be satisfied simultaneously. Indeed, the odd Ising gauge theory on the triangular lattice is dual to a frustrated transverse field Ising model on the honeycomb lattice. As the honeycomb lattice is bipartite, it needs to be frustrated "by hand," by choosing exactly one bond on each hexagon to be antiferromagnetic; such a model is known as fully frustrated. In the limit of low temperatures for $\gamma \gg \kappa$, minimizing the number of frustrated bonds together with the odd constraint yields the hardcore dimer condition: each site has one, and only one, bond with $\sigma_{ij}^x = -1$. In this limit, the Rokhsar–Kivelson quantum dimer model and the fully frustrated Ising model on the honeycomb lattice are equivalent. The RVB liquid of the former thus corresponds to a quantum paramagnetic phase of the latter.

We close the discussion of these dualities with two parenthetical remarks. First, if one demanded that not one, but exactly three, bond variables have $\sigma_{ij}^x = -1$, the corresponding topological liquid phase is more extended (Balents et al., 2002). Second, the fully frustrated version of the square lattice Ising magnet, which is dual to the odd Ising gauge theory on the square lattice, supports no topological liquid phase. This can be derived following the discussion of the transverse field Ising model on the triangular lattice (itself dual to the quantum dimer model on the honeycomb lattice) (Box 2.3). This now maps onto an XY-model with a fourfold clock term, which is relevant in $d = 2$ at the KT transition, so that in the phase diagram in Figure 2.5, the two KT transitions coincide, and the intervening KT phase vanishes. This reflects the general absence of RVB liquids in quantum dimer models on bipartite lattices.

6.2 Ising Gauge Theory with Matter

The study of gauge theories is a vast field of various branches of physics. Our specific aim here is to make a connection between deconfinement phenomena in gauge theory and the physics of topological phases. In this section, we study the Ising gauge theory in more detail, in order to display its important features.

The next step is to introduce matter fields to the Ising gauge theory. As we see below in analogy to the case of the U(1) gauge theory familiar from Maxwell electromagnetism, Eq. 6.2 contains terms corresponding to the energies of electric and magnetic fields; what is missing are charges, that is to say, sources and sinks of field lines. With the electric fields residing on the links of the lattice, the corresponding charges live on the lattice sites. For the Ising gauge theory, absence/presence of a charge is denoted by an Ising operator τ^x with eigenvalues $+1/-1$. To capture the notion that these charges are sources and sinks of the electric field lines, an appropriate form of Gauss's law, defined for each site i, now reads

$$\tau_i^x \prod_+ \sigma_{ij}^x = \pm 1 \; . \tag{6.13}$$

Here, the $\prod_+$ stands for the product over the set of (four) neighbors of site i as in Eq. 6.12. The ± 1 on the right-hand side refers to the (standard) even Ising gauge theory that we have been discussing so far for the case $+1$. The odd case of -1 applies to frustrated Ising models, or rather, their dual, the gauge theories related to quantum dimer models as discussed at the end of the previous section.

These charges, with creation energy ("rest mass") $2\gamma_m$ can chosen to be static; if dynamical charges are desired, one needs to add a kinetic term obeying Gauss's law; in the case of Ising matter with its parity, rather than number-conserving Gauss's law, the simplest such term reads $\tau_i^z \sigma_{ij}^z \tau_j^z$. Given τ^x encodes the presence/absence of an Ising charge, this can be understood as toggling two adjacent sites between absence and presence of a charge. The σ^z in turn toggles the link joining them between the absence and presence of a flux line.

The resulting Hamiltonian, obtained as for the case without matter fields, reads

$$H_{\text{IGT}}^Q = -\kappa \sum_{\square} W_p^Q - \gamma \sum_{-} \sigma_{ij}^x - \gamma_m \sum_{\cdot} \tau_i^x - J \sum_{-} \tau_i^z \sigma_{ij}^z \tau_j^z . \tag{6.14}$$

Like in the pure case, this in turn corresponds to a classical Ising gauge theory with matter:

$$\frac{H_{\text{IGT}}}{T} = -K \sum_{\square} W_p - J \sum_{-} \tau_i \sigma_{ij} \tau_j . \tag{6.15}$$

The phase diagram of this Ising gauge theory in $d = 2 + 1$ is shown in Figure 6.2 (Fradkin and Shenker, 1979). The phase structure of the pure gauge theory, corresponding to the lower horizontal boundary of the phase diagram, with its phase transition dual to that of the Ising model, was already discussed in Section 6.1.

Another easily analyzed boundary is the vertical one on the right-hand side: for $K \to \infty$, the gauge fields get pinned to $\sigma \equiv 1$ (or a gauge-equivalent configuration) and effectively drop out as dynamical degrees of freedom. One can then analyze the remaining model involving only the matter field; upon the gauge-fixing, $\sigma = 1$, the pure matter theory turns into a simple Ising model, $H_{\text{IM}} = -J \sum_{-} \tau_i \tau_j$, without the need for a duality transformation. As a function of J this exhibits an Ising transition. This transition is between a deconfined "paramagnet" at small J, and a "ferromagnet" at large J, where the kinetic energy of the matter field dominates. The latter is known as the Higgs phase, and in this model, it is continuously connected to the confining phase at small K and J.

At the other side of the transition, the charges are sparse yet freely mobile: this is the deconfined phase. It is continuously connected to the dual phase of the paramagnet of the pure gauge theory. Our next stop is to consider the properties of this phase in more detail. Ultimately, we want to move from the gauge theory, defined

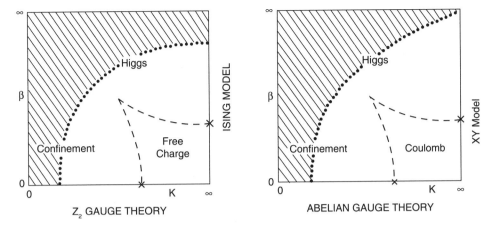

Fig. 6.2 (Left) Phase diagram of the classical Ising gauge theory with matter in $d = 3$, corresponding to the quantum IGT in $d = 2 + 1$. The deconfined phase, delineated in the bottom right quadrant, corresponds to the deconfined phase of $\mathbb{Z}_2$ models such as the RVB liquid and the toric code. The β on the vertical axis corresponds to J in Eq. 6.15. The bottom boundary of the phase diagram corresponds to the pure gauge theory with Wegner's dual Ising transition, while the right boundary corresponds to an Ising ferromagnet. (Right) Analogous phase diagram and U(1) gauge theory with matter. Here, $d = 3$ is always confining, so that this only applies to $d \geq 4$. In particular, this rules out a deconfined U(1) theory in a quantum model in $d = 2 + 1$, and instead necessitates a system in $d = 3 + 1$ dimensions like quantum spin ice with its deconfined Coulomb phase. From Fradkin and Shenker (1979). Reprinted with permission by *Physical Review*.

on the lattice scale, to the topological field theory, which is impervious to any physics on finite lengthscales.

6.2.1 Wegner–Wilson Loop

As emphasized at the outset, the phase transition in the Ising gauge theory cannot be diagnosed by a local order parameter. Instead, it is the expectation value of a many-body operator having support on a large number, $L \to \infty$, of links which can diagnose the transition in the IGT without matter. This is known as the (Wegner–) Wilson loop, $W_w = \prod \sigma_{ij}^z$, where the product is taken around, say, a square region of linear size L. It can be thought of as the product of all elementary plaquette flux operators, $\prod W_\square^Q$, enclosed by the loop.

Loosely speaking, if the value of an Ising variable on each plaquette is a random quantity taking on values ± 1, the expectation value of the W_w loop, the overall parity of their product, is *exponentially* small in the number of plaquettes involved. It turns out that the two phases of the IGT differ in whether the number of plaquettes effectively involved in the product scales with the perimeter, $4L$, of the loop, or with its area, $A \sim L^2$. The former is known as a perimeter law, $W_w \sim \exp(-aL)$, while the latter is the area law $W_w \sim \exp(-bL^2)$.[1]

The simplest setting to derive perimeter and area laws is to do so perturbatively around the points $\gamma/\kappa = 0, \infty$ of H_{IGT}^Q in $d = 2 + 1$ at $T = 0$. Starting with the former, we note that the ground state is the state with $W_\square^Q = 1$ for all plaquettes: no magnetic charges (visons) are present. On top of that ground state, visons can only be created virtually, in pairs on spatially adjacent plaquettes sharing a link, by application of $\gamma\sigma_{ij}^x$ on that link. They cost excitation energy 2κ, so that this process is suppressed by a factor $\gamma/\kappa \ll 1$ in perturbation theory. The members of the pair can also be separated from each other by further applications of $\gamma\sigma_{ij}^x$. While this does not necessarily create further visons, the energy denominator persists as one is moving between excited states throughout the process. The consequence is that the probability of encountering well-separated visons is exponentially suppressed in a length ξ which is small in $\gamma/\kappa \ll 1$.

Now, the only virtual vison pairs which degrade the value of the Wilson loops are those which have one partner inside it, and the other outside. As the vison pairs are localized on a lengthscale ξ, this means that only plaquettes within that distance of the perimeter are relevant for W_w. Therefore W_w is exponentially small in ξL: this is just the perimeter law.

At the other edge of the phase diagram, $\kappa/\gamma = 0$, the ground state has $\sigma_{ij}^x = 1$ for all the links. Written in the σ^z basis, this amounts to completely random combinations of σ^zs, and hence completely random values of the flux plaquette operators. In the vicinity of that point, where the probability distribution of $W_\square^Q$ gets slanted toward $+1$ for all plaquettes, this still means that the total parity expectation value is evaluated over a product of L^2 random variables. Hence, one obtains the area law.

6.2.2 Deconfinement and Its Diagnostics

The phase with the perimeter law is known as the deconfined phase. To see the origin of this nomenclature, let us first recall the conventional statement of deconfinement of a pair of charges. This states that if two charges are deconfined, then there is a finite energy cost, $E_2(L)$, to separate them to a large separation: $\lim_{L\to\infty} E_2(L) < \infty$. This can again straightforwardly be checked near the two corners of the phase diagram. For $\kappa/\gamma \ll 1$, the cost of a flux line joining the two charges, which consists of links with $\sigma^x = -1$, is linear in its length L: the flux line has a tension set by γ, and the charges are therefore confined.

---

[1] More recently, the term area law has become commonly used in quantum information theory to describe the behavior of the entanglement entropy between a region and its surroundings (see Chapter 11). There, it distinguishes low-entanglement states from high-entanglement ones, which are said to obey a volume law. These two meanings are obviously quite distinct.

By contrast, in the phase with well-defined σ^z, the σ^xs are random, and there-fore, the system consists of a soup of interpenetrating flux lines, which are now tensionless (see also the discussion around Figure 5.16): rather than having to cre-ate a flux line when separating a charge pair, one can reorient a preexisting one as needed. Charges can therefore be separated at a finite energy cost $E_2(\infty)$. They are deconfined.

Note that this diagnostic is only useful for the pure gauge theory, as it ceases to work in the presence of dynamical matter. Rather than having to extend an increas-ingly long flux string as the two charges are separated to $L \rightarrow \infty$, it becomes cheaper to pop a new charge pair out of the vacuum. The energy cost for this $4\gamma_m$, which no longer depends on the separation L. This phenomenon is known as string breaking, and is quite familiar from the case of confinement in QCD: when separat-ing a quark-antiquark pair $q, \overline{q}$, another $q\overline{q}$ pair is created, so that one ends up with two pions, that is, bound states of $q\overline{q}$, which are readily separated from one another.

In this situation, the Wilson loop loses its value as a diagnostic, as it always yields a perimeter law, regardless of (de)confinement. However, a variant of the Wilson loop introduced in Fredenhagen and Marcu (1986) allows to distinguish if the perimeter law arises because of the absence of a string tension (deconfinement), or just the absence of a string (string breaking and confinement).

6.2.3 Stability of Topological Order at Finite T

So far, the discussion of the IGT has centered on two dimensions at temperature $T = 0$. We now address what happens for $T > 0$.

The somewhat disappointing answer is that in $d = 2$, topological order is destroyed immediately for any $T > 0$. The above argument for the perimeter/area law distinction can be adapted to show this as follows. For the demonstration of the perimeter law, the crucial ingredient was the suppression of well-separated virtual vison pairs in the ground state. However, in order to evaluate a thermal expecta-tion value, one sums over *all* states, with a Boltzmann factor determined by their respective excitation energies. Here, it is important that the energy of a (real) pair of visons does not arbitrarily grow with their separation. One therefore sums over all distributions of visons, with the Boltzmann factor essentially just given by the vison number. Thus, each plaquette inside the Wilson loop, and not just those straddling its perimeter, contributes with a nonzero vison probability, and we are back to an area law–the deconfined phase with its topological order is destroyed. This means in particular that a two-dimensional topological memory for a quantum computer (Chapter 9) is not strictly stable at nonzero temperature (Castelnovo and Chamon, 2007).

The situation in $d = 3$ dimensions is different. The crucial new ingredient is that magnetic fluxes are no longer zero-dimensional in nature, like the plaquettes hosting a vison. Rather, magnetic fluxes now form loops, that is, one dimensional objects. (This is of course familiar from the closed flux lines of Maxwell electromagnetism in $d = 3$. The analogous U(1) theory in $d = 2$ also has pointlike magnetic fluxes.)

In the present context for a lattice gauge theory, this can be seen by considering the action of $\gamma \sigma_{ij}^x$ on a single link of the cubic lattice in $d = 3$. This flips the value of σ_{ij}^z on that link. Therefore, it inserts fluxes $W_\square^Q = -1$ on all four plaquettes which share this link. This can be pictorially captured by drawing a circular loop through the midpoints of these four plaquettes, with the flipped bond piercing the plane of that loop at its centerpoint. Further action of $\gamma \sigma_{ij}^x$ now lengthens the loop, that is, creates further fluxes. This results in a tension of energetic origin for the flux loop. Now, in $d = 3$, the Wilson loop measures not the expectation value of the parity of the number of visons in the loop, but that of the number of flux loops linking the Wilson loop, that is, the number of flux loops which cannot be contracted to a point without crossing the Wilson loop. The result is that the tension for the flux loops imparts a characteristic lengthscale to the flux loops, and one retains a perimeter law as in $d = 2$ at $T = 0$: the topologically ordered deconfined phase of the IGT persists to finite temperature.

6.3 Kitaev's Toric Code

We next take a shortcut toward a discussion of the purely topological content of the IGT by considering a very special point in the phase diagram in which all correlations which are not of topological nature are ultra-short range, that is, the correlation length vanishes. Therefore, topological information can be read off directly from the microscopic model, while nonuniversal information is "absent": the Wilson loop does not decay at all as L is increased, thereby not distinguishing between area and perimeter laws. The model corresponds to the Ising gauge theory with matter for the parameter values at which the two lines mentioned above meet. It is arguably the simplest lattice model exhibiting topological order, and has proven to be tremendously useful. It is known as Kitaev's (2003) toric code.

The Hamiltonian of the toric code follows from Eq. 6.14 by first setting two terms to zero, $J = \gamma = 0$, and using Gauss's law $\tau_i^x \prod_+ \sigma_{ij}^x = 1$ to replace the matter τ^x operator at site i by a "star term" comprising four gauge fluxes $W_s^Q = \prod_+ \sigma_{ij}^x$ on the links emanating from that site:

$$H_{\text{TC}} = -\kappa \sum_\square W_\square^Q - \gamma_m \sum_+ W_s^Q . \qquad (6.16)$$

This Hamiltonian has a very useful feature: all of its terms commute with each other. This follows as the plaquette and star terms commute trivially amongst themselves as they only contain Pauli matrices of the same component; while they commute with one another as noncommuting terms always appear in pairs. Pictorially, acting with $W_\square^Q$ on a plaquette flips the σ^x on *two* bonds for each $+$ at its corners, yielding two factors of -1 in the commutators.

As the terms individually square to 1, being combinations of commuting Ising variables, their eigenvalues can be fixed independently to be ± 1 for each site, and for each plaquette, respectively. On top of the vacuum, denoted by 1, there are therefore three different quantum numbers for possible excitations: $W_p^Q = +1$, $W_s^Q = -1$, an electric charge denoted e; $W_p^Q = -1$, $W_s^Q = +1$, a magnetic charge/vortex (vison), denoted m; and $W_p^Q = -1$, $W_s^Q = -1$, a combination of the two, denoted $\epsilon = e \times m$. These are the particles we have encountered for the quantum dimer model in the previous chapter – the spinon (monomer), the vison, and the bound state of spinon and vison.

Having more than one excitation of each type does not lead to further distinct sectors because the relevant excitations can be annihilated locally in pairs. Such operations are encoded by the fusion rules which describe how the different excitations combine. For this case, they are

$$e \times e = 1; \quad m \times m = 1; \quad e \times m = \epsilon$$
$$\epsilon \times e = m; \quad \epsilon \times m = e; \quad \epsilon \times \epsilon = 1 . \tag{6.17}$$

This is a particularly simple set of rules, essentially only encoding the Ising pair annihilation of the electric and magnetic excitations. For more complex topological models, in particular those involving non-Abelian anyons, such fusion rules are richer, and are one of the ingredients to topological quantum computation discussed in Chapter 9.

The quantum statistics of the e and m particles can straightforwardly be determined to be bosonic. This follows from the fact that exchanging a pair of es, that is, two plaquettes with $W_\square^Q = -1$, involves a closed loop of σ_{ij}^x operators. This loop also commutes with all the terms in the Hamiltonian, and therefore no nontrivial phase is generated. The argument for the ms is analogous.

The *relative* statistics of these excitations can be obtained explicitly as well for the toric code. These encode the Aharonov–Bohm phase that is picked up when one particle encircles the other. This is the counterpart of the idea of the exchange statistics for identical particles (Box 5.2). For distinguishable particles, however, a simple exchange leads to a distinguishable state, so that one needs to consider two successive exchanges, which just correspond to the encirclement of one particle by the other. For the case of e and m, the mutual statistics is fermionic. This is the same result as the factor -1 picked up by a monomer encircling a vison in the case of the quantum dimer model (Fig. 5.12).

The remaining question then concerns the quantum statistics of the ϵ particle. Exchanging two $\epsilon = e \times m$ pairs requires first moving one e past an m, and then one m past an e. Together, these two operations amount to an m encircling an e, which yields the Aharonov–Bohm phase of -1. Therefore, the ϵ is a fermion.

Generally speaking, for the toric code on a torus, there are thus four sectors: $1, e, m, \epsilon$. These correspond to the system having globally no excitations; an electric excitation; a magnetic excitation; or both. Any further excitations can be locally pair-annihilated. By contrast, the action of *any* local Hamiltonian does not connect these sectors with one another–they are therefore known as superselection sectors. This is a reference to the idea that selection rules for, say, transitions between atomic states are subject to selection rules which make some transition matrix elements vanish on account of symmetries; here, the vanishing is more strongly protected, not by symmetry but by topology.

The presence of a topological ground-state degeneracy signaling topological order also follows transparently, by noting that in fact any operator $\prod_C \sigma_{ij}^z$ defined on a closed loop C commutes with H_{TC}. This holds true in particular for any noncontractible loop, say one winding around a cylinder. For each independent noncontractible loop, there is hence a twofold degeneracy, according to the sign of $\prod_C \sigma_{ij}^z$. In Section 6.8 we will also see how the mutual statistics of e and m charges can be encoded in a continuum topological quantum field theory, BF theory.

6.4 Maxwell Electromagnetism

The naming of electric and magnetic charges in the above incarnations of Ising gauge theories parallels a much more familiar case of a gauge theory – the U(1) theory of Maxwell electromagnetism, with the electric and magnetic fields ubiquitous in our daily lives.

In Section 5.4.2, on spin ice, we have introduced an emergent gauge structure in the form of magnetostatics. This included pointlike defects charged under the emergent gauge field, which were christened magnetic monopoles on account of the fact that they *also* carried magnetic charge of conventional Maxwell electromagnetism. It would therefore seem natural to extend this emergent gauge structure to full Maxwell electrodynamics, by also introducing the complementary electric sector of the theory.

Before we turn to this, we make the side remark that in the language of gauge theories used above for the case of the Ising gauge theory, the link variables are the electric fields; this term would then naturally apply to the spins in spin ice when considered as link variables on the diamond lattice; we therefore denote the microscopic **M** field as oriented Ising link variable $E_{jk} = -E_{kj}$. (Such an orientation is possible as the diamond lattice is bipartite.) The sources and sinks of such electric flux would in this convention be called electric charges. For the remainder of this chapter, we stick to this convention, for the sake of internal consistency of the gauge theory treatment, at the price of referring to the magnetic monopoles as electric charges (of the emergent gauge field) rather than as the magnetic charges (of Maxwell electromagnetism). This somewhat annoying complication arguably reflects the otherwise intriguing subtlety of the phenomenon of fractionalization in this compound.

6.4.1 Quantum Spin Ice Hamiltonian

Microscopically, spin ice admits a simple quantum kinetic term very much like the one presented for the quantum dimer model in Eq. 5.42, pictorially represented as

$$H_{\text{SI−res}} = -\Gamma \sum_{\bigcirc} \left[|\circlearrowright\rangle\langle\circlearrowleft| + |\circlearrowleft\rangle\langle\circlearrowright| \right]. \tag{6.18}$$

The basic constraint on possible kinetic terms within the ice manifold is that they must not lead to the creation of electric charges. Reversing a single field line is ruled out by this constraint, but it is possible to reverse a closed electric flux loop – corresponding to spins arranged head-to-tail on the shortest closed loop, a hexagon, $\bigcirc$, of the diamond lattice.

Microscopically, in order to turn spin ice into a quantum theory, a variable conjugate to the electric field E_{jk} is introduced, the vector potential A_{jk} with $[A_{jk}, E_{jk}] = i$. Since A_{jk} needs to act on a loop of E_{jk}s, the resonance term Eq. 6.18 corresponds to the (lattice) curl $H_{\text{SI−res}} = \cos(\nabla \times \mathbf{A})$ (Hermele et al., 2004). Coarse-graining this expression, along the lines of the $\mathbf{M}$ field in the previous chapter, and expanding the cosine, yields the final desired form of the familiar Maxwell Hamiltonian:

$$H_{\text{U(1)}} = \int d^3x \left[\frac{k_E}{2} \mathbf{E}^2 + \frac{k_B}{2} \mathbf{B}^2 \right], \tag{6.19}$$

with the stiffness constants k_E, k_B, like the permittivity and permeability of an effective medium, not determined quantitatively from the coarse-graining process.

6.4.2 Emergent Photons

At this stage, results from Maxwell electromagnetism can be directly transferred thanks to its identical form to the Hamiltonian, equation 6.19. Particularly notable from a phenomenological perspective is the existence of emergent photons implied by $H_{\text{U(1)}}$. These well-known excitations have a linear dispersion and two transverse polarizations.

Quantum spin ice is therefore an instance of a topological spin liquid with gap-less excitations. Its effective description in this sense is not a topological field theory, as the gapless photons are not – in contrast to Chern-Simons theory discussed further down – insensitive to aspects of the local geometry of the space they inhabit. Note that the existence of photons adds one further inhabitant to our zoo of quasi-particles. It is not fractionalized, unlike the electric charges. However, it is very distinct in origin from the linearly dispersing magnons encountered in conventionally ordered *Heisenberg* antiferromagnets. Indeed, the fact that gapless excitations exist in an Ising magnet with only a discrete spin inversion symmetry is in itself a non-trivial manifestation of the emergent U(1) gauge structure: conventionally ordered Ising magnets tend to have gapped excitations, as gapless modes by Goldstone's theorem generally require the breaking of a continuous symmetry. In distinction to the topological order of a gapped system, the stable gapless excitations of the emergent gauge field are said to reflect what is known as quantum order.

6.4.3 Emergent U(1) Gauge Theory and Coarse-Graining

The fact that an emergent magnetostatics leads to Maxwell electromagnetism may not be all that much of a surprise. Indeed, much of the U(1) structure was already evident on the lattice scale, where the spin ice constraint can be formulated as the requirement that the net number of electric flux lines emanating from each vertex be conserved; number conservation corresponds to a microscopic U(1) constraint. A similar U(1) constraint applies to the dimer models discussed above: each vertex has one and only one dimer impinging on it.

Why then does one not get deconfined U(1) gauge theories with emergent photons in the case of the square and triangular lattice quantum dimer models as well? For the latter, the reason is that the U(1) constraint gets reduced to an Ising parity constraint under coarse-graining, as it turns out to be impossible to orient the dimers on the nonbipartite triangular lattice consistently, unlike in the case of the bipartite diamond and square lattices. For the former, the absence of a photon is a dimensionality effect: a celebrated result states that the relevant U(1) gauge theory in $d = 2$ is in fact confining (Polyakov, 1987) and does not exhibit a phase analogous to the Coulomb phase described by Maxwell electromagnetism in $d = 3$.

6.4.4 Gauge-Charged Excitations

This discussion has been restricted to the case of the pure U(1) gauge theory. The addition of matter is quite analogous to the case of the IGT, or indeed for a gauge theory with any other (Abelian) gauge group. Things are more complicated in detail, however. For the IGT, all variables are $\mathbb{Z}_2$ in nature, that is, magnetic as well as electric charges and fluxes. For the U(1) case, by contrast, angular variables are conjugate to a number operator. As a consequence, the phase diagram in Figure 6.2 is now delineated at the right-hand side by an XY model. A detailed microscopic treatment of quantum spin ice as an instance of such a theory is provided in Hermele et al. (2004).

For the present purposes, we focus on the nature of the gauge-charged objects. First, note that the defect tetrahedra (also known as spinons, or magnetic monopoles in the convention of Chapter 5) represent electric charges. So, what do the magnetic charges look like? Like for the visons of the $\mathbb{Z}_2$ gauge theory, Eq. 5.53, the insertion of a magnetic charge generates nontrivial phases in the wavefunction.

These phases are not particularly easy to visualize. Perhaps the simplest way of doing so is as follows. Consider a wavefunction containing the spin ice configurations $|c\rangle$ and $|c'\rangle$ which differ by flipping spins arranged head-to-tail around a closed loop. (The simplest example of this are two configurations differing by inverting the spins around a hexagon as effected by the resonance term in Eq. 6.18.) One can then define a cone-like object by drawing a line between the location of the magnetic charge and each point on this (oriented) loop. The intersection of the interior of this object with the surface of a sphere around the location of the monopole defines a solid angle $0 \leq \Omega \leq 4\pi$.

The insertion of a magnetic charge will then change the relative phase of $|c\rangle$ and $|c'\rangle$ in the wavefunction by $\varphi \propto \Omega$. The constant of proportionality is fixed by Dirac's original condition, Eq. 5.64, linking the size of the emergent electric and magnetic charges. This happens because, as in the case of conventional quantum electrodynamics, there has to be a Dirac string linking the magnetic charge to a point at infinity (or an oppositely charged such Dirac monopole). An electric charge encircling this string must pick up an unobservable Aharonov–Bohm phase $2n\pi$, so that the amount of flux emanating from the magnetic charge, equal to that entering it via the Dirac string, is inversely proportional to the emergent electric charge.

This prescription can be used to fix the relative phases between all components of the wavefunction by transforming one into the other by consecutive loop flips. The fact that the solid angle is used in the definition, rather than the loop length, automatically accounts for the fact that the strength of the field falls off as $1/r^2$ in $d = 3$ dimensions.

The existence of both electric and magnetic charges in the emergent U(1) gauge theory of quantum spin ice thus drops out in a straightforward manner, which perhaps makes the absence of the latter in conventional quantum electrodynamics all the more remarkable.

6.4.5 *The Fine Structure Constant of Quantum Spin Ice*

One may ask what has been gained by constructing artificial electromagnetism in spin ice: it would seem to be a rather roundabout way of recreating in a magnetic material the universe that we have already lived in forever. Besides the point that magnetic monopoles have not been found in our universe yet, another observation is that the Maxwell electromagnetism in quantum spin ice also differs quantitatively.

To see this, consider the fine structure constant, the value of which is close to $\alpha_{\text{QED}} \approx 1/137$. This is a dimensionless measure of the strength of interactions in Maxwell electromagnetism; and it is small. Indeed, this smallness underpins the success of perturbative calculations in QED.

In quantum spin ice, the fine structure constant is found to be considerably larger – by more than an order of magnitude, $\alpha_{\text{QSI}} \approx 0.1$. This value in itself is not fixed. Rather, by adding additional spin interactions, the value can be increased further, and also tuned all the way down to zero. There is a certain irony to the fact that a topological magnet allows the tuning of the fine structure constant, given the fact that it was the immutability of α_{QED} in an imperfect semiconductor that was taken to be so remarkable in von Klitzing's experiment on the integer quantum Hall effect.

In addition, there are reasons to believe that there is a global instability to the gauge theory at around $\alpha = 0.2$, so that α_{QSI} is actually quite close to the maximal possible coupling. Quantum spin ice, if ever realized in a magnetic material or a quantum computer, may therefore turn into a laboratory quantum simulator of a strongly coupled gauge theory.

6.5 Tensor Gauge Theories and Fractons

With the mathematical structure of gauge and field theories in hand, this structure is open to generalization; one can then ask about the physical content, and possible physical realizations, of the generalized structure afterward. In the case of gauge theories, several routes of generalization are known (Savit, 1980).

One such generalization which has proven to be physically interesting is a family of tensor gauge theories, which go under the name of fractonic models (Vijay et al., 2016; Pretko et al., 2020). Here, the gauge field in question is no longer given by a vector – such as in the case of Maxwell electromagnetism – but rather a tensor. By "simply" adding an index to the electric and magnetic fields, one is led to the Hamiltonian

$$H_{\text{tensor}} = \int d^3x \left[\frac{k_E}{2} E_{jk} E^{jk} + \frac{k_B}{2} B_{jk} B^{jk} \right] . \qquad (6.20)$$

This needs to be supplemented by a definition of the relation of E and B, alongside a Gauss's law. Such a Gauss's law again defines the physical sector of the theory, and it also plays a role in generating the gauge transformations which leave the theory invariant. The gauge transformations act on the gauge field, that is, the field A_{jk} conjugate to E_{jk}.

One possible choice is a gauge field that is a symmetric tensor, $A_{jk} = A_{kj}$. The simplest choice would be the so-called scalar charge theory with gauge transformations given by a scalar function Υ

$$A_{jk} \to A_{jk} + \partial_j \partial_k \Upsilon , \qquad (6.21)$$

with the magnetic tensor defined as

$$B^{jk} = \epsilon^{jmn} \partial_m A_n^{\ k} , \qquad (6.22)$$

and Gauss's law

$$\partial_j \partial_k E^{jk} = \rho . \qquad (6.23)$$

Another possibility is to add a vector index to the charges ρ^j also,

$$A_{jk} \to A_{jk} + \partial_j \Upsilon_k + \partial_k \Upsilon_j \qquad (6.24)$$

$$B^{jk} = \epsilon^{jmn} \epsilon^{kpq} \partial_m \partial_p A_{nq} \qquad (6.25)$$

$$\partial_j E^{jk} = \rho^k . \qquad (6.26)$$

Both of these have charge conservation as in conventional electromagnetism, as seen from putting Gauss's law into integral form over a surface A (with oriented surface elements a_k) and using Stokes's theorem to yield

$$Q = \int_V d^3x \rho = \oint_A \partial_j E^{jk} da_k \text{ and } Q^j = \int_V d^3x \rho^j = \oint E^{jk} da_k , \qquad (6.27)$$

respectively.

These theories, however, exhibit additional conservation laws. For the scalar theory, this is straightforward dipole moment conservation: the total dipole moment of the system is conserved via

$$P^j = \int_V d^3x \, x^j \, \rho = \oint_A \left(x^j \partial_l E^{lk} - E^{jk} \right) da_k . \qquad (6.28)$$

For the vector theory, the quantity in question is an angular moment density,

$$L^j = \int_V d^3x (\mathbf{x} \times \boldsymbol{\rho})^j = \oint_A \epsilon^{jkl} x_k E_{lm} da^m . \qquad (6.29)$$

Such constraints have immediate consequences for the correlations of the fields. Recall that, in the case of spin ice, implementing charge conservation led to a transverse/"dipolar" projector, which manifests itself in the correlations in reciprocal space (Figure 1.3), as a bow tie with a pinch-point motif at its center, with an angular dependence $\cos 2\phi$ around the pinch-point. Analogously, the "quadrupolar" projector at work here provides a nonanalytic structure $\cos 4\phi$, reminiscent of a Maltese cross.

These conservation laws endow the theories with distinctive kinematic restrictions. For the scalar theory, dipole moment conservation simply implies that charges cannot move on their own, but need to take along an oppositely charged partner, as only the motion of charge-neutral objects leaves the dipole moment invariant. (Two like charges moving in opposite directions will of course also do, but this is harder to arrange for, as described below.) Analogously, for the vector model, charges ρ^j can only move longitudinally, that is, along the spatial direction corresponding to their index j, as the angular momentum $(\mathbf{x} \times \boldsymbol{\rho})$ does not depend on this component. Both of these constrain the charges to subdimensional motion, $d_m = 0$ and $d_m = 1$, respectively; theories with $d_m = 2$ are also available. Collectively, particles with such mobility restrictions are known as fractons.

One of course needs not to know about tensor gauge theories or fractons in order to be study interesting kinetically constrained models. Indeed, such studies have in part been motivated by the desire to find mechanism, beyond "conventional" glassiness, naturally providing slow dynamics in many-body systems. Thus, models in this general class have been around for longer than the general mathematical framework provided by fractonic models. For instance, there is a lattice model devised in the context of quantum glassiness (Chamon, 2005), the effective description of which is in terms of such a theory. With the general formulation, one can now construct further lattice models, following the reverse route taken from spin ice to electromagnetism: one needs to construct a lattice with geometric objects (sites, links, plaquettes) to place the appropriate physical fields and terms in the Hamiltonian.

A known physical system exhibiting a fractonic phenomenology (Pretko and Radzihovsky, 2018) is already familiar from two-dimensional elasticity theory. For an introduction to the theory of topological defects in ordered media in the language of homotopy theory, see Chapter 2.8 and also the review by Mermin (1979). The objects in questions are lattice defects known as dislocations and disclinations. In a hexagonal (triangular) lattice, a pentagonal plaquette is an instance of a disclination. This type of defect is immobile; or rather, moving it leads to a trail of dislocations. The disclinations thus behave as the fractonic charges, while a dislocation can be thought of as a dipole, that is, an oppositely charged pair of disclinations. This can be constructed from a combination of a pentagonal and a heptagonal plaquette. Such dislocations are indeed mobile, but only in the direction perpendicular to the Burgers vector which defines the dipolar axis. This is just the subdimensional mobility of the fractonic dipoles.

There is a particularly simple way of enforcing the dipole conservation underpinning fractonic physics by applying an external field. This is based on Wannier–Stark localization: consider a single particle on a one-dimensional lattice (i.e., a chain) with nearest-neighbor hopping matrix element t, so that its energy band in the absence of the field has bandwidth $E_b = 4t$. Adding an electric field along the chain direction yields a potential difference between adjacent sites of $V = e E a$, given by the product of charge, field strength and lattice constant of the chain. This means that an eigenstate of the particle can at most be delocalized over n lattice sites such that $n = E_b/V$, otherwise the kinetic energy cannot compensate for the tilt in potential energy.

This however does not stop a particle pair from delocalizing. The electric field couples to the sum of their coordinates, that is, their center of mass, while it is indifferent to their relative coordinate. As the dipole moment (for simplicity, assume a neutralizing background makes the system charge-neutral overall) is proportional to the center of mass coordinate, this implies dipole conservation as promised.

Another twist awaits the particle pair, though. If a particle cannot hop in isolation, neither can two isolated particles. To convince yourself of this, consider the pair-hopping process in second-order perturbation theory, with one particle hopping left and one right. Depending on the order the hops are executed, the intermediate state has opposite energy difference (hopping left is downhill, and right uphill, say); the two processes thus precisely cancel, preserving the notion of locality. It is only when the particles interact that the intermediate state energies are shifted (say, on account of an attraction) and pair hopping becomes possible. Nature of the resulting many-particle states, and the structure of the energy spectrum in this disorder-free localization problem, are only now beginning to be investigated.

6.6 Long-Wavelength and Topological Field Theories

We now return to the discussion of the topological content of gauge theories. We recall that the set of three excitations (e, m, ϵ), their statistics, and the topological ground-state degeneracy, were identically identified for the quantum dimer model in Section 5.3.2, as well as for the Ising gauge theory and the toric code in this chapter. Indeed, from a topological perspective, these models are the same – they differ only in short-distance details. A topological field theory for this hence only needs to encode this ground-state degeneracy, and nature and statistics of the excitations.

In an ideal world, the low-energy description of a given phase will be obtained in two steps. First, just like in the case of the derivation of the nonlinear sigma model, or the description of spin ice, one needs to coarse-grain the theory from the lattice scale in order to be able to take a continuum limit. Subsequently, one needs to identify the low-energy sector of the continuum theory. However, carrying out these steps in practice is in general not just a mechanically straightforward exercise.

The more promising approach is to identify some key properties of the phase in question – such as the nature and mutual statistics of its excitations and ground-state degeneracy obtained above for the Kitaev toric code model – and to identify promising candidate theories. For the Kitaev model, this leads to one kind of topological field theory, the BF theory, which is discussed in the last section of this chapter. Our first major topic is the field-theory description of (Abelian) quantum Hall states via Chern–Simons theories, which are arguably the simplest – and particularly well understood – examples of topological field theories appearing in condensed matter. Before launching into the detailed discussion of Chern–Simons theory, however, it seems worthwhile to see how a continuum theory can capture topological aspects in a simple toy model, the Jackiw–Rebbi model (Box 6.2).

Box 6.2 Bound States of the Dirac Equation: Jackiw–Rebbi Model

We would like to show, in the simplest possible continuum model (Jackiw and Rebbi, 1976), how bound states can emerge for topological reasons at boundaries between domains of different character. This serves as a conceptual warm-up for the discussion of edge states from Chern–Simons theory in this chapter and for the discussion of bound states in certain vortices of superconductors in Chapter 9. Our starting point is the one-dimensional Dirac equation,

$$i\gamma^0 \partial_t \psi + i\gamma^1 \partial_x \psi - m(x)\psi(x) = 0 \qquad (6.30)$$

where the γ matrices can be chosen to be, in terms of the Pauli matrices,

$$\gamma^0 = \sigma_x, \quad \gamma^1 = i\sigma_z, \qquad (6.31)$$

and the speed of light is unity.

One of the most basic topological questions we might ask of a function is the number of times it passes through zero. We explain that an isolated point where the mass passes through zero always traps a bound state of the fermion with zero energy. A zero-energy solution satisfies $\partial_t \psi = 0$ and hence, if we write the components of ψ as

$$\psi = \begin{bmatrix} \psi_1 \\ \psi_2 \end{bmatrix}, \qquad (6.32)$$

then

$$-\partial_x \psi_1(x) = m(x)\psi_1(x),$$
$$\partial_x \psi_2(x) = m(x)\psi_2(x). \qquad (6.33)$$

The solutions to these equations, if any, must take the form

$$\psi_1(x) \propto e^{-\int^x m(x')dx'}, \quad \psi_2(x) \propto e^{\int^x m(x')dx'} \qquad (6.34)$$

Now, if we have $m(-\infty) < 0 < m(\infty)$, then a zero mode is given by

$$\psi(x) = \begin{bmatrix} e^{-\int^x m(x')dx'} \\ 0 \end{bmatrix} \qquad (6.35)$$

whereas if $m(-\infty) > 0 > m(\infty)$,

$$\psi(x) = \begin{bmatrix} 0 \\ e^{\int^x m(x')dx'} \end{bmatrix} \qquad (6.36)$$

We see that there is one trapped mode with amplitude falling off exponentially with distance away from the zero crossing. The scale of the decay is determined by the magnitude of the mass, and if there are multiple zero crossings but they are farther apart than this length scale for bound state, we would expect the bound states to modify each other only slightly.

This simple model is a useful picture to keep in mind of how edge or surface states decay into the gapped bulk in more complex models. Indeed, while in this book we have tried not to stray too far from the microscopic basis in condensed matter, there are many more analogies between the effects we discuss and simple continuum models in field theory and high energy physics (see, e.g., Witten, 2016).

6.6.1 Chern–Simons Theory

We will now start the process of developing a more abstract description of the fractional quantum Hall effect that will help us understand what type of order it has. For example, this will define precisely what it means to say that the physical state is adiabatically connected to the Laughlin wavefunction. Our main tool will be Chern–Simons theory; we briefly encountered the Chern–Simons term of the electromagnetic gauge potential when we discussed quantum Hall layers at the surface of the strong topological insulator in Section 4.5, which will reappear in a moment. However, a more fundamental use of the Chern–Simons theory is to describe the internal degrees of freedom of the quantum Hall liquid. In other words, we will have both an internal Chern–Simons theory (topological, i.e., metric-independent) describing the quantum Hall liquid and a Chern–Simons term induced in the (nontopological) electromagnetic action.

Since that sounds complicated, let's start by understanding why a Chern–Simons theory might be useful. To begin, we come up with a picture for the Laughlin state by noting that, since the filled lowest Landau level has one magnetic flux quantum per electron, the Laughlin state at $m = 3$ (i.e., $\nu = 1/3$) has three flux quanta per electron. To get a picture for how the Laughlin state is connected to the $\nu = 1$ state, we imagine attaching two of these flux quanta to each electron; this approach is pursued further in Section 8.7. The resulting composite fermion still has fermionic statistics, by the following counting. Interchanging two electrons gives a -1 factor. The Aharonov–Bohm factor from moving an electron all the way around a flux quantum is $+1$, but in this exchange process, each electron moves only half-way around the flux quanta attached to the other electron. So when one of these objects is exchanged with another, the wavefunction picks up three factors of -1 and the statistics is still fermionic.

We should note in passing that this intuitive composite fermion picture was used by Jain to motivate explicit model wavefunctions for many fractional quantum Hall states, which turn out to be rather accurate in comparison to exact diagonalization of small systems (Section 8.7). As with the Laughlin state, we still expect that in a sufficiently large system the overlap between any simple model wavefunction and the actual many-body wavefunction goes to zero, and one motivation for the continuum limit description is to capture the essential physics that the model and real wavefunctions should have in common.

These composite fermions now can form the integer quantum Hall state in the remaining field of one flux quantum per composite fermion, leading to a $\nu = 1/3$ incompressible state in terms of the original electrons. More generally, the phase picked up by a particle of charge q moving completely around a flux Φ is

$$e^{i\theta} = e^{iq\Phi/\hbar}. \tag{6.37}$$

We will now see how the Chern–Simons term lets us carry out a flux attachment related to the above composite fermion idea: in fact, by attaching three flux quanta rather than two to each electron, we would obtain bosons moving in zero applied field, and the Laughlin state can be viewed as a Bose–Einstein condensate of these composite bosons (Read, 1989; Zhang et al., 1989).[2]

The Abelian Chern-Simons theory we will study is described by the Lagrangian density in 2+1 dimensional Minkowski spacetime

$$\mathcal{L} = 2\gamma \epsilon^{\mu\nu\lambda} a_\mu \partial_\nu a_\lambda + a_\mu j^\mu \tag{6.38}$$

where γ is a numerical constant that we will interpret later, a is the Chern–Simons gauge field, and j is a conserved current describing the particles of the theory. Under a gauge transformation $a_\mu \to a_\mu + \partial_\mu \chi$, the Chern–Simons term (the first one) transforms as

$$\epsilon^{\mu\nu\lambda} a_\mu \partial_\nu a_\lambda \to \epsilon^{\mu\nu\lambda} a_\mu \partial_\nu a_\lambda + \epsilon^{\mu\nu\lambda} \partial_\mu \chi \partial_\nu a_\lambda, \tag{6.39}$$

---

[2] One feature of the composite fermion picture that is preferable to the composite boson picture is that the former is naturally described as topological order, while the latter would lead to a picture of the phase in terms of the symmetry-breaking order of a BEC.

where the term with two derivatives of χ drops out by antisymmetry. The new term can be written as

$$\delta S = 2\gamma \int d^2x dt \epsilon^{\mu\nu\lambda} \partial_\mu (\chi \partial_\nu a_\lambda), \qquad (6.40)$$

where again the term with two derivatives acting on a gives zero by antisymmetry. So the variation in the action is a total derivative and hence can be moved to the spatial boundary; *if we can neglect the boundary*, the Abelian Chern–Simons term is gauge-invariant.[3] Later on we will actually consider a system with a boundary and see how the boundary term leads to physically important effects.

Consider the equation of motion from varying this action. We get

$$4\gamma \epsilon^{\mu\nu\lambda} \partial_\nu a_\lambda = -j^\mu. \qquad (6.41)$$

where the 4 rather than 2 appears because the Chern-Simons term has nonzero derivative with respect to both a and ∂a. For a particle sitting at rest, the spatial components of the current vanish, but there must be a flux: writing in components,

$$\int d^2x (\partial_1 a_2 - \partial_2 a_1) = -\frac{1}{4\gamma} \int d^2x \, j^0. \qquad (6.42)$$

Hence a charged particle in the theory gains a flux of the a field (since the left term is just the integral of a magnetic field). If the charge is localized, then the flux is localized as well.

What good is this? Well, we know that when one charged particle with respect to the a field moves around another, it will now pick up an Aharonov–Bohm phase from the attached flux in addition to any statistics factor. The additional statistics factor is

$$\theta = \frac{1}{8\gamma}, \qquad (6.43)$$

where the 1/2 here results because the particles only move halfway around each other in an exchange. In other words, if we started with $\theta = 0$ bosonic particles but added a $\gamma = \frac{1}{8\pi}$ Chern–Simons term, we would obtain fermions, and vice versa. But so far nothing constrains γ, suggesting that in two dimensions, braiding statistics is not constrained to be bosonic or fermionic. We have thus obtained a field-theoretic description of particles in two dimensions (anyons) that may be neither bosonic nor fermionic, similar to the electric and magnetic charges of the toric code introduced earlier in this section.

---

[3] Side remark: as we discussed previously in the context of magnetoelectric polarizability, the non-Abelian Chern–Simons term is not gauge-invariant, because large (non-null-homotopic) gauge transformations change the integral; this is related to the third homotopy group of SU(N). A consequence is that the theory is not acceptable classically but is acceptable quantum-mechanically as long as the *exponential* $\exp(iS/\hbar)$ is gauge-independent, which imposes a quantization condition on the prefactor. We will find below that for different reasons we take the prefactor γ to be quantized even in the Abelian theory, where the need to do so is less obvious.

Why is two spatial dimensions so special? It turns out that an argument about why generalized statistics are possible for point particles in two spatial dimensions but not higher dimensions was given long ago (Leinaas and Myrheim, 1977), as detailed in Box 5.2. The key observation is that an exchange path that takes one particle around another and back to its original location is not smoothly contractible in 2D without having the particles pass through each other, while in higher dimensions, such a path is contractible.

Integrating of Gauge Fields to Obtain Electromagnetic Coupling

Aside from the composite fermion/composite boson pictures, why might the Chern–Simons theory with Lagrangian density given by (6.38) describe quantum Hall states? Without working through a detailed derivation starting from nonrelativistic quantum mechanics of many interacting electrons in a magnetic field, we can note the following. A conserved electromagnetic current in 2+1D can always be written as the curl of a gauge field:

$$J^\mu = \frac{1}{2\pi}\epsilon^{\mu\nu\lambda}\partial_\nu a_\lambda. \tag{6.44}$$

(Note that this electromagnetic current might in general be distinct from the particle current above.) Here a is automatically a gauge field since the U(1) gauge transformation does not modify the current. Gauge invariance forbids the mass term $a^\mu a_\mu$, so the lowest-dimension possible term is the Chern–Simons term, which we write for future use with a different normalization than above:

$$\mathcal{L}_{CS} = \frac{k}{4\pi}\epsilon^{\mu\nu\lambda}a_\mu\partial_\nu a_\lambda. \tag{6.45}$$

The point of the new normalization $k = 8\pi\gamma$ compared to (6.38) is that the boson-fermion statistics transformation above now corresponds just to $k = 1$. It will turn out that k should be an integer for the electron to appear somewhere in the spectrum of excitations of the theory.

Does this term need to appear? No, for example, in a system that has P or T symmetry, it cannot appear. However, if it does appear, then since there is only one spatial derivative, it dominates the Maxwell term at large distances. Effectively we define the quantum Hall phase as one in which $\mathcal{L}_{CS}$ appears in the low-energy Lagrangian; for example, this is true in both the Laughlin state and the physical state with Coulomb interactions, even though the overlap between those two ground-state wavefunctions is zero in the thermodynamic limit.

What if we added the $a_\mu J^\mu$ coupling and integrated out the gauge field? Well, the main reason not to do that is that we obtain a nonlocal current-current coupling. Since the original action is quadratic in the fields, this integration is not too difficult, but an alternate, equivalent way to do it is to solve for a in terms of J. Here our treatment follows the review of Wen (1992). Given a general Lagrangian

$$\mathcal{L} = \phi \mathcal{Q} \phi + \phi J, \tag{6.46}$$

where $\mathcal{Q}$ denotes some operator, we have the formal equation of motion from varying ϕ

$$2\mathcal{Q}\phi = -J \tag{6.47}$$

which is solved by

$$\phi = \frac{-1}{2\mathcal{Q}} J. \tag{6.48}$$

Then substituting this into the Lagrangian (and ignoring some subtleties about ordering of operators), we obtain

$$\mathcal{L} = \frac{1}{4} J \frac{1}{\mathcal{Q}} J - J \frac{1}{2\mathcal{Q}} J = -J \frac{1}{4\mathcal{Q}} J. \tag{6.49}$$

So for the Chern–Simons term we need to define the inverse of the operator $\epsilon^{\mu\nu\lambda}\partial_\nu$ that appears between the a fields. This is a bit subtle because there is a zero mode of the original operator, related to gauge invariance: for any smooth function g, $\epsilon^{\mu\nu\lambda}\partial_\nu(\partial_\lambda g) = 0$. To define the inverse, we fix the Lorentz gauge $\partial_\mu a_\mu = 0$. In this gauge, we look for an inverse using

$$(\epsilon^{\mu\nu\lambda}\partial_\nu)(\epsilon^{\lambda\alpha\beta}\partial_\alpha a_\beta) = \epsilon^{\mu\nu\lambda}\epsilon^{\lambda\alpha\beta}(\partial_\nu \partial_\alpha a_\beta). \tag{6.50}$$

We can combine the ϵ tensors by noting that $\epsilon^{\mu\nu\lambda} = \epsilon^{\lambda\mu\nu}$, so there are two types of nonzero terms in the above: either $\mu = \alpha$ and $\nu = \beta$ or vice versa, with a minus sign in the second case. From the first type of term, we obtain $\partial_\alpha(\partial_\beta a_\beta)$ which is zero by our gauge choice. From the second type, we obtain

$$-\partial_\nu^2 a_\mu. \tag{6.51}$$

So the inverse of the operator appearing in the Chern–Simons term in this gauge is $-\epsilon^{\mu\nu\lambda}\partial_\nu/\partial^2$, and the Lagrangian (6.38) with the gauge field integrated out is just

$$\mathcal{L} = \frac{1}{8\gamma} j_\mu \left(\frac{\epsilon^{\mu\nu\lambda}\partial_\nu}{\partial^2} \right) j_\lambda. \tag{6.52}$$

Aside from showing another interesting difference between the Chern–Simons term and the Maxwell term, we can use this inverse to couple the Chern–Simons theory to an external electromagnetic gauge potential $\mathcal{A}_\mu$. We will set $e = \hbar = 1$ except as noted. We do not include the Maxwell term to give this field dynamics, but rather view it as an imposed field *beyond the magnetic field producing the phase*. For example, we could use this additional field to add an electrical field, and we should find a Hall response. Let's try this:

$$\mathcal{L} = \frac{k}{4\pi}\epsilon^{\mu\nu\lambda}a_\mu\partial_\nu a_\lambda - \frac{1}{2\pi}\epsilon^{\mu\nu\lambda}A_\mu\partial_\nu a_\lambda = \frac{k}{4\pi}\epsilon^{\mu\nu\lambda}a_\mu\partial_\nu a_\lambda - \frac{1}{2\pi}\epsilon^{\mu\nu\lambda}a_\mu\partial_\nu A_\lambda,$$
(6.53)

where in the second step we have dropped a boundary term and used the antisymmetry property of the ϵ tensor. Note that to obtain the second term we have just rewritten $A_\mu J^\mu$ using (6.44).

Now we can integrate out a_μ using (6.52) above, recalling $\gamma = k/(8\pi)$, and obtain

$$\mathcal{L}_{\text{eff}} = \frac{\pi}{k}J_\mu\epsilon^{\mu\nu\lambda}\partial_\nu\frac{1}{\partial^2}J_\lambda = \frac{1}{4\pi k}\epsilon^{\mu\alpha\beta}\partial_\alpha A_\beta\epsilon^{\mu\nu\lambda}\partial_\nu\frac{1}{\partial^2}\epsilon^{\lambda\gamma\delta}\partial_\gamma A_\delta.$$
(6.54)

where in the second step we have used the rewritten Lagrangian in (6.53) to identify $J^\mu = \frac{1}{2\pi}\epsilon^{\mu\nu\lambda}\partial_\nu A_\lambda$. As above, the nonzero possibilities for the indices and resulting sign factors are $\alpha = \nu$ and $\beta = \lambda$ (+1) or vice versa (−1), and also $\gamma = \mu$ and $\delta = \nu$ (+1) or vice versa (−1). Working through these, one is left with the $\gamma = \nu$ and $\delta = \mu$ terms,

$$\mathcal{L}_{\text{eff}} = \frac{1}{4\pi k}\epsilon^{\mu\nu\lambda}A_\mu\partial_\nu A_\lambda.$$
(6.55)

This is the *electromagnetic* Chern–Simons term. The electromagnetic current is obtained by varying A:

$$J^\mu = -\frac{\delta\mathcal{L}_{\text{eff}}}{\delta A_\mu} = \frac{1}{2\pi k}\epsilon^{\mu\nu\lambda}\partial_\nu A_\lambda.$$
(6.56)

where the factor of 2 is obtained because the variation can act on either A.

We can see immediately that this predicts a Hall effect: in response to an electrical field along x, we obtain a current along y. What about the factor $1/(2\pi)$? That is here just so that the response, once we restore factors of e and $\hbar$, is

$$\sigma_{xy} = \frac{e^2}{(2\pi)k\hbar} = \frac{1}{k}\frac{e^2}{h}.$$
(6.57)

Here we get a clue about the physical significance of k. Another clue is to consider the electromagnetic charge J^0 induced by a change in the magnetic field δB (i.e., an additional field beyond the one producing the FQHE):

$$J^0 = \delta n = \frac{1}{2\pi k}\delta B.$$
(6.58)

where we have written $J^0 = \delta n$ to indicate that this electromagnetic density describes the change in electron density from the ground state without the additional field. For the IQHE, a change of one flux quantum corresponds to one additional electron, while we can see that the $k = 3$ Chern–Simons theory predicts a change in density $e/3$, consistent with the quasihole and quasiparticle excitations.

To summarize what we have learned so far, we now see that Chern–Simons theory predicts a connection between the Hall quantum, the statistics of quasiparticles in the theory (from the previous section), and the effective density induced by a local change in the magnetic field. Here the term quasiparticle, which we will discuss later, means whatever particle couples to the Chern–Simons theory as in the preceding section, which need not be an electron.

Bulk Spectrum and Gapless Edge Excitations

One obvious respect in which the Chern–Simons theory is topological is that, because ϵ rather than the metric tensor g was used to raise the indices, there is no dependence on the metric. In Tony Zee's elegant language, it describes a world without rulers or clocks. Since the stress-energy tensor in a relativistic theory is determined by varying the Lagrangian with respect to the metric, the stress-energy tensor is identically zero.

How can a theory be interesting if all its states have zero energy, as in the pure Chern–Simons theory? Well, one interesting fact is that the number of zero-energy states is dependent on the manifold where the theory is defined. We will not try to compute this in general but will solve the theory for the case of the torus. It is quite surprising that we can solve this 2+1-dimensional field theory exactly; the key will be that there are very few physical degrees of freedom once the U(1) gauge invariance is taken into account.

We wish to solve the pure Chern–Simons theory with action

$$\mathcal{L}_{CS} = \frac{k}{4\pi}\epsilon^{\mu\nu\lambda}a_\mu\partial_\nu a_\lambda \qquad (6.59)$$

on the manifold $\mathbb{R}$(time) $\times T^2$(space). The gauge invariance is under $a_\mu \to a_\mu + \partial_\mu\chi$, χ an arbitrary scalar function. Given an arbitrary configuration of the gauge field a_μ, we first fix $a_0 = 0$ by the gauge transformation $a_\mu \to a_\mu + \partial_\mu\chi$ with $\chi = -\int a_0\,dt$. The Lagrangian is then

$$\mathcal{L} = -\frac{k}{4\pi}\epsilon^{ij}a_i\dot{a}_j, \qquad (6.60)$$

where $i, j = 1, 2$. The equation of motion from varying the original Lagrangian with respect to a_0 now gives a constraint

$$\epsilon_{ij}\partial_i a_j = 0. \qquad (6.61)$$

There is still some gauge invariance remaining in a_1, a_2: we can add a purely spatially dependent χ, so that a_0 remains 0, to make $\partial_i a_i = 0$. Then $(a_i(t), a_j(t))$ have zero spatial derivatives and hence are purely functions of time. The

Lagrangian (6.60) is now just the minimal coupling of a particle moving in a position-dependent vector potential; thinking of (a_1, a_2) as the coordinates of a particle moving in the plane, and noting that a constant magnetic field can be described by the vector potential $(By/2, -Bx/2) = (Ba_2/2, -Ba_1/2)$, we see that this is the interaction term of a particle in a constant magnetic field.

So far, using gauge invariance we can reduce the degrees of freedom from a 2+1-dimensional field theory to the path integral for the quantum mechanics of a particle moving in two dimensions. There is one last bit of gauge invariance we need to use. This will reduce the space on which our particle moves, which so far is $\mathbb{R}^2$ because the gauge fields are noncompact, to the torus T^2 on which the theory is defined. We consider a gauge transformation of the form $a_j \to a_j - iu^{-1}\partial_j u$, where u is purely a function of space. Note that if we can write $u = \exp(i\theta)$, this becomes a conventional gauge transformation $a_j \to a_j + \partial_j\theta$. This gauge transformation will not break the previous two gauge constraints if $\nabla^2\theta = 0$.

However, the periodicity of the torus means that we might not be able to define θ periodically, even if u is defined globally and the gauge transformation is indeed periodic. Taking the torus to be $L_1 \times L_2$, the following θ has zero Laplacian everywhere and gives rise to a periodic u and hence a periodic gauge transformation, even if θ is not itself periodic:

$$\theta = \frac{2\pi n_1 x}{L_1} + \frac{2\pi n_2 y}{L_2}. \tag{6.62}$$

The effect of this gauge transformation is that we can shift the particle's trajectory by an arbitrary constant integer multiple of L_1 in the x direction and L_2 in the y direction. To make the torus equivalent to the unit torus, we can rescale $a_i(t) = (2\pi/L_i)q_i(t)$. So finally we have shown

$$S = \int d^2x\, dt\, \frac{k}{4\pi}\epsilon^{\mu\nu\lambda}a_\mu\partial_\nu a_\lambda = -\frac{kL_1L_2}{4\pi}\int dt\, \frac{(2\pi)^2}{L_1L_2}\epsilon^{ij}q_i\dot{q}_j. \tag{6.63}$$

Here one L_1L_2 factor is from the spatial integrals and one is from the change of variable from a_i to q_i. We still haven't done anything quantum-mechanical to solve the path integral. However, we can temporarily add a term $m\dot{q}_i^2/2$ to the Lagrangian and recognize it as the path integral for a particle moving on the torus in a constant magnetic field. The gauge potential is $A_i = k\pi\epsilon_{ij}q_j$, which corresponds to a magnetic field $B = 2\pi k$ (this factor of 2 always appears in the rotational gauge). This is in our theorist's units with $\hbar = e = 1$; it means that there are a total of k flux quanta through the torus.

The limit that gives a description of pure CS theory is $m \to 0$, which takes all states not in the lowest Landau level to infinite energy. This makes sense because in a topological theory there can be no energy scale; the states either have some constant energy (the lowest Landau level here), which can be taken to zero, or infinite energy (the other Landau levels here). A quick calculation shows that there are exactly k states in the lowest Landau level on the torus pierced by k flux quanta; if we had instead worked on the sphere, there would be an additional state, which is an example of the (Wen-Zee) "shift". For example, the lowest Landau level with one flux quantum through the *sphere* corresponds to the coherent-state path integral for a $s = 1/2$ particle, with 2 degenerate states.

The conclusion is that the parameter k also controls the ground-state degeneracy on the torus. An argument (see Eq. 5.17) based on the magnetic translation group (Wen and Niu, 1990) (a direct calculation would be challenging for general manifolds) shows that the general degeneracy of the pure Abelian CS theory on a 2-manifold of genus g is k^g. So for a topological theory, the physical content of the model is determined not just by explicit parameters in the action, such as k, but also by the topology of the manifold where the theory is defined. In this sense topological theories are sensitive to global or "long-ranged" properties, even though the theory is massive/gapped. (Of course, in the pure CS theory there is no notion of length so the distinction between local and global doesn't mean much, but adding a Maxwell term or something like that would not modify the long-distance properties; it would just mean that the other Landau levels are no longer at infinite energy.)

Bulk-Edge Correspondence

We noted above that the Chern-Simons term has different gauge-invariance properties from the Maxwell term: in particular, in a system with a boundary, it is not gauge-invariant by itself because the boundary term, Eq. 6.40, need not vanish. Our last goal in this section is to see that this gauge invariance leads to the free massless chiral boson theory at the edge,

$$S_{\text{edge}} = \frac{k}{4\pi} \int dt \, dx \, (\partial_t + v\partial_x)\phi \partial_x \phi. \tag{6.64}$$

Here k is exactly the same integer coefficient as in the bulk CS theory, while v is a nonuniversal velocity that depends on the confining potential and other details. Note that the kinetic term here is topological in the sense that it does not contribute to the Hamiltonian, because it is first-order in time. The second term is not topological and hence shouldn't be directly obtainable from the bulk theory.

The theory of the bulk and boundary is certainly invariant under restricted gauge transformations that vanish at the boundary: $a_\mu \to a_\mu + \partial_\mu \chi$ with $\chi = 0$ on the boundary. From (6.40) above, the boundary term vanishes if $\chi = 0$ there. This constraint means that degrees of freedom that were previously gauge degrees of freedom now become dynamical degrees of freedom. We will revisit this idea later.

To start, choose the gauge condition $a_0 = 0$ as in the previous section and again use the equation of motion for a_0 as a constraint.[4] Then $\epsilon^{ij} a_j = 0$ and we can write $a_i = \partial_i \phi$. Substituting this into the bulk Chern–Simons Lagrangian

$$
\begin{aligned}
S &= -\frac{k}{4\pi} \int \epsilon^{ij} a_i \partial_0 a_j \, d^2 x \, dt = -\frac{k}{4\pi} \int \left(\partial_x \phi \partial_0 \partial_y \phi - \partial_y \phi \partial_0 \partial_x \phi \right) d^2 x \, dt \\
&= -\frac{k}{4\pi} \int \left(\partial_x (\phi \partial_0 \partial_y \phi) - \partial_y (\phi \partial_0 \partial_x \phi) \right) d^2 x \, dt \\
&= -\frac{k}{4\pi} \int (\nabla \times \mathbf{v})_z \, d^2 x \, dt = -\frac{k}{4\pi} \int \mathbf{v} \cdot d\mathbf{l} \, dt,
\end{aligned}
\tag{6.65}
$$

where $\mathbf{v}$ is the vector field

$$
\mathbf{v} = (\phi \partial_0 \partial_x \phi, \, \phi \partial_0 \partial_y \phi).
\tag{6.66}
$$

(You might wonder why this doesn't let us transform the action simply to zero in the case of the torus studied earlier. The reason is that using Stokes's theorem in the second line, we have assumed the disk topology–since the torus has nontrivial topology, we are not allowed to use Stokes's theorem to obtain zero.) So at the boundary, which we will assume to run along x for compactness, the resulting action is, after an integration by parts,

$$
S_{\text{edge}} = \frac{k}{4\pi} \int \partial_t \phi \, \partial_x \phi \, dx \, dt.
\tag{6.67}
$$

This action describes a topological edge theory determined by the bulk physics; this edge theory is topological in that the Hamiltonian is identically zero. However, in order to obtain an accurate physical description we need to include nonuniversal, nontopological physics arising from the details of how the Hall droplet is confined. One approach to this is to start from a hydrodynamical theory of the edge and then recognize one term in that theory as S_{edge} above (Wen, 1992). The other term in that theory is a nonuniversal velocity term, and the combined action is

$$
S_{\text{edge}} = \frac{k}{4\pi} \int (\partial_t \phi - v \partial_x \phi) \, \partial_x \phi \, dx \, dt.
\tag{6.68}
$$

Here the nonuniversal parameter v clearly has units of a velocity, and in the correlation functions of the theory discussed below indeed appears as a velocity. The Hamiltonian density is

$$
\mathcal{H} = \frac{kv}{4\pi} (\partial_x \phi)^2
\tag{6.69}
$$

---

[4] Here and before we are assuming that the Jacobians from our gauge-fixings and changes of variables are trivial. That this is the case is argued in Elitzur et al. (1989). Another nice discussion in this paper is how, for the non-Abelian case, the bulk can be understood as providing the Wess–Zumino term that keeps the edge theory gapless.

Note that for the Hamiltonian to be positive definite, the product kv needs to be positive: in other words, the sign of the velocity is determined by the bulk parameter k even though the magnitude is not, and the edge is indeed chiral. (The density at the edge is found from the hydrodynamical argument to be proportional to $\partial_x \phi/(2\pi)$, so the above interaction term corresponds to a short-ranged density-density interaction; as usual, we neglect the differences that arise if the long-ranged Coulomb interaction is retained instead.)

Observables from the Edge Theory

We give a quick overview of how the above theory leads to detailed predictions of several edge properties. The general approach to treating one-dimensional electronic systems via free boson theories is known as bosonization, and is the subject of several books, such as Stone (1994). The resulting phase of one-dimensional electrons is known as a Luttinger liquid (Haldane, 1981) to contrast it from the Fermi liquid appearing in higher dimensions. Hence the edge state found above in Eq. 6.64 is known as a chiral Luttinger liquid. In fact, it is possible to work backward and build up two-dimensional quantum Hall phases through coupling one-dimensional Luttinger liquids (Teo and Kane, 2014). While we will not calculate the main results in detail, it turns out that there is a close similarity between the 1-dimensional free (chiral or nonchiral) boson Lagrangian and the theory of the algebraic phase of the XY model studied previously.

The reason such a connection exists is simple: the Euclidean version of the nonchiral version of the above free boson theory is just the 2D Gaussian theory. However, we know from the study of the XY model that subtleties such as the Berezinskii–Kosterlitz–Thouless transition arise when the variable appearing in the Gaussian theory is taken to be periodic, as when it describes an angular variable in that model. One of the surprising results we found was a power law phase with continuously variable exponents: the correlations of spin operators $S_x + i S_y = \exp(i\theta)$ go as a power law with the coefficient depending on the prefactor of the Gaussian.

The connection between the edge theory above and physical quantities is that the electron correlation function is represented in the bosonized theory as $e^{ik\phi}$: effectively ϕ describes a single quasiparticle and k quasiparticles make up the electron. The electron propagator in momentum space is likewise here found to have an exponent that depends on k: there is a factor of k^2 from the ks in the electron operator, and a factor of k^{-1} from the quasiparticle propagator since k appears as a coefficient in the Lagrangian. The result is

$$G(q, \omega) \propto \frac{(vq + \omega)^{k-1}}{vq - \omega}. \qquad (6.70)$$

This describes an electron density of states $N(\omega) \propto |\omega|^{k-1}$, and this exponent can be measured in tunneling exponents: $dI/dV \propto V^{k-1}$. As a sanity check, the $k = 1$ case describes a constant density of states and the predicted conduction is Ohmic: $I \propto V$.

For readers with an interest in field theory, the manner in which the edge properties determine the tunneling exponent is an interesting example of how topology and Luttinger liquid physics connect. Part of the magic of Luttinger liquids is that the correlation functions of a fermion operator, such as the electron creation/annihilation operator, have variable scaling dimensions. Mathematically this happens in the free boson theory because the electron creation operator, say, is represented by a vertex operator $e^{i\alpha\phi}$ times a formal factor to generate fermionic statistics, and the power law in correlation functions of this operator is a continuous function of α. However, in a simple chiral edge such as that of the Laughlin state at $\nu = 1/m$, the value of α is effectively fixed by the chirality and the fractional charge $\pm e/m$ of the edge excitations. In nonchiral edges with modes moving in both directions, the theory of the tunneling behavior is considerably richer and covered in several of the quantum Hall references mentioned in the appendix.

Experimental agreement is reasonable albeit by no means perfect; at $\nu = 1/3$ the observed tunneling exponent $I \propto V^\alpha$ is $\alpha \approx 2.7$, which is far from the Ohmic value ($\alpha = 1$) but reasonably close to the predicted value $\alpha = 3$. The tunneling exponent also does not appear to be perfectly constant when one is on a Hall plateau, as the theory would predict. Other measurements include noise measurements that attempt to see the quasiparticle charge directly (see Section 5.1.4), and in recent years interferometry measurements (Figure 9.3), that try to check more subtle aspects of the theory.

6.7 Mutual Statistics and the Quantum Hall Hierarchy

The above Chern–Simons and edge theories can be generalized without too much work to many more complicated (but still Abelian) quantum Hall states. These states, as suggested by the hierarchy picture, have multiple types of particles, and two particles can have nontrivial statistics whether or not they belong to the same species. These statistics are defined by a universal integer matrix, the K matrix (Wen, 1992), which can be taken as a fundamental aspect of the topological order in the state.[5] The resulting Chern–Simons theory is

$$\mathcal{L} = \frac{1}{4\pi} K^{IJ} a_\mu^I \partial_\nu a_\lambda^J \tag{6.71}$$

[5] Information must also be provided about the allowed quasiparticle types through the compactification conditions of the fields, which is what makes it meaningful that K is integral since the fields cannot be arbitrarily rescaled.

The ground-state degeneracy is now $|\det K|^g$. This effective theory works for all but a few of the many quantum Hall states observed in experiment. In particular, the various fractional quantum Hall states constructed by the elegant composite fermion or hierarchy constructions (Section 8.7) are described by various K matrices.

The exceptions, such as the Moore–Read (Moore and Read, 1991) or Pfaffian state believed to describe the quantum Hall plateau observed at $\nu = 5/2$, have non-Abelian statistics and are a major subject of Chapter 9. Their field theory description by non-Abelian Chern–Simons theories is more complicated in some ways (Nayak et al., 2008), but together with other techniques, such as microscopic wavefunctions or second-quantization treatment of their quasiparticle excitations, has led to great advances in understanding their properties. In fact these states are closely connected in many ways to topological superconductors, and hence we discuss them together with the latter in Chapter 9.

The statistics of e and m charges in the $\mathbb{Z}_2$ spin liquid might seem like a good candidate for a two-by-two K matrix description. The actual description is by the related but distinct BF theory introduced in the next section. Another reason why BF theories are of interest is that, unlike Chern–Simons theories which only exist in odd space-time dimension, BF theory has a direct generalization to 3+1 spatial dimensions, whose physical import for actual materials is not yet fully clear.

6.8 *BF* Theory

To motivate another topological field theory with the nonilluminating name BF theory, we start with the simplest spin liquid in two dimensions, the $\mathbb{Z}_2$ spin liquid appearing in the quantum dimer model and toric code. From earlier in this chapter, we know that there are electric and magnetic charges, and the key statistical fact is that moving an electric charge around a magnetic charge or vice versa gives a nontrivial phase. Working in the basis of these charges, we could guess from the Chern–Simons case that we should represent the statistics of two statistical gauge fields, sourced by e and m particles, by a K matrix of the form

$$K = \begin{pmatrix} 0 & 2 \\ 2 & 0 \end{pmatrix}, \qquad (6.72)$$

if the Laughlin state at $\nu = 1/m$ is represented by $K = (m)$. This correctly reproduces the ground-state degeneracy on the torus, $|\det K| = 4$.

We can encode the relative statistics in a single-term Lagrangian

$$L_{BF} = \frac{1}{\pi}\varepsilon^{\mu\nu\lambda}(a_\mu \partial_\nu b_\lambda). \qquad (6.73)$$

To understand the overall prefactor, note that an integration by parts moves the derivative from a to b. The name BF theory arises for this Lagrangian simply because, after doing so, the term is of the form $b\,da$ or $b\,F$, since $F = da$. The bulk properties of this Lagrangian are very closely related to Chern–Simons theory: again, one can add particle sources for e and m particles that attach a and b flux to them, generating their relative statistics via the BF term.

It might be tempting, looking just at the form of the action and K matrix, to reorient the fields by $\pi/4$ so that the K matrix became

$$K' = \begin{pmatrix} 2 & 0 \\ 0 & -2 \end{pmatrix}. \tag{6.74}$$

This has the same determinant and hence the same ground-state degeneracy. It is actually a different theory, sometimes called the double semion theory: it could describe two $\nu = 1/2$ fractional quantum Hall states with opposite senses of time-reversal-breaking, for example. One way to see that this probably is not the right description is that K' naturally has a pair of oppositely propagating edge states, which the $\mathbb{Z}_2$ spin liquid does not. The subtle distinction arises because of the compactification of the fields (Wen, 1992), which we have so far neglected in our mostly classical treatment: we cannot treat the a and b fields as noncompact, as they are compact in order to give rise to discrete particle charges.

An exciting feature of BF theory is that it has a generalization to 3+1 space-time dimensions, which Chern–Simons theory does not. The key is that we can generalize b to a two-index tensor field,

$$L_{BF} = \frac{k}{4\pi}\varepsilon^{\mu\nu\lambda\rho}a_{\mu}\partial_{\nu}b_{\lambda\rho} - j^{\mu}a_{\mu} - \frac{1}{2}\Sigma^{\mu\nu}b_{\mu\nu}. \tag{6.75}$$

This time we have included source terms in order to stress that the b term must be sourced by a tensor, which could be viewed as a density of linelike objects. This theory can describe fractional statistics, which are possible in three dimensions between pointlike and linelike objects.

Its most direct relevance is to the $\mathbb{Z}_2$, or more generally $\mathbb{Z}_n$, spin liquid in three dimensions. It is a fun exercise to generalize the toric code model to three dimensions and obtain an explicit construction of a spin liquid, now with both pointlike and linelike excitations. This theory has also been discussed in the context of superconductors (Hansson et al., 2004) and topological insulators (Cho and Moore, 2011). There does not yet seem to be a complete understanding in the literature of the possible compactifications of this bulk action and what surface states they could lead to, unlike in the two-dimensional case where things are relatively clear.

The physics of fractional statistics in three dimensions remains an active and poorly understood subject, at least in comparison to the two-dimensional case.

Finding plausible physical realizations of systems with relative statistics between pointlike and linelike objects has been challenging, and even purely theoretical progress is difficult without the connection to two-dimensional conformal field theory, which was essential to unravel many aspects of fractionalized matter in two spatial dimensions. There are a number of recent ideas about fractional 3D states, however, so perhaps future books in this area will have exciting news to report.

7

Topology in Gapless Matter

The original examples of topological phases occurred in systems that in bulk had a unique ground state, or perhaps a small number of ground states as in the fractional quantum Hall effect on a torus. Excited states lay above a nonzero energy gap, which could be the gap between Landau levels in the integer Hall effect, or the gap between energy bands in a Chern insulator. The existence of this gap meant that dynamical processes could be adiabatic, and hence protected from losses, for simple reasons. The Laughlin pumping picture of conductance in the Hall effect, for example, is valid because a slow process will not create any transitions out of the ground state.[1]

What about gapless systems such as metals? Our conclusion in this chapter will be that geometry and topology are, perhaps surprisingly, as important in metals as they are in insulators. It may be worthwhile to give an example of what we mean by geometry and topology in this context. Transport in magnetic metals such as iron was already measured by Hall and others in the nineteenth century, and more recent work has clarified that a piece of the measured Hall effect is determined by the geometry (i.e., the Berry phase) of the electronic wavefunctions of the *pure* material (Nagaosa et al., 2010).

This piece, which can dominate other contributions in some regimes of temperature and material quality, is referred to as intrinsic as it is independent of the details of impurities in the material, unlike most transport processes in a metal. Understanding this piece makes clear how wavefunction geometry modifies even the simple semiclassical picture of how electrons move in a metal. The intrinsic

---

[1] Note that in most experimental quantum Hall systems, there are localized states at all energies and no true energy gap or singularities in the density of states. There is nevertheless a mobility gap (an energy window in which there are no extended states that could degrade transport), and theoretical analyses of transport start from the adiabatic limit.

contribution is closely related to the integer quantum Hall effect in insulators from Chapter 3, which can be written as

$$\sigma_{yx} = \frac{e^2}{h} \sum_{n \text{ occ}} \int \frac{\mathcal{F}_z^n}{2\pi} d^2k. \tag{7.1}$$

Here the sum is over occupied bands: each occupied band contributes an integer to the Hall effect, and that integer is given by the topologically invariant Chern number of the band's wavefunctions.

In a metal, the intrinsic piece comes from the same integrand (the Berry curvature), but now the region of integration is not the whole Brillouin zone but only the occupied portion, the Fermi sea. There are additional contributions from ordinary metallic transport processes of electrons scattering off impurities (skew-scattering and side-jump processes; Nagaosa et al. 2010), so we write schematically

$$\sigma_{yx} = \int_{FS} \frac{\mathcal{F}_z^n}{2\pi} d^3k + \sigma_{\text{extrinsic}}, \tag{7.2}$$

where the first integral is over all occupied states in both filled and empty bands, and the second term depends upon impurity concentration. We have made the integral three-dimensional in order to describe bulk metals, which gives a conductivity with one more power of inverse length, as it should. One could argue for something like this formula by continuity: starting from a 2D Chern insulator with $\sigma_{xy} = e^2/h$, for example, and removing a few electrons from the filled band, it might be surprising if the Hall conductivity immediately dropped to zero. We will see in a little while how understanding σ_{xy}, in materials as familiar as iron, led to a significant change in the intuitive picture of how electrons move in metals.

Equation 7.2 implies that the intrinsic contribution to the Hall effect is not quantized. One might worry that therefore, while geometry might be important in metals, topological invariants and quantization might not appear. Fortunately such is not the case, as amply confirmed by recent work on topological *semimetals* in three dimensions. These were discovered experimentally in 2013-2015, long after they were first discussed as a theoretical possibility in the 1930s by Conyers Herring. Gapless spin systems also show consequences of geometry and topology, and gapless spin liquids in particular are a challenging state of matter to describe that may have recently been observed in experiment.

The following subsection explains how the familiar semiclassical picture of electrons in a free-electron metal needs to be augmented with geometrical concepts, at least in materials that break either inversion or time-reversal symmetries. We then introduce topological semimetals and their unique properties. The family of Kitaev models as prime examples of gapless spin liquids is discussed in the final section.

7.1 Geometric Quantities in the Semiclassical Theory of Metals

The semiclassical equations of motion for the center of a wave packet provide a very useful approach to understanding how the crystalline background modifies the motion of electrons in a metal. Suppose that an electron moves in a nondegenerate band of energy $\epsilon(\mathbf{k})$, and assume for the moment that spin can be ignored (the band is SU(2) symmetric). Then many solid-state textbooks will explain that a wave packet with central position $\mathbf{r}$ and central momentum $\hbar\mathbf{k}$ moves according to

$$\dot{r}_i = \frac{1}{\hbar}\frac{\partial\epsilon(\mathbf{k})}{\partial k_i}, \qquad (7.3)$$

$$\hbar\dot{k}_i = -e\left(E_i(\mathbf{r}) + \left[\dot{\mathbf{r}}\times\mathbf{B}(\mathbf{r})\right]_i\right). \qquad (7.4)$$

The second equation is just the Lorentz force law on an electron located at the center of the wavepacket in electric and magnetic fields $(\mathbf{E}, \mathbf{B})$. The first equation states that the proper velocity for the wavepacket is the group velocity, which captures physics such as the change in effective mass of an electron at the bottom of a quadratic band.

These equations require the applied fields to be both weak and slowly varying in space and time, but there is a deeper problem that causes the set of equations (7.3,7.4) to be incorrect in many materials even in those limits. In materials that break either inversion or time-reversal symmetry, there is an additional term in the semiclassical velocity (7.3) that results from the same Berry curvature $\mathcal{F}(\mathbf{k})$ studied in Chapters 3 and 4. Recall that the Berry curvature is determined[2] from the periodic part of the Bloch wavefunctions via (Eq. 2.9)

$$\mathcal{F}(\mathbf{k}) = \nabla_{\mathbf{k}}\times\mathcal{A}(\mathbf{k}), \qquad \mathcal{A}(\mathbf{k}) = \langle u_{\mathbf{k}}|i\nabla_{\mathbf{k}}|u_{\mathbf{k}}\rangle. \qquad (7.5)$$

The quantity $\mathcal{F}$ has units of inverse wavevector squared, that is, area. Then the anomalous velocity is the second term in the improved semiclassical velocity equation for a single band,

$$\dot{r}_i = \frac{1}{\hbar}\frac{\partial\epsilon(\mathbf{k})}{\partial k_i} - \left[\dot{\mathbf{k}}\times\mathcal{F}(\mathbf{k})\right]_i. \qquad (7.6)$$

Why is this anomalous velocity term present? It has the appealing feature of making the velocity equation look similar in form to the Lorentz equation, and for this reason the Berry curvature is sometimes called a magnetic field in momentum space. We learned from the modern theory of polarization in Chapter 4 that, in insulators, the Berry connection and Berry curvature must contain some information about the electron's location in space, as they control the electric polarization and changes in polarization must require changes in that position. Note that bands

---

[2] As mentioned earlier in this book, some authors choose to define $\mathcal{A}$ and $\mathcal{F}$ with the opposite sign, which leads to a sign change in the expression for the anomalous velocity.

with no Berry curvature do not have an anomalous velocity. A simple example is a tight-binding model with only one atom per unit cell, where at every momentum the orbital composition of the wavefunction within the unit cell is the same.

A heuristic way to think about the anomalous velocity is as follows: view this term as, ignoring spatial indices and gauge invariance to focus on dimensional analysis,

$$\dot{\mathbf{k}} \times \mathcal{F} = \frac{\partial k_j}{\partial t} \frac{\partial \mathcal{A}_i}{\partial k_j} \approx \frac{d\mathcal{A}_i}{dt}. \tag{7.7}$$

This looks like the velocity associated with a polarization current (a current induced by a change in polarization). As the electron's momentum changes, the Bloch wavefunction changes also, and it might evolve between two different distributions in the unit cell (Figure 7.1). As we can follow that change continuously in time, there is a unique velocity we can associate with the change: the anomalous velocity originates in the electron's motion within the unit cell. Clearly this has a different origin than the group velocity, which is the effective velocity of a point particle. The anomalous velocity exists because the electron's wavefunction, rather than being a point, has structure that changes with momentum.

We can sketch a more serious derivation that gives the correct form of the anomalous velocity. Consider the gauge-invariant operator

$$\mathbf{r} = -i \nabla_{\mathbf{k}} + \mathcal{A}, \tag{7.8}$$

where $\mathcal{A} = \langle u | i \nabla_{\mathbf{k}} | u \rangle$ is again the Berry connection, to be thought of as a vector potential in momentum space. This has the form of a minimal coupling derivative

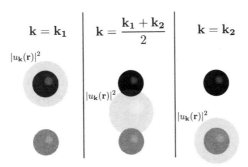

Fig. 7.1 The anomalous velocity depicted as motion within the unit cell. Suppose that at one momentum k_1, the electron wavefunction $u_{\mathbf{k}}(\mathbf{r})$ is located in one region of the unit cell, say around the black atom, and that another momentum k_2 the electron is around the gray atom. As momentum changes, the electron center of mass moves continuously. One possibility is shown, with the electron moving downward; a physically different scenario would be to have the electron move from black to gray by crossing the boundary of the unit cell upward, with velocity in the opposite direction.

in momentum space, known as a Peierls substitution. We will assume that $\mathbf{r}$ is connected to the position of the electron relative to its Wannier center, that is, the center of mass of its Wannier orbital (Vanderbilt, 2018). Then the time derivative of $\mathbf{r}$ is

$$\mathbf{v} = \dot{\mathbf{r}} = -i[\mathbf{r}, H] = -i[\mathbf{r}, \epsilon_{\mathbf{k}} + V(\mathbf{r})]. \tag{7.9}$$

Let the potential $V = Er_x = E(-i\partial_{k_x} - \mathcal{A}_x)$, in order to describe a constant electric field in the x direction.

Now note that different components of $\mathbf{r}$ do not commute with each other; indeed their commutator gives the Berry connection $\mathcal{F}$. As a result, in the simplest case of two dimensions, we obtain an anomalous Hall effect with a current along y proportional to $E_x \mathcal{F}_{yx}$. One might worry about the validity of the Peierls substitution in this argument. Panati et al. (2003) develops a rigorous approach to the semiclassical equations at first order in applied fields including the anomalous velocity. Extensions of the semiclassical formalism to higher order in field and spatial variation are discussed in the review by Xiao et al. (2010).

Before turning to applications of the semiclassical formalism, we mention two additional corrections that occur in the full treatment. A desirable feature of the semiclassical equations of motion is that they should be Hamiltonian, in order that Liouville's theorem be satisfied and an initial volume of phase space evolve to the same final volume; a change in volume would be difficult to reconcile with the Pauli exclusion principle for electrons. That holds for the modified semiclassical equations provided we define the volume of phase space with an additional factor in situations where both magnetic field and Berry curvature are present. The usual phase space volume element $d\mathbf{V} = d\mathbf{r}\, d\mathbf{k}$ becomes

$$d\mathbf{V} = \frac{d\mathbf{r}\, d\mathbf{k}}{1 + (e/\hbar)\mathbf{B}(\mathbf{r}) \cdot \mathcal{F}(\mathbf{k})}. \tag{7.10}$$

(The amount of phase space volume h^3 corresponding to one quantum state cannot be determined without building in information from underlying quantum physics, of course.) Consequently we need to correct the density of states that appears in integrals over phase space from simply $(2\pi)^{-3}$ to

$$D(\mathbf{r}, \mathbf{k}) = \frac{1}{(2\pi)^3}\left(1 + \frac{e}{\hbar}\mathbf{B}(\mathbf{r}) \cdot \mathcal{F}(\mathbf{k})\right). \tag{7.11}$$

An instructive example of how this correction is necessary to give correct physical results in a metal is given in Box 7.1. In an insulator, the density of states correction means that a two-dimensional band of nonzero Chern number must, when a $\mathbf{z}$-directed magnetic field is applied, either take in holes or fill a nearby band with electrons. This is nothing other than the response of an integer quantum Hall state to a magnetic field: a magnetic field applied to a filled Chern band or

Landau level changes the density of states locally and increases or decreases the charge, depending on the sign of the field. One can understand this density change as resulting from inward or outward Hall currents, driven by the in-plane electric fields generated from Maxwell's equations as the magnetic field changes in time.

The second important correction also appears in a magnetic field. An electron in an atom has an orbital magnetic moment proportional to its angular momentum, in addition to its spin magnetic moment. What should we expect in a solid? Well, overall angular momentum is no longer simple to define in the absence of rotational invariance, but we should certainly expect that a superposition of atomic orbitals each with positive L_z, for example, would have a different energy in a z-directed magnetic field than a superposition of the time-reversed versions of those orbitals, which would have negative L_z.

A consequence is that when we make a wavepacket out of Bloch states around a particular value of momentum, it should not be too surprising if that wavepacket has a rotational sense to its motion as well as a translational sense. The intrinsic part of this motion is independent of the details of wavepacket shape and contributes an orbital magnetic moment. It turns out that the total orbital magnetic moment of a material even in the independent-electron approximation is fairly complicated (Thonhauser et al., 2005), but the correction to the semiclassical velocity (7.6) is simple to state: the energy $\epsilon(\mathbf{k})$ that appears in the group velocity should be replaced by

$$\tilde{\epsilon}(\mathbf{k}) = \epsilon(\mathbf{k}) - \mathbf{m}(\mathbf{k}) \cdot \mathbf{B}(\mathbf{r}), \tag{7.12}$$

where $\mathbf{m}(\mathbf{k})$ is the orbital magnetic moment of a Bloch electron, given by

$$m_i(\mathbf{k}) = -\frac{ie}{2\hbar} \varepsilon_{ijk} \left\langle \frac{\partial u_{\mathbf{k}}}{\partial k_j} \middle| H_{\mathbf{k}} - \epsilon(\mathbf{k}) \middle| \frac{\partial u_{\mathbf{k}}}{\partial k_k} \right\rangle. \tag{7.13}$$

Here ε_{ijk} indicates the fully antisymmetric tensor and we trust the reader to distinguish between wavevector $\mathbf{k}$ and index k. Subtracting out the band energy from the Bloch Hamiltonian $H_{\mathbf{k}}$ has the effect of removing a piece that would otherwise be proportional to the Berry curvature $\mathcal{F}$. Indeed, the orbital magnetic moment has the same odd symmetry under time-reversal and even symmetry under inversion as the Berry curvature. Hence the variation in magnetic moment with $\mathbf{k}$ gives an extra contribution to the group velocity in a magnetic field, and we will discuss later in this section how this affects the optical response of materials at low frequencies.

Incorporating the spin degree of freedom, or generalizing the above semiclassical equations to multiple bands, requires using the non-Abelian Berry connection that was previously introduced in Section 4.5 for the magnetoelectric effect in insulators. Even in the simple case of a single band, deriving the intrinsic contribution from the semiclassical equations of motion is subtle: at first glance, the externally applied electric field to produce a current is cancelled in steady state by the impurity electric fields that produce scattering, in order for the longitudinal acceleration of electrons to stop on average and σ_{xx} to be finite. We refer the reader to Nagaosa et al. (2010) for a review of quantum and semiclassical calculations of σ_{xy}. Perhaps the strongest evidence for the intrinsic contribution being dominant in some materials requires remembering that σ in a 3D material is a tensor: the anisotropy of the antisymmetric component of this tensor is found to agree well between experiment and microscopic theory in materials such as hexagonal close-packed cobalt (Roman et al., 2009).

Box 7.1 Semiclassical Equilibrium

To understand how the phase space correction and the anomalous velocity are linked in practice, let us consider a metal in an applied static magnetic field. We would like to understand if these terms give an additional current response, in addition to standard Pauli paramagnetism and Landau diamagnetism. An extensive transport current induced by a static magnetic field would contradict the expectation, credited to Bloch, that while an infinite solid does not reach equilibrium in an electric field (and even in an insulator there is some degree of tunneling), it can reach equilibrium in a magnetic field.

We look for a spatially uniform equilibrium of the semiclassical equations of motion and work to leading order in the magnetic field $\mathbf{B} = B\hat{\mathbf{z}}$. The gradient form of the group velocity means that there will be zero total current from this term in the ground state made by filling all states up to the Fermi energy, that is, with $\tilde{\epsilon} \leq \epsilon_F$. The dynamics of how a system moves toward this equilibrium, starting from the zero-field ground state, are discussed later in this chapter and turn out to control an important piece of the optical response.

What about the anomalous velocity? At first glance this does not need to vanish. At any momentum where the group velocity is nonzero along $\mathbf{x}$ or $\mathbf{y}$, there can be an anomalous velocity term from the magnetic field and Berry curvature:

$$\dot{\mathbf{r}} = -\dot{\mathbf{k}} \times \mathcal{F}(\mathbf{k}) = \frac{e}{\hbar}\left(\mathbf{v}_k \times \mathbf{B}\right) \times \mathcal{F}(\mathbf{k}), \qquad (7.14)$$

where we use $\mathbf{v}_k$ to denote the group velocity $\hbar^{-1}\nabla_{\mathbf{k}}\tilde{\epsilon}$. Simple algebra leads to an apparent ground-state current, dropping the $\mathbf{k}$ arguments of $\mathbf{v}$ and $\mathcal{F}$ for simplicity,

$$j_z = (-e)\frac{-eB}{\hbar}\int_{\tilde{\epsilon}_{\mathbf{k}} \leq \epsilon_F} \frac{d\mathbf{k}}{(2\pi)^3}\left(v_x\mathcal{F}_x + v_y\mathcal{F}_y\right), \qquad (7.15)$$

but there is an extra piece that comes from expanding the density of states correction (7.11) to first order in B:

$$j_z' = (-e)\frac{-eB}{\hbar}\int_{\tilde{\epsilon}_{\mathbf{k}} \leq \epsilon_F} \frac{d\mathbf{k}}{(2\pi)^3}\, v_z\mathcal{F}_z. \qquad (7.16)$$

So we obtain a nonzero velocity along the magnetic field *unless* the sum over all occupied states of $\mathbf{v}(\mathbf{k}) \cdot \mathcal{F}(\mathbf{k})$ is zero.

Indeed this is true, for topological reasons that are interesting in themselves, and this handy little fact appears in other contexts as well. At first this is surprising since the group velocity and Berry curvature have different physical origins, but note that the former is a gradient and the latter is a curl, at least if we can ignore singularities and have a globally defined Berry connection $\mathcal{A}$. We can thus write

$$\int_{\tilde{\epsilon}_\mathbf{k} \leq \epsilon_F} \frac{d\mathbf{k}}{(2\pi)^3} (\mathbf{v} \cdot \mathcal{F}) = \int_{\tilde{\epsilon}_\mathbf{k} \leq \epsilon_F} \frac{d\mathbf{k}}{(2\pi)^3} (\nabla_\mathbf{k} \tilde{\epsilon} \cdot \mathcal{F}) \tag{7.17}$$

$$= \int_{\tilde{\epsilon}_\mathbf{k} \leq \epsilon_F} \frac{d\mathbf{k}}{(2\pi)^3} \nabla_\mathbf{k} \cdot [\tilde{\epsilon}\mathcal{F}] = \epsilon_F \int_{\mathrm{FS}} d\mathbf{S} \cdot \mathcal{F}. \tag{7.18}$$

The last expression used Green's theorem to make an integral over the Fermi surface of the Berry curvature. Based on our previous experience with Chern number, we expect this to be a topological integer, and to be zero under our condition that $\mathbf{A}$ is globally defined. What if we had a Fermi surface with nonzero Chern number? It turns out that if we take more care and sum over all Fermi surfaces properly, then even with monopoles, the integral is still zero (Zhou et al., 2013). But the possibility of a metal with monopoles of Berry curvature, giving rise to Fermi surface sheets that separately have nonzero Chern number, provides a nice lead-in to the discussion of topological semimetals.

7.2 Dirac and Weyl Semimetals

An ideal semimetal can be defined as one in which the locus of points with gapless excitations has lower dimension than a normal metallic Fermi surface, which has dimension $d - 1$ in d spatial dimensions.[3] Undoped graphene provides a ready example of an ideal semimetal in two dimensions; in three dimensions we could have either point or nodal line semimetals. In practice the term semimetal is also used for materials like bismuth, which at stoichiometry have small but nonsingular electron and hole Fermi surfaces. A recent review with background on experimental developments, including the discovery of 3D ideal semimetals starting around 2013, is Armitage et al. (2018).

The focus of this section and the following one is on points of various types (Dirac and Weyl) where bands come together in three-dimensional materials. In an ideal semimetal, the Fermi surface would consist precisely of such crossings. We still use the term semimetal when the Fermi surface passes nearby but not exactly through such crossings, because many properties are still sensitive to the

---

[3] A semiconductor, on the other hand, is an insulator that can be doped to create a small density of metallic electrons or holes.

crossings. Similarly the electronic properties of graphene, or of the metallic surface of a 3D topological insulator, can remain distinctive even when the Fermi surface is not precisely at a Dirac point. Examples in the TI case include the half-integer quantum Hall effect at the surface of a 3D TI, and the existence of surface Majorana fermions when the surface is proximitized by a superconductor, neither of which is sensitive to the precise value of the chemical potential. While graphene is not topological in the same way, we shall give an example below in the discussion of optical properties of an elegant effect in the optical transmission of a single sheet of graphene that is not too sensitive to chemical potential.

Soon after Dirac invented his celebrated equation as a relativistic description of the electron in three dimensions, Hermann Weyl[4] pointed out a limit in which the equation simplifies. Suppose that in one frame of reference, the electron is in an eigenstate of helicity: it is spinning clockwise around its direction of motion, for example. The electron is a massive particle, and it is possible to invert the helicity by transforming to a frame moving faster than the particle. However, for a massless particle, there is no such frame: a particle can have a definite helicity, and in this case Dirac's equation of four-dimensional matrices separates into left-handed and right-handed pieces.

An implication for solids is that the massive 2D electrons of graphene have two natural generalizations to three dimensions. Dirac semimetals involve four bands coming together at a point crossing; the Dirac point can split into two two-band crossings, near which the electrons are effectively Weyl fermions. Materials realizing both possibilities have been found in recent years, but a first question is whether such 3D semimetals have any topological stability. Graphene, for example, is expected to be unstable to disorder that mixes electrons at its two Dirac points (valleys), and it is also unstable to a variety of crystalline perturbations. Going from graphene to single-layer boron nitride, which can be created as a 2D material with the same honeycomb lattice, opens up a large bandgap. We start by giving some intuition for why Weyl points in 3D Weyl semimetals have a higher degree of stability than the Dirac points of graphene, even if in principle the Weyl points are also unstable to extremely strong disorder that means momentum is ill defined.

A model Hamiltonian for a 2D Dirac point, such as in graphene, can be taken to be, setting velocity equal to 1 (see Chapter 2):

$$H = k_x \sigma_x + k_y \sigma_y. \qquad (7.19)$$

Here the Pauli matrices act within a two-band subspace (pseudospin) and we ignore real spin. The degeneracy is broken by a perturbation proportional to σ_z: in graphene, for example, one can add spin-independent perturbations that produce

---

[4] One of Einstein's original colleagues at the Institute for Advanced Study.

either an integer quantum Hall phase (a Chern insulator) or a band insulator (see Chapter 3). The most direct generalization of this two-band crossing to 3D is a Weyl point of two bands, such as

$$H = k_x \sigma_x + k_y \sigma_y + k_z \sigma_z. \tag{7.20}$$

This 3D version is robust to perturbations within the two-band subspace. For example, adding a constant σ_z term would now just displace the k_z location of the Weyl point. This picture suggests that a single Weyl point in isolation might be rather stable, and that mixing with another Weyl point of opposite topological charge might be necessary to eliminate the two-band crossing.

We can use the Berry curvature to see how the robustness of Weyl points results from a more general topological consideration, namely, a Chern number that is ± 1 in the simplest case. Recall that the Berry curvature is a two-form with units of inverse wavevector squared, that is, we need to integrate it over a surface, not a volume, in order to get a dimensionless and potentially topological quantity. In two dimensions the surface of integration was the Brillouin zone. What could be such a surface in three dimensions? Well, we can form a tiny sphere S^2 around the Weyl point (Figure 7.2) and compute the topological charge of the point as the flux of Berry curvature through that surface:

$$n_w = \frac{1}{2\pi} \int_{S^2} d\mathbf{S} \cdot \mathcal{F}, \quad n_w \in \mathbb{Z}. \tag{7.21}$$

In other words, a simple Weyl point as in Eq. 7.20 is a unit monopole of Berry flux (Murakami, 2007). Note that the Berry flux calculation from (7.20) is just the momentum-space version of the previously studied problem of an $s = 1/2$ spin in a time-varying magnetic field, described in Chapter 2.

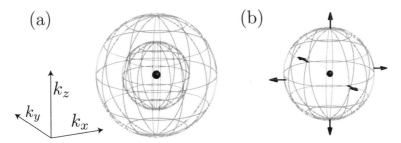

Fig. 7.2 The energy spectrum (a) and Berry curvature (b) around a simple Weyl point. In (a) the two spheres around the Weyl point show two constant energy surfaces $v|\mathbf{k} - \mathbf{k}_0| = \Delta E$ and $v|\mathbf{k} - \mathbf{k}_0| = 2\Delta E$, where $\mathbf{k}_0$ is the momentum of the Weyl point. The energy surfaces are spherical close to the Weyl point. The nontrivial topological charge of the Weyl point appears as (positive or negative) flux of Berry curvature through an enclosing sphere in momentum space.

This association of an integer invariant to a band crossing generalizes easily to more than two bands, as it is based on exactly the same Chern numbers as in the integer quantum Hall case. A nice feature of two-band crossings with linear dispersion is that they must essentially reduce to Eq. 7.20 or its topological opposite, that is, they cannot have zero charge. A theorem of Nielsen and Ninomiya concludes that the total charge of all crossings in a lattice system, at all energies, must be zero; consequently, while in principle one could write a single Weyl fermion in the continuum, we do not believe that this can arise for electrons in a solid. In the simplest case, a material has multiple isolated Weyl points, at locations consistent with the symmetries of the crystal.

In a crystal with both time-reversal and inversion symmetry, Weyl points must come together in pairs and form 3D Dirac points; to see this, note that time-reversal symmetry means that a Weyl point at k must be paired with one at $-k$ with the same Chern number. A center of inversion pairs a Weyl point at k with one with the *opposite* Chern number at $-k$. Hence isolated Weyl points are forbidden when both time-reversal and inversion are present as the above implies that the Chern number must be zero. The Hamiltonian at a Dirac point is a four-by-four matrix as now four bands are involved. These Dirac points are not topologically protected as their total Chern number is zero, and under some circumstances the four-band crossing can be protected by crystalline symmetries. The Weyl materials studied most intensely so far have time-reversal symmetry and break inversion. TaAs is a well-characterized example: it has 24 Weyl points related by a variety of crystalline symmetries other than inversion. Mirror symmetries act similarly to inversion, in that they invert Weyl charge between one point at k and one point at the mirror image.

We end this section on the band-structure properties of Weyl materials by mentioning a consequence of the topological nature of Weyl points that has received striking experimental confirmation. Consider the projection of two Weyl points of opposite charge to the surface Brillouin zone (Figure 7.3). The surface Fermi surface (i.e., the Fermi surface on the surface) turns out to have a Fermi arc (Wan et al., 2011) that connects the two Weyl point projections. This unique surface state helped greatly in the use of ARPES to confirm Weyl semimetal candidates. In other words, Weyl semimetals, like 3D topological insulators, include a surface metal that partitions a normal 2D metal between the top and bottom surfaces. In a 3D topological insulator, each surface Fermi surface has half as many closed sheets as usual, while in a Weyl semimetal, the surface Fermi surface is not closed; only when the top and bottom real-space surfaces are taken together as a single 2D material (Figure 7.3) is there a closed Fermi surface.

These Fermi arcs might seem paradoxical. When a metal is put in a large magnetic field for quantum oscillation measurements, electrons move along the Fermi surface, semiclassically speaking. Where does an electron go when it reaches the

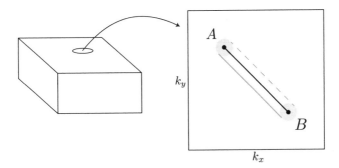

Fig. 7.3 Surfaces and Fermi arcs in a Weyl semimetal, for the ideal case of two Weyl points *A* and *B* of opposite topological charge. The Fermi arc (solid black and gray lines, corresponding to two different energies) on the top surface of the slab connects the projections of bulk Weyl points *A* and *B* into the surface Brillouin zone. Here it is assumed that the Weyl points are at the same energy, and the black line is the Fermi arc at that energy. As the energy changes, more bulk states become available near the Weyl points (gray circles), and the Fermi arcs move. The gray arcs shown with dotted lines are states present on the bottom surface, assumed related by symmetry, rather than the top, and will have opposite Fermi velocity and hence opposite motion measured in quantum oscillations.

end of a surface Fermi arc? The resolution is that these Fermi arc ends are precisely the points where the *bulk* band structure is gapless: an electron in principle executes a complicated orbit in a finite slab that moves from one surface Fermi arc to the other by moving through the bulk. Strong evidence for such orbits has been obtained by nanopatterning surfaces in a Dirac semimetal (Moll et al., 2016). It should be noted that there are many different kinds of topological semimetals known to date, for example with multifold Dirac or Weyl points, or nodal rings rather than points, beyond the simplest case of isolated Dirac/Weyl points treated here, and new kinds seem to be found with regularity.

7.3 Electromagnetic Response of Topological Semimetals

A considerable challenge has been to identify unique response properties resulting from the existence of Weyl points. Such properties would be analogous to the quantized Hall effect in 2D materials or to the quantized magnetoelectric polarizability in 3D topological insulators, which define those states beyond one-particle physics. The Hall effect in a Weyl semimetal that breaks time-reversal symmetry has a topological contribution that follows directly from the quantized charge of Weyl points, but it is less clear what unusual response there might be in topological semimetals that preserve time-reversal. Emerging belief, based on a combination of experiment and theory, is that nonlinear optics is surprisingly interesting in $\mathcal{T}$-invariant Weyl semimetals, which also allows us to give a brief introduction into the microscopic origins of such responses and how topology might emerge. *Nonlinear* responses, including standard nonlinear optical properties, are already observed in experiment to be exceptionally strong in Weyl semimetals (Wu et al., 2017; Orenstein et al., 2020).

We have seen that in 2D insulators the integer quantum Hall effect probes the Chern number of occupied bands. Think of the simplest case of a $\mathcal{T}$-breaking Weyl semimetal: there are two Weyl points of opposite charge ± 1, located at different points in the Brillouin zone. Assume for simplicity that these are the only two points on the Fermi surface. Any plane with periodic boundary conditions, which we take for definiteness to be an xy plane, in the Brillouin zone that does not pass through either Weyl point could be the band structure of an insulating 2D material, and we can define its Chern number by summing over occupied bands.

What happens when the plane we are looking at passes through one of the Weyl points in the Brillouin zone? The effect of the monopole of Berry flux is that the Chern number of the xy planes jumps by precisely the Weyl charge. The Chern number jumps back at the opposite Weyl point. Hence a consequence of this topological charge is that the Weyl material cannot have zero Chern number in all of its xy planes in the Brillouin zone. Instead, even if we take the Chern number to be zero in some xy planes, there will be a fraction of planes that have a nonzero value, which leads to a Hall effect in this example that is not quantized but probes one component of the *separation* $\mathbf{k}^W$ between Weyl points in momentum space:

$$|\sigma_{xy}| = \frac{e^2}{h} k_z^W. \tag{7.22}$$

Note that this is a three-dimensional transverse conductivity and hence should differ by one power of inverse length from a two-dimensional conductivity, which would have the same units as e^2/h. That extra power of length comes from the Weyl point separation. We have essentially just computed the anomalous Hall effect in this $\mathcal{T}$-breaking metal from (7.2) by taking advantage of its particularly simple band structure. If one chose to define the separation using the other direction around the Brillouin zone, that would effectively add an amount to σ_{xy} that is compatible with the Hall effect in a 3D time-reversal-breaking insulator, where the Weyl node separation vector is replaced by a reciprocal lattice vector.

Time-reversal symmetric case: Most Weyl materials discovered so far are nonmagnetic and preserve time-reversal symmetry, which forces σ_{xy} to be zero. First, though, consider linear response in a single electromagnetic field, meaning responses that originate from the bulk of the material and involve a single power of the vector potential $\mathbf{A}$. Linear response is believed not to lead to unique behavior in Weyl materials, at least those with time-reversal symmetry. This is perhaps not too surprising since the quantum Hall effect involves both **E** and **B** fields, as does the quantized magnetoelectric effect. The linear response that has been most studied involves the currents induced by time-varying, low-frequency electric and magnetic fields, which is connected to the expansion of the conductivity or dielectric constant to first order in wavevector q (Zhong et al., 2016) We start from the current density in a time-reversal symmetric material induced by a monochromatic optical field $\mathbf{A}(t, \mathbf{r}) = \mathbf{A}(\omega, \mathbf{q})e^{i(\mathbf{q}\cdot\mathbf{r}-\omega t)}$ at first order in $\mathbf{q}$,

$$j_i(\omega, \mathbf{q}) = \Pi_{ijl}(\omega)A_j(\omega, \mathbf{q})q_l + \dots . \tag{7.23}$$

The time-reversal symmetric part of Π can be expressed in terms of a rank-two tensor

$$\Pi_{ijl} = i\varepsilon_{ilp}\alpha_{jp} - i\varepsilon_{jlp}\alpha_{ip} \tag{7.24}$$

This piece of Π, which is called either natural optical activity or gyrotropy in the optics literature, gives the rotation of light polarization on transmission through low-symmetry materials such as quartz or selenium that have a chirality or handedness. Hence it is analogous to the better-known Faraday effect in magnetic materials.[5] Understanding this response at low frequencies in a metal, where it is called the gyrotropic magnetic effect, provides an example of the use of the semiclassical equations as inputs to a Boltzmann approach and of how low-frequency optics is often controlled by the same wavefunction properties that determine transport.

We will assume from the outset $\hbar\omega \ll \Delta$, where Δ is the energy gap to other bands, which is required for the semiclassical description of transport in metals to hold. The key ingredient is the correction to the band velocity (as opposed to the Berry-curvature anomalous velocity) in the presence of a magnetic field discussed earlier:

$$\mathbf{v}_{\mathbf{k}n} = \frac{1}{\hbar}\frac{\partial\tilde{\epsilon}_{\mathbf{k}n}}{\partial\mathbf{k}}, \quad \tilde{\epsilon}_{\mathbf{k}n} = \epsilon_{\mathbf{k}n} - \mathbf{m}_{\mathbf{k}n}\cdot\mathbf{B}. \tag{7.25}$$

In a static magnetic field, the conduction electrons will eventually, assuming some relaxation which we model in a moment, reach a new equilibrium state with distribution function $f^0_{\mathbf{k}n}(\mathbf{B}) = f(\tilde{\epsilon}_{\mathbf{k}n})$, where f is the zero-temperature Fermi function. The current in this state vanishes according to Box 7.1. We need to consider oscillating fields $\mathbf{E}, \mathbf{B} \propto e^{i(\mathbf{q}\cdot\mathbf{r}-\omega t)}$. Under these fields, the electrons are in an excited state with a time-dependent distribution function $g_{\mathbf{k}n}(t, \mathbf{r})$ which we find by solving the Boltzmann equation in the relaxation-time approximation.

$$\partial_t g_{\mathbf{k}n} + \dot{\mathbf{r}}\cdot\frac{\partial g_{\mathbf{k}n}}{\partial\mathbf{r}} + \dot{\mathbf{k}}\cdot\frac{\partial g_{\mathbf{k}n}}{\partial\mathbf{k}} = -\frac{\left[g_{\mathbf{k}n} - f^0_{\mathbf{k}n}(\mathbf{B})\right]}{\tau}, \tag{7.26}$$

---

[5] The order q^2 term in the expansion of Hall conductivity is also interesting: it is closely connected to the nondissipative Hall viscosity in systems with broken time-reversal symmetry (Hoyos and Son, 2012).

where τ is the relaxation time to return to the instantaneous equilibrium state described by $f_{kn}^0(\mathbf{B}(t, \mathbf{r}))$, assuming a slow spatial variation of $\mathbf{B}$.

Using the semiclassical equations for $\dot{\mathbf{r}}$ and $\dot{\mathbf{k}}$, the distribution function to linear order in $\mathbf{E}$ and $\mathbf{B}$ is $g_{kn}(t, \mathbf{r}) = f_{kn}^0(\mathbf{B}(t, \mathbf{r})) + f_{kn}^1(t, \mathbf{r})$ with

$$f_{kn}^1 = \frac{\partial f/\partial \epsilon_{kn}}{1 - \frac{\mathbf{q}}{\omega} \cdot \mathbf{v}_{kn} + \frac{i}{\omega\tau}}\left[\mathbf{m}_{kn} \cdot \mathbf{B} + (ie/\omega)\mathbf{E} \cdot \mathbf{v}_{kn}\right]. \qquad (7.27)$$

As the current associated with $f_{kn}^0(\mathbf{B})$ vanishes, the current induced by an oscillating $\mathbf{B}$-field is obtained by integrating the perturbation f^1 in the distribution function against the velocity. The result in the long-wavelength limit is

$$\mathbf{j} = \frac{ie\omega\tau}{1 - i\omega\tau}\sum_n \int d\mathbf{k}\,(\partial f/\partial \epsilon_{kn})\,\mathbf{v}_{kn}\,(\mathbf{m}_{kn} \cdot \mathbf{B}). \qquad (7.28)$$

Hence the first-order-in-q correction to the conductivity defined in (7.29), describing optical rotation, takes a simple form in the clean limit $\omega\tau \to \infty$ involving only the orbital magnetic moment on the Fermi surface and the Fermi velocity $\mathbf{v}_F$:

$$\alpha_{ij} = \frac{e}{(2\pi)^2 h}\sum_{n,a}\int_{S_{na}} dS\,\hat{v}_{F,i}m_{kn,j}. \qquad (7.29)$$

Spin can be incorporated straightforwardly and the same result is obtained from fully quantum-mechanical calculations (Ma and Pesin, 2015; Zhong et al., 2016) that are similar in spirit to the TKNN calculation of the integer quantum Hall effect, outlined in Chapter 4. Just as in TKNN, a standard perturbation-theory expression can be transformed, taking advantage of the fact that frequency is smaller than energy differences, by bookkeeping and the use of a topological identity[6] into a simple expression with no energy denominators. This expression also shows that Weyl materials are not especially unique as far as their linear optical response is concerned, except for having particularly simple band structures: the magnitude of the response depends on the separation in energy between oppositely charged Weyl points and other material-dependent properties. In that sense, the gyrotropic magnetic effect is similar to the anomalous Hall effect and not quantized.[7]

Another optical property where semiclassics can be used to obtain low-frequency behavior, now controlled by the Berry curvature (Moore and Orenstein, 2010), is the circular photogalvanic effect or CPGE in inversion-breaking materials. This is the chiral part of photocurrent, which switches sign depending on the

---

[6] The topological identity is actually the same one as in Box 7.1: the integral over occupied momenta of the dot product of group velocity and Berry curvature, treated as a vector, is zero. This identity is also crucial in the full quantum-mechanical treatment of the GME.

[7] Hence one linear response that was conjectured for Weyl materials, sometimes called the chiral magnetic effect (but distinct from the chiral anomaly class of $\mathbf{E} \cdot \mathbf{B}$ effects, possibly visible in nonequilibrium states; Kharzeev, 2014), turns out to be related to natural optical activity or gyrotropy and to be controlled by the orbital magnetic moment of Bloch electrons rather than by band topology.

sense of circular polarization of incident light. A semiclassical calculation similar to the above gives a Berry phase contribution. In a time-reversal symmetric material that breaks inversion, the Berry curvature is nonzero and odd in momentum k, and it is the magnitude of this variation with k that determines the strength of the CPGE in either 2D (Moore and Orenstein, 2010) or 3D (Sodemann and Fu, 2015). Hence the strength of the effect is not quantized in this low-frequency limit, which can be viewed as a nonlinear version of the Hall effect. A surprise is that, specifically in Weyl semimetals with low spatial symmetry and at *nonzero* frequency, there is a quantization of the rate of CPGE current injection (de Juan et al., 2017): in phasor notation

$$\frac{\mathrm{d}j_i}{\mathrm{d}t} = i\pi \frac{e^3 C}{h^2} \hat{\beta}_{ij}(\omega)[\mathbf{E}(\omega) \times \mathbf{E}^*(\omega)]_j \qquad (7.30)$$

where $\mathrm{Tr}\hat{\beta}_{ij} = 1$ and C is the monopole charge.

This result cannot be obtained from semiclassics as the key process is the optical transition that crosses a Weyl node. Since it is somewhat surprising to find quantization of an observable quantity in a metal (leaving aside ratios, such as the approximately quantized Wiedemann–Franz ratio), it may be worth recalling a simpler kind of quantization observed in optics on graphene, a two-dimensional semimetal. The transmission of light in the visible range through a single layer of graphene is approximately $T = 1 - \pi\alpha$, where $\alpha = \frac{1}{4\pi\epsilon_0}\frac{e^2}{\hbar c}$ is the fine structure constant.

An observation of photocurrents consistent with (7.30) was recently reported in the multifold Weyl system RhSi, where the quantum is enhanced by a factor of 4 (Rees et al., 2020). The magnitude is not directly measured in that experiment, but the frequency dependence clearly shows the anticipated pattern of an approximate plateau starting at low frequencies but turning off once contributions from oppositely charged Weyl transitions become allowed. Additional quantitative observation of the magnitude would represent the first example of quantization in a nonlinear effect, and of the quantum e^3/h^2. As of the time of writing, this area is developing quite rapidly, but it already seems clear that several topological semimetals have truly unique nonlinear optical responses, while others have unexpectedly long mean free paths in transport.

7.4 Kitaev Honeycomb Model

Fermion band structures are particularly simple settings in which to observe the kind of topological phenomena described in this chapter: noninteracting electrons provide a simple starting point through a one-body hopping analysis, which turns into a nontrivial many-body problem thanks to the Pauli principle allowing access to special points in the dispersion relation by varying the chemical potential. For magnetic systems with local moments, there is no similarly simple starting point. Superficially, conventionally ordered magnets are promising, as their excitation spectrum also assumes a simple single-particle form in the form of a spin-wave dispersion spectrum. However, spin waves are bosons, removing the convenience provided by the tunable chemical potential. While the study of topological magnons is an interesting subject in its own right, we here take a different route to generalizing the above discussions to topological magnets.

This route involves the identification of a set of topological spin liquids which themselves exhibit *emergent* fermionic degrees of freedom. This is a priori a tall order. As outlined in Chapter 5, spin liquids are in fact rather hard to come by, certainly experimentally, but also theoretically. While we now know of recipes for creating spin liquids, the possibility of devising specifically gapless fermionic spin liquids was, for a long time, out of reach of any controlled treatment of a reasonable Hamiltonian in dimension $d > 1$.

This situation changed completely with the advent of Kitaev's honeycomb model (Kitaev, 2006). This model is part of a family of spin-orbit coupled compass models (Nussinov and van den Brink, 2015), in which the nature of a spin interaction depends on the spatial orientation of the bond connecting the spins. The Hamiltonian reads

$$\mathcal{H}_{\mathrm{Kh}} = \sum_{\langle ij \rangle} J_{ij,a} \sigma_i^a \sigma_j^a \ . \tag{7.31}$$

Here, σ_i^a denote the Pauli matrices, and the interactions are anisotropic as indicated in Figure 7.4. In the following, we concentrate on the case where all nonzero couplings $J_{ij,a}$ have the same strength J.

The model looks innocuous enough, and its solvability, lacking from any number of similar models, is very remarkable. Kitaev's solution of this model proceeds via a representation of the spin operators in terms of Majorana fermions. We defer a discussion of the exact solution to Box 9.4 and concentrate on the physical insights which flow from the solution in the remainder of this account.

Indeed, the exact solution is all the more remarkable as it transparently conveys a number of physical insights. The most immediate one is the identification of the effective low-energy degrees of freedom of the Kitaev model. These come in two flavors. First, a set of Ising fluxes, and second, a set of fermionic excitations. The nondynamical nature of the former degrees of freedom underpins the solvability in terms of effectively free fermions. It is the latter which provide the connections to the above ideas about gapless band structures.

The Ising fluxes are defined on each plaquette and are given by

$$W_p = \sigma_1^x \sigma_2^y \sigma_3^z \sigma_4^x \sigma_5^y \sigma_6^z \ , \tag{7.32}$$

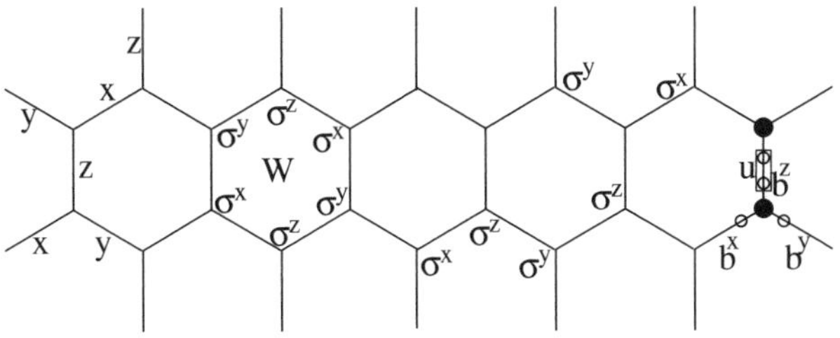

Fig. 7.4 The Kitaev honeycomb model. Conventions, from left to right. Pairwise Ising interactions $J_{ij,a}\sigma_i^a\sigma_j^a$ involve spin components $a = x, y, z$ according to bond direction, as indicated. The emergent Ising plaquette flux variables, W, are products of six spin operators as indicated. Applying a magnetic field perturbatively induced three-spin interactions of size h^3/J^2. The left triplet gives rise to a hopping term in the fermionic description which breaks time-reversal symmetry, thereby gapping the Dirac cone and inducing a gapped topological phase with non-Abelian quasiparticles (see Chapter 9). The right triplet, by contrast, spoils integrability, and is thus typically ignored. The mapping to fermions proceeds by defining four Majorana fermions, c, b^a with $\sigma_a = ib^a c$ at each site (Eq. 9.25). The c "matter" Majoranas are indicated by a solid circle. Two b-Majoranas combine to form a bond variable u (indicated by a rectangular box) which have no dynamics of their own and combine into the fluxes W when multiplied around the six bonds of a plaquette.

again as indicated in Figure 7.4. These commute with each other, and with the Hamiltonian,

$$[W_p, W_q] = [W_p, H_{\text{Kh}}] = 0 \,. \tag{7.33}$$

As $W_p^2 = 1$, each Pauli matrix appears twice in the product, and commutes with all other pairs, the eigenvalues of the fluxes take on the Ising values ± 1.

H_{Kh} is therefore block-diagonal, with each block labeled by the choice of eigenvalues of $\Phi_\bigcirc = \{W_p\}$. The individual blocks can be mapped onto a hopping problem of conventional (complex) fermions, the nature of which is simply stated: it is given by a honeycomb lattice subject to the flux distribution $\Phi_\bigcirc$. However, this "is" not a simple hopping problem–rather, the connection to physical observables in the spin system requires a backward mapping of the insights from the flux-fermion problem.

The spectrum of the two is perhaps the most directly connected feature: in a given flux sector, the excitation energies of the two are identical. This does come with some small print, however. While a simple fermionic hopping problem would have low-energy excitations in the form of particle-hole excitations, here, there is no fermion number conservation, but only a fermion parity conservation.

In addition, there are excitations involving the flux variables. Their energy is obtained by comparing the zero-point energies of the hopping problems with different flux distributions. It is known that the zero-point energy is minimal for the flux-free configuration (Lieb, 1994), $W \equiv 1$ for all plaquettes, which implies that the ground state has an excitation spectrum exhibiting the Dirac dispersion characteristic of the hopping problem on the honeycomb lattice without flux, as detailed in Section 2.5.

It is on this level that a connection to the above topological band structures appears. The Dirac cones are the magnetic representatives of gapless topological band structures. In fact, they are by no means the only ones. It has been realized that Kitaev's solution strategy can be applied to a wide range of lattices. They have in common that they are at most threefold coordinated, with each vertex hosting interactions of different components for different bonds. This class of models is large enough to allow for the construction of a wide variety of band structures. For instance, in three dimensions, one finds not only the usual Fermi surface, but also Weyl nodes as well as nodal rings, as summarized in Hermanns et al. (2018).

It is worth reemphasizing that the mapping between fermionic band structures and observables is not entirely simple. Besides the conservation laws mentioned above, another feature lies in the role of symmetries, which are important, for example, for stability considerations of the band structures. Since the original spin degrees of freedom have been decomposed in a pair of fractionalized degrees of freedom, the relevant symmetries are now represented projectively. The analysis of the resulting projective symmetry group representations to determine under what conditions these band structures are stable, and what instabilities they naturally exhibit, is in its infancy and much remains to be discovered here.

One straightforward way of replacing the gapless Dirac point with a gapped dispersion arises when time-reversal symmetry is broken. This can be arranged for by simply applying a magnetic field,

$$H_{111} = -\sum_{j} \left(h_x \sigma_j^x + h_y \sigma_j^y + h_z \sigma_j^z \right). \tag{7.34}$$

Kitaev argued already in his original publication that a magnetic field in the [111] direction, $h_x = h_y = h_z = h$, would lead to an effective three-spin term of the form

$$H_3 \sim -\frac{h^3}{J^2} \sum_{jkl} \sigma_j^x \sigma_k^y \sigma_l^z, \tag{7.35}$$

whose strength $K \sim h^3/J^2$ to leading order in an expansion in h/J. This is depicted in Figure 7.4. Unfortunately, the field term induces two such terms at this order in h/J. One of them spoils the exact solvability of the model, but for the purposes of the following argument, one can restrict the consideration to the three-spin interaction which does not, and omit the physical motivation of a magnetic field being at its origin.

In the ground-state flux sector, such a time-reversal symmetry-breaking term translates into an additional hopping term for the Majorana problem. This next-nearest-neighbor hopping connects sites on the same sublattice. In the flux-free ground-state sector of the Kitaev-honeycomb model, this appears to mimick the hopping of the Haldane model (see Section 3.2) although its actual manifestation is to generate a Bogoliubov–de Gennes form of the effective complex Fermion problem (see Box 9.4). It thus opens a superconducting gap at the Dirac cones. We will return to this feature in the context of non-Abelian anyons in Chapter 9 on topological quantum computing.

Returning to the discussion of the gapless spin liquid, the reader may wonder to what extent we have built a purely theoretical edifice with little connection to experiment. In particular, a construction involving three bonds on the *two*-dimensional honeycomb lattice labeled with the *three* Cartesian coordinate axes labels x, y and z would seem to be rather unnatural. However, the versatility of condensed matter physics comes to the rescue (Jackeli and Khaliullin, 2009). Consider the right panel of Figure 5.13, which shows a cube and its body $\langle 111 \rangle$ diagonals (dashed lines). Now, let us label the orthogonal cube axes emanating from a given corner by their respective x, y and z directions. Next, project that cube corner into a two-dimensional plane perpendicular to the body diagonal incident at that corner. The result is just the threefold vertex with the labels x, y and z attached to the three lines forming 120° angles as in Figure 7.4. A compound featuring two-dimensional layers stacked in a $\langle 111 \rangle$ direction can therefore provide a natural realization of Kitaev's honeycomb model.

Nonetheless, the question of an actual materials realization is still open, even though there is by now a respectable number of magnetic compounds as candidate hosts of a Kitaev quantum spin liquid (Takagi et al., 2019). A great deal of attention has been paid to a quasi two-dimensional magnetic compound, α-RuCl$_3$, which has been identified as a candidate for realizing this kind of Kitaev spin liquid physics based on an original quantum chemical analysis in the aforementioned work (Jackeli and Khaliullin, 2009) and backed up by first principles calculations (Winter et al., 2017). These suggest that the Hamiltonian describing this system may be dominated by a Kitaev term H_{Kh}. Things are of course not straightforward; in particular, this system exhibits magnetic order at low temperatures, in clear contradiction with the behavior of the pure Kitaev honeycomb model; and the nature of the ordering even depends on the stacking of the two-dimensional layers in the third dimension. However, at temperatures above such ordering, in a proximate spin liquid regime, one can try one's luck with an analysis based on H_{Kh}.

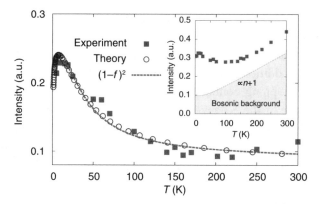

Fig. 7.5 Signature of fermionic excitations in $RuCl_3$: the Raman scattering signal is well fitted by an ansatz based on weakly interacting modes with a thermal Fermi occupation factor, $(1 - f)^2$ (dotted line). Both agree well with the exact solution of $\mathcal{H}_{Kh}$ with a single fitting parameter. From Nasu et al. (2016). Reprinted with permission by Nature Publishing Group.

The result of such an analysis of the temperature dependence of the Raman scattering response is shown in Figure 7.5. These experiments involve the scattering of light by the magnetic system, mediated by their coupling to the lattice degrees of freedom. Like in other scattering experiments, the energy difference between incoming and outgoing photons provides information on the excitation spectrum of the magnetic system. The details of the scattering process, including the polarization dependence of the matrix elements, can be rather involved, and the background subtraction of nonmagnetic contributions is a considerable complicating factor. With all these caveats, the response calculated from the pure Kitaev model agrees reasonably well with the experimental signal shown in Figure 7.5. There, a single fitting parameter is used to fix an overall energy scale of the magnetic exchange, and the signal is unnormalized.

Quite remarkably, a simple phenomenological model also fits the data very well. This is based on the idea that the simplest scattering process involves a photon either the scattering off a single fermionic excitation; or the creation of a pair of them. Both processes leave the fermion parity unchanged, but have different temperature dependences, as they are subject to Fermi–Dirac occupancy factors $f(\omega_1/T)[1 - f(\omega_2/T)]$ for the former, and $[1 - f(\omega_1/T)][1 - f(\omega_2/T)]$ for the latter. These factors simply encode Pauli blocking: the state into which is supposed to be occupied by the scattered excitation must initially be empty, just as the state being scattered out of must initially be occupied. A fit involving a typical frequency scales ω of the excitations does indeed reproduce the temperature dependence of the relevant regimes. It is left to the reader to decide whether to accept this as convincing evidence for the existence of fermionic excitations.

8

Disorder and Defects in Topological Phases

So far, we have largely stayed clear of the discussion of disorder. The first justification one might attempt is that for a gapped phase like the $\mathbb{Z}_2$ dimer liquid, disorder should not have too much of an effect as long as it is weak enough that states cannot move across the gap. However, this imposes a restriction on the strength of disorder that is unnecessarily strong and not satisfied by many experimental systems; for example, integer quantum Hall samples, at least for idealized noninteracting electrons, do not have a true energy gap but only a gap to spatially extended states.

More pragmatically, disorder comes in so many guises – be it vacancy sites in a spin liquid, magnetic ions coupling the two channels of a helical edge, or strong fields of ionized donors in a semiconductor heterostructure – that it would seem hard to say all that much that has a significant degree of universality to it. Nonetheless, there are a number of ways in which topology and disorder interact in a very characteristic way, and this chapter presents a selection of those.

Indeed, quite generally in condensed matter physics, disorder can broadly be said to play three roles. It can be a nuisance, an asset, or a source of new physics; these roles are not mutually exclusive. The nuisance aspect is that disorder may place a veil over interesting phenomena by adding complicated artifacts to an experimental signal. Once such artifacts are systematically understood, they can become useful as experimental probes, as used, for example, in scanning tunneling spectroscopy, where a tunneling microscope supplies an electronic wavepacket to a sample and then probes its interference with secondary waves resulting from scattering off static impurities, thereby revealing properties of the underlying many-body system. Much of this chapter is devoted to the third aspect, with the very existence of quantum Hall plateaux, which would not be there in the absence of disorder, the most striking aspect. We use this as justification to discuss the physics of a disordered two-dimensional electron gas in more detail. This is a rewarding exercise as it allows us to visit a set of topics with interesting connections to field theories and geometric phase transitions. The role of defects is also central for topological quantum computing, and will be discussed in the chapter dedicated to this topic. At the same time, we here provide a few examples of the other two aspects, as these play such an important role for the practicing condensed matter physicist.

In most of our discussions of topological phases so far, we have considered systems with no randomness, such as ideal crystals with a repeating unit cell. Real solids have imperfections of various kinds, and it is natural to regard our previous discussions as incomplete without some consideration of how topological properties can still exist with sufficiently strong disorder to modify significantly the underlying one-electron states. Here we assume for now that the disorder is quenched, that is, constant in time, rather than thermally induced, and work at zero temperature.

It turns out that the connections between disorder and topological behavior are profound and include some remarkably general results, at least for the case of independent particles. These connections could well justify an extended treatment of their own, and we can only hope to scratch the surface here, but we can at least indicate the basic questions that arise and pictures of some key phenomena. The first challenge is to understand how topology appears when we look at an ensemble of Hamiltonians obtained from a statistical weighting of random potentials, which means that physical observables are averages over that ensemble.[1] We will then visit a number of other topics which are particularly enlightening, for instance how to glean information about the nature of a topological phase by analyzing its response to the presence of disorder, be it deliberately introduced or not. This includes the physics of vortices in conventional superconductors as well as in quantum Hall states, where we will also encounter composite fermions. Another subject is the physics of defects in integer topological phases. Finally, we present a collection of topics concerning the interplay of disorder and spin liquids, including the generation of quantum dynamics from static disorder; the role of strain in the creation of a random synthetic gauge field; the capacity of a spin liquid to provide information about the level of disorder in the lattice is resides on; and finally, the response of a gapless spin liquid to dilution and distortion, which will make contact to parallel work on graphene, as well as to a different localization problem which goes under the name of random bipartite hopping.

---

[1] Fortunately, experimentally relevant observables are often found to be self-averaging, meaning that their fluctuations become relatively small as the system size becomes large, or more precisely that the value of the observable in a single realization lies in a shrinking window around the mean value over all realizations with probability close to 1.

8.1 Introduction to Disorder and Localization

The idea of a metal embodies the notion of delocalization: an electron can be added on one side of a sample, and extracted on the other side, by applying only an infinitesimal potential bias between the two. Without wanting to belittle the complexity of what transport in metals actually does involve in the presence of interactions and disorder, a simple cartoon picture for this is that a wavepacket can propagate through the sample as if it obeyed the Schrödinger equation of free space.

The alternative option is for the electron to be stuck in a finite portion of the sample. This possibility – localization – is taken for granted nowadays, such a phenomenon having been observed in all sorts of settings involving, for example, sound, matter (cold atoms) and light waves. However, when it was first raised by Anderson in 1958, this proposal was so revolutionary that its importance was not appreciated until much later. As the Nobel laureate himself famously remarked in his lecture in 1977 (Anderson, 1977), "among those who failed to fully understand [the importance of localization] at first was certainly its author."

The basic idea is that a wave scatters off impurities, that is, objects which break the translational symmetry of space, so that plane waves cease to be solutions of the Schrödinger equation. The incident and scattered waves are then subject to interference. This degrades the ballistic motion of the wavepacket, and replaces it by diffusive or even localized behavior. At its most extreme, the interference between incident and (multiply) scattered waves can be so destructive at long distances as to restrict the wave to a finite portion of space, the size of which is used to define what is called the localization length.

Localization is a subtle phenomenon, as attested by the fact that there have been literally thousands of publications devoted to this topic. The simplest symmetry class arises in considering the nonrelativistic quantum mechanics of a particle in a random potential $V(\mathbf{r})$. We study the electronic wavefunctions that solve the nonrelativistic time-independent Schrodinger equation

$$H\psi = E\psi, \quad H = -\frac{\hbar^2}{2m}\nabla^2 + V(\mathbf{r}). \tag{8.1}$$

A simple guess is that at low energy E, there might be trapped states that decay at least exponentially at large distances (for example, if E is less than the typical values of V), while at high energy, there are scattering states in which on long times the electron executes a random walk, and the wavefunction extends to spatial infinity. In other words, at low energy, there should be some bound states trapped near minima of the potential, while at high energy, we expect that there should be some free states, where the electron scatters occasionally off bumps in the potential but is unbound. A mathematical distinction can be made between localized eigenstates whose magnitude falls off exponentially at spatial infinity, and extended eigenstates which fall off more slowly. Often the term critical is used for wavefunctions that fall off algebraically, and extended reserved for wavefunctions like plane waves that do not fall off at all.

In three dimensions this picture is correct: in fact there is a special energy, known as the mobility edge, that separates extended from localized states. An argument due to Mott for the existence of the mobility edge is that having extended and localized states coexist at the same energy is only possible for rare disorder potentials, as a small perturbation will mix the two types (as the energy denominator is zero) and give rise to extended states. In one and two dimensions, however, it is now believed that all states are localized by a random potential, although the localization length (the length scale on which the localized wavefunctions decay exponentially) becomes extremely long at high energies, especially in 2D.

How is localization of high-energy particles by a weak potential even possible? Recall that a random walk in 1D or 2D returns the same points over and over again, while in 3D and higher the mean number of returns to a given point is finite. As a result of this repetition, even a small bump in the random potential is amplified by being visited a large number of times: the result of constructive quantum interference is to lead to localization by even weak disorder. In two dimensions, the number of returns of a random walk to a particular point diverges only logarithmically in the number of steps, so we might expect the localizing tendency to be feeble in a weakly disordered 2D system, which is indeed the case.

More mathematically, diffusive spreading as in a random walk means that the mean squared distance after time t scales as $\langle R^2 \rangle \sim Dt$ for some diffusion constant D. We can think of the electron density at time t as concentrated in a sphere of radius proportional to $\sqrt{Dt}$. (More realistically, of course, the probability distribution of the density would be Gaussian.) Then, normalizing the overall density to 1, we have that the probability for the particle to be near the origin at time t goes as $(Dt)^{-d/2}$, where d is the spatial dimensionality, since this is the reciprocal of the sphere's volume. Now we can ask, how many times is the electron expected to have returned to the origin by time T? The expected number of returns is thus, up to a factor of order unity,

$$N = \int_{t_0}^{T} \frac{1}{(Dt)^{d/2}} dt. \tag{8.2}$$

Here we ignore any possible singularity at the origin (since we know that at short times our assumption of a Gaussian spread breaks down) and focus on the long-time behavior. The integral converges for $d > 2$. So for $d > 2$, the electron returns only a finite number of times to any particular fluctuation; if the fluctuations are weak enough, then the electron will not be localized since different fluctuations are independent for a random potential with only short-ranged correlations in space.

Why should the interference be constructive when the electron returns to the same point? We give two arguments. One is a heuristic picture of the relevant physics appearing in the diagrammatic approach to computing electronic properties (weak localization). This picture gives a clue that the behavior in a magnetic field may be quite different than in zero field, which indeed is the case as the integer quantum Hall effect must somehow appear. The other approach, in Box 8.1, is a nice example of renormalization group ideas that some readers may have previously seen in statistical physics. This one-parameter scaling theory is a famous example of how a simple RG argument can be used to get a qualitative picture for a complicated system where an exact treatment is extremely difficult.

Box 8.1 One-Parameter Scaling Approach to Anderson Localization

We can give a scaling argument that supports the result stated above, that two dimensions is the marginal dimension for localization, and is much simpler than a serious calculation (Abrahams et al., 1979). (The four authors of this paper are widely referred to as the Gang of Four, a reference to Chinese politics of that era coined by Patrick Lee.) Be forewarned that magnetic fields or strong spin-orbit coupling give alternative behavior to what we find here; in the case of magnetic fields, for example, one needs to consider scaling as a function of two parameters (Khmelnitskii, 1984), which can be taken to be the diagonal and Hall conductivities (σ_{xx}, σ_{xy}), as we will touch on below; see also Figure 8.4.

The formal version of this argument is made using the renormalization group, but the basic idea is quite simple. Let us try to understand the behavior of the function $g(L)$, which gives the conductance (not conductivity) of some material in a (hyper)cube of side L. Our goal will be, given an initial value $g(L_0)$ at a short length scale L_0, to understand what happens when we go to larger scales.

The choice of $g(L)$ as a coupling constant that can be subjected to an RG treatment may at first sight be surprising. Indeed, as a dimensionful quantity to do with electronic transport, it does not look like much of a coupling constant to start with. Its use – as a dimensionless parameter obtained by expressing it in units of the conductance quantum e^2/h which we have encountered extensively in the context of the quantum Hall effects – can nonetheless be motivated as follows (see, e.g., John Chalker's article in Comtet et al., 1999).

Imagine what happens to the states of the small hypercube of linear size L_0 when 2^d of them are combined into a larger one of size $2L_0$. From the point of view of one level in the small hypercube, there are two interesting energy scales. One is the energy window ϵ over which it hybridizes with states from adjacent hypercubes; and the other is the distance to other states in its own hypercube, the mean level spacing Δ. Their ratio ϵ/Δ therefore encodes how easily states from adjacent hypercubes can mix: if $\epsilon/\Delta \ll 1$, there is likely no state in a neighboring hypercube available with which to hybridize, so that each state will remain localized in its own hypercube.

The connection to $g(L)$ then proceeds as follows. Let us return to the idea of particle diffusion, Eq. 8.2, where a displacement scales as $(Dt)^{d/2}$. This allows the conversion of a length scale to a timescale, and in turn an energy scale. For a diffusive conductor of size L, this is known as the Thouless energy $E_T = \hbar D/L_0^2$. We identify ϵ with this Thouless energy. At the same time, the mean level spacing is given by $\Delta = 1/(L_0^d n)$ for a hypercube with density of states n, while an Einstein relation connects the conductivity with the diffusion constant as $\sigma = e^2 n D$. Putting these together makes g pop out as desired: $g(L_0) \sim \epsilon/\Delta = (h/e^2)\sigma L_0^{d-2}$.

The RG flow is conventionally parameterized in terms of the β-function defined as

$$\beta = \frac{d \log g}{d \log L}. \tag{8.3}$$

Suppose first of all that a scattering picture is correct: noninteracting electrons in the material move diffusively (rather than being localized or ballistic), and Ohm's law is satisfied. Then the conductance, once the cube is larger than the mean free path l, should go as

$$g(L) \approx \sigma L^{d-2} \tag{8.4}$$

where σ is the conductivity. Already we can see that $d = 2$ is marginal – a power 0 usually translates into a logarithmic flow. As an aside, we note that it seems possible only in $d = 2$ to have a scale-invariant conductance. From the quantum Hall effects, we are certainly familiar with the quantization of the Hall conductance in units of e^2/h.[a]

Suppose now that instead of having diffusive electron motion, all electrons are in localized states. How then should g behave? Well, if the longest localization length is $\xi \gg l$ (the localization length is always longer than the mean free path), then we expect for $\xi \ll L$

$$g(L) \sim \exp(-L/\xi). \tag{8.5}$$

Here we are ignoring possible power-law factors which will be dominated for large L by the exponential. For a realization of disorder characterized by a particular strength, we expect the microscopic conductance to flow from its initial value $g(L_0)$ with increasing L until reaching one of the above two asymptotic regimes. The challenge is now to justify this picture and understand the importance of dimensionality.

The main conjecture by Abrahams et al. (1979) was that $\beta = \frac{d \log g}{d \log L}$ can be taken to be a function of g *only*. We might think that other properties such as L, the details of disorder, and so on would be important, but at least in the long-length-scale limit, it seems that β is indeed a function of g alone: this is known as one-parameter scaling. We write $\beta(g)$ henceforth to emphasize this. The idea behind one-parameter scaling is that the properties of the system on scale $2L$ are determined by the effective level of disorder at scale L, and that the dimensionless conductance g is a sufficient measure of this disorder: two types of microscopic disorder that give rise to the same conductance g at a large scale L then are predicted to give the same conductance at all larger length scales. This is still an assumption that needs to be tested, but one-dimensional calculations support this picture. It is also possible to justify some of the above using perturbation theory in the disorder strength, which gives the weak localization theory described in the main text.

Returning to our above guesses for the asymptotic form of β, we now have deep in the diffusive regime (high conductance, $g \gg 1$, $\log g > 0$)

$$\beta(g) = \frac{d \log g}{d \log L} = (d - 2). \tag{8.6}$$

Deep in the localized regime (low conductance, $g \ll 1$, $\log g < 0$), we have

$$\beta(g) = \frac{d \log g}{d \log L} = (-L/\xi) \approx \log g, \tag{8.7}$$

which is negative in any dimension.

The point of one-parameter scaling is that now we can make a plot of $\beta(g)$ versus $\log g$, as in Figure 8.1, and argue based on continuity that $d = 1$ and $d = 2$ are very different from $d = 3$. If there were more dimensions to the plot, as occurs in a strong magnetic field, then the situation would be more complicated. In $d = 1$ and $d = 2$, the simplest continuity assumption is that $\beta(g)$ is always negative for finite g, since it is negative at $g = 0^+$ and zero or negative at $g = \infty$. (Regarding the latter, a controlled perturbation theory for $g \gg 1$ in $d = 2$ does indeed yield a negative rather than zero value.) In $d = 3$, we have a more complicated situation because $\beta(g)$ must have a zero.

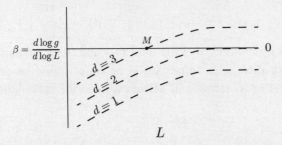

Fig. 8.1 One-parameter scaling flow in various dimensions and mobility edge in $d = 3$. At low conductivity, when the system is localized, the β function describing the (logarithmic) change of conductance is negative in all dimensions. At high conductance, Ohm's law in Eq. 8.6 determines $\beta > 0$ in $d = 3$, so the conductance increases under rescaling, while in $d = 1, 2$ the conductance does not increase since $\beta \leq 0$. This leads to a critical point or mobility edge in $d = 3$ at the point where $\beta = 0$, labeled M.

Above this critical point, the flow is to the diffusive regime; below this critical point, the flow is to a localized regime. This picture corresponds roughly to our intuitive idea of a mobility edge separating extended and localized states in three dimensions. So the simplest guess for how g evolves with L by interpolating between these two asymptotic regimes leads to the conclusion that in 3D there are two regimes separated by an unstable fixed point, while in 1D and 2D, the only stable fixed point is at $g = 0$.

---

[a] In $d = 1$, however, there is also a sort of conductance quantization in units of e^2/h. The resolution to this paradox is that the 1D finite conductance results purely from the contacts; transport is quasi-ballistic in the bulk of the system.

To understand where localization comes from and to present one famous prediction of weak-localization theory, consider the probability that an electron initially at spatial location A in the plane propagates to another point B. We give a heuristic argument that is borne out by summing over a series of diagrams in perturbation theory in the disorder potential (Lee and Ramakrishnan, 1985). In the path-integral picture of quantum mechanics, we would expect the amplitude to be a given by a weighted sum over possible real-space paths from A to B.

The probability is the squared magnitude of the amplitude and will include cross terms between these paths. There are multiple paths to get from A to B, and some of those will have self-intersections. The relevant cross terms are those involving two paths, say 1 and 2, with the same starting and ending point, but which trace an intermediate loop in opposite directions, for example, clockwise for path 1 but counter-clockwise for path 2. In the absence of a magnetic field (i.e., because of time-reversal symmetry), these paths interfere constructively so that their cross terms are positive (Figure 8.2):

$$\langle |\Psi|^2 \rangle = \langle |\psi_1 + \psi_2|^2 \rangle = \langle |2\psi_1|^2 \rangle = 4\langle |\psi_1|^2 \rangle > \langle |\psi_1|^2 \rangle + \langle |\psi_2|^2 \rangle. \qquad (8.8)$$

Now we know by unitarity that the total probability of the particle to get from A to any final location after some period of time should be unity. The above argument says that cross terms will act to increase the probability of paths with

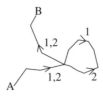

Fig. 8.2 Constructive interference of time-reversed paths in weak localization is destroyed by application of a magnetic field.

self-intersections, and the number of self-intersections is larger for paths where A is not too far from B. This suggests that the first quantum correction to random-walk behavior is in the localizing direction.

A test of this scenario is to consider what happens when a uniform orbital magnetic field normal to the plane of Fig. 8.2 is added to the system. The magnetic field modifies the cross terms by generating Aharonov–Bohm phases for the two directions around the internal loop. Hence constructive interference is no longer guaranteed, with loops of different area causing various phases of the cross term. The prediction is thus that a metal showing signs of localization should behave differently in a magnetic field than classical expectations would suggest: normally a metal's diagonal resistivity increases because the magnetic field bends the electron trajectories, but in a strongly disordered metal, applying a magnetic field famously *decreases* the resistivity by reducing the strength of the localizing corrections. This negative magnetoresistance is one of the major predictions of weak localization theory.

8.2 A Semiclassical Model of Quantum Hall Transitions

One of the central features of the quantum Hall effect is its absence in disorder-free systems: for a translationally invariant system, $\rho_{xy} = \frac{h}{\nu e^2}$ for any value of the filling factor ν, not just for integer or simple rational numbers. This can be seen by a simple thought experiment involving a Lorentz transformation of a system at rest in the laboratory frame, now viewed in a frame moving with velocity $\mathbf{v} \perp \mathbf{B}$. Then, to leading order in $\beta = v/c$, the density n along with the magnetic field $B\hat{z}$ remain unchanged, while an electric field $\mathbf{E} = -\mathbf{v} \times \mathbf{B}$ appears, along with a current density $\mathbf{J} = -ne\mathbf{v}$.

Collecting these together gives

$$\mathbf{E} = \frac{B}{ne}\mathbf{J} \times \hat{B} \ . \tag{8.9}$$

Defining the resistivity tensor as $E^{\mu} = \rho_{\mu\nu}j^{\nu}$, and its inverse, the conductivity, as $j^{\mu} = \sigma_{\mu\nu}E^{\nu}$, we find

$$\rho = \frac{B}{ne}\begin{pmatrix} 0 & +1 \\ -1 & 0 \end{pmatrix} \qquad \sigma = \frac{ne}{B}\begin{pmatrix} 0 & -1 \\ +1 & 0 \end{pmatrix} . \tag{8.10}$$

Note that, for densities corresponding to rational occupancy of the Landau levels, $n = \nu/2\pi\ell^2 = \frac{\nu eB}{2\pi\hbar}$, we obtain the Hall plateau values for $\nu = p/q$: $\sigma = \frac{\nu e^2}{h}$. However, there actually is no plateau at all: σ_{xy} is a continuous function of ν.

One is therefore required to relax the original assumption of translational invariance and ask what happens in its absence. To do this, consider the most natural form of translational symmetry breaking in a semiconductor, which is due to the disordered electric field set up by the ionized donor or receptor atoms that provide the charge carriers for the 2DEG.

It may at first sight seem surprising that there is any conduction in the presence of disorder at all, given the celebrated result detailed in the previous section, that in two dimensions, disorder localizes all electric states. However, the decrease of the resistance upon applying a magnetic field already holds the promise of things "improving" as far as delocalization is concerned. Nonetheless, it is somewhat discouraging that the simplest theory of conduction in the presence of disorder – Drude theory – while giving a nonzero ρ_{xx}, yields precisely the same value for $\rho_{xy} = \frac{-B}{ne}$ as for the clean system. Drude theory supplements Newton's equations of motion by a relaxation term, which in the absence of external driving leads the electrons to return to rest with a timescale τ

$$\frac{d\mathbf{v}}{dt} = \frac{\mathbf{F}}{m^*} - \frac{\mathbf{v}}{\tau}. \tag{8.11}$$

In the presence of crossed $\mathbf{E}$ and $\mathbf{B}$ fields, in the steady state, $\frac{d}{dt} \equiv 0$, the Lorentz force counter balances the relaxation, so that the steady-state drift velocity

$$\mathbf{v}_0 = -\mathbf{J}/ne = -\frac{e\tau}{m^*}\left[\mathbf{E} + \mathbf{v}_0 \times (B\hat{z})\right]. \tag{8.12}$$

This in turn implies

$$\mathbf{E} = \frac{m}{ne^2\tau}\mathbf{J} \times \hat{z} \Longrightarrow \rho = \begin{pmatrix} \frac{m}{ne^2\tau} & -\frac{B}{ne} \\ \frac{B}{ne} & \frac{m}{ne^2\tau} \end{pmatrix} : \tag{8.13}$$

the off-diagonal term ρ_{xy} is entirely independent of the scattering processes supposed to be captured by τ.

The role of localization together with the origin of a conducting channel can be explained elegantly from a totally different point of view by a semiclassical theory, and an analogous quantum theory of the Chalker–Coddington network model (Chalker and Coddington, 1988).

The semiclassical theory connects the behavior of an electron in a strong magnetic field subject to a disorder potential to ideas of classical percolation. The basic ingredient is the observation that an electron in the xy-plane subject to an uniform crossed field $\mathbf{E} = E\hat{x}, \mathbf{B} = B\hat{z}$ executes a motion combining circular motion with a drift in a direction $\mathbf{v}_0||(\mathbf{E} \times \mathbf{B})$ perpendicular to both $\mathbf{E}$ and $\mathbf{B}$, since

$$\dot{x} = a_0\cos(\omega t + \alpha_0), \ \dot{y} = a_0\sin(\omega t + \alpha_0) - \frac{eE}{m^*\omega_c}, \tag{8.14}$$

where a_0, α_0 depend on initial conditions.

Now, in the limit of strong fields, when the difference in electrostatic energy across a Larmor radius $\sim \sqrt{n}\ell$ (where n is the Landau level index) is much less than the cyclotron energy, this motion approximates a fast circular cyclotron motion superimposed on a slow drift in the plane, that is, perpendicular to $\hat{B}$, and along an equipotential of the disorder potential, that is, perpendicular $\hat{E}$.

A study of the single-particle states and their localization properties hence reduces to a study of equipotentials of the disorder potential. Before we consider the disorder potential, note that for a *confining* potential, this picture immediately yields the existence of edge states. At the boundary of each sample, think of a field pointing inward which stops the electron from escaping. As the magnetic field always points in the same direction perpendicular to the plane, the electron's drift is thence always in the same sense (clockwise, say) around the edge, since the direction of the field follows the normal of the sample surface. In the semiclassical picture, there is one edge state per Landau level, that is, for each value of n.

Let us now turn to the disordered problem. For the simplest case, assume disorder that is symmetric about $V = 0$. As shown in Figure 8.3, these equipotentials either enclose local maxima or minima, which are encircled in opposite senses. However, for $V = 0$, a single contour emerges which percolates, and which can therefore support delocalized motion. The resulting density of states is sketched in Figure 8.4.

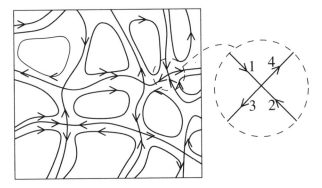

Fig. 8.3 (left) Semiclassical motion of electrons in a disordered potential landscape. The trajectories form closed contours along equipotentials of the disorder potential. On these, the motion is either clockwise or counterclockwise, depending on whether the contour encloses a maximum or a minimum of the disorder potential. These closed contours are separated by a percolating contour. The Chalker–Coddington network model focuses on the properties near the percolating contour. The saddle points are abstracted to nodes (right) of a square network. Each node has two ingoing and two outgoing channels. Quantum interference is taken into account in a random scattering matrix between these (Eq. 8.17).

In the semiclassical picture, there is one such contour for each Landau level, that is, for each value of n. It is the states near the contour that are responsible for the nonzero conductivity of the 2DEG. In particular, at the plateau, the chemical potential is close to the center of a Landau level where the delocalized state resides, and therefore a spike in ρ_{xx} appears (see Figure 1.1).

In this process, the localization length is infinite at the delocalized contour, and finite otherwise (Figure 8.4). It is thus a natural question how the localization length diverges as the energy of the critical equipotential is reached. A standard scaling ansatz from the theory of critical phenomena would be a scaling form like

$$\xi(E) = |E - E_c|^\nu \,, \tag{8.15}$$

where ν is the critical exponent for the correlation length.

At this stage, the alert reader may worry why we are pretending this to be a continuous transition, even though the Hall resistivity ρ_{xy} is discontinuous in the limit of zero temperature. The answer is that such transport coefficients are not thermodynamic quantities such as a free energy, and therefore their discontinuities do not make the quantum phase transition first order. This is akin to the case of the Berezinskii–Kosterlitz–Thouless transition (Box 2.3), which certainly is not first order, but where the superfluid stiffness exhibits a discontinuity, in the form of a "universal jump."

The localization length being given by the size of the equipotentials at a given energy, we are thus interested, in this semiclassical picture, in the critical properties of this geometric object. Its critical exponent is known from the theory of percolation, which predicts

$$\nu = \nu_P = 4/3 \,. \tag{8.16}$$

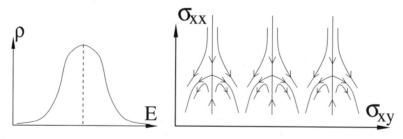

Fig. 8.4 (left) Density of states, $\rho(E)$, of a disordered Landau level. The localization length ξ diverges, Eq. 8.15, near the center of the band, denoted by the dashed line. The tails at large E are known as Lifshits tails. (right) Two-parameter scaling plot of the diagonal and Hall conductivities (Khmelnitskii, 1984). The plateau transitions correspond to the unstable fixed points; the plateaux themselves correspond to attractive fixed points with quantized σ_{xy} and vanishing σ_{xx}.

8.3 Adding Quantum Mechanics: Network Models

The semiclassical model provides an attractive picture of the plateau transition. Alas, the value of ν_P turns out to be incorrect. One basic reason is that the semiclassical model does not take into account any quantum interference between electrons which have traversed different paths along the backbone of the delocalized state.

In disorder problems, it is an entirely normal state of affairs not to include all possible sorts of randomness, if only because there are so many of them. Nor indeed have we considered any interactions between the electrons. Universality provides a justification for at least the former: as long as the disorder does not have any special properties (such as preserving certain symmetries), this should not matter. However, it turns out that quantum interference does alter the universality class of the plateau transition, so that a more elaborate model is needed after all.

The next step in complexity, and the next step historically, is provided by the Chalker–Coddington model (Chalker and Coddington, 1988), reviewed in Huckestein (1995). The network model adds two further ingredients, namely, that of scattering between different trajectories; and randomness in the phase of the wavefunctions. It is in that sense based on a different cartoon of the electrons' motion, treating the behavior of the electrons "more" quantum mechanically.

The first item that is clearly missing from the description of the electronic motion given above is the scale of the cyclotron radius itself: once the guiding center motions of two trajectories approach to within a magnetic length, their wavefunctions overlap in real space, and they cannot be considered as independent any more.

The network model assumes that this situation can be treated as a simple scattering problem, with two incoming and two outgoing channels at each node of the network. The semiclassical model simplifies this to two pairs of channels which do not intercommunicate at all, unless they are right at the critical energy. Two pairs of channels meet at the saddle points of the potential; these thus form the nodes of the network model, and the links are the (directed) channels connecting them (see Figure 8.5).

The scattering problem at a given node is normally cast in terms of a complex number, Z_i on each link of the four incident links, with $|Z|^2$ denoting the flux on that link, and $\arg(Z)$ the phase of the wave incident on the node:

$$\begin{pmatrix} Z_1 \\ Z_3 \end{pmatrix} = M \begin{pmatrix} Z_2 \\ Z_4 \end{pmatrix}. \tag{8.17}$$

Charge conservation then demands the incoming and outgoing fluxes to be equal, $|Z_1|^2 + |Z_2|^2 = |Z_3|^2 + |Z_4|^2$. This restricts the form of M to

$$M = \begin{pmatrix} \exp(i\phi_1) & 0 \\ 0 & \exp(i\phi_2) \end{pmatrix} \begin{pmatrix} \cosh\theta & \sinh\theta \\ \sinh\theta & \cosh\theta \end{pmatrix} \begin{pmatrix} \exp(i\phi_3) & 0 \\ 0 & \exp(i\phi_4) \end{pmatrix}. \tag{8.18}$$

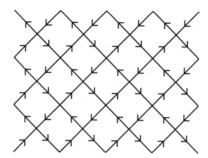

Fig. 8.5 The Chalker–Coddington network model is abstracted from the semiclassical model (Figure 8.3) by replacing the equipotential contours with a bipartite square network at the nodes of which the random matrices M encoding the energy-dependent scattering properties (Eq. 8.18) reside.

For a single node, the values of the phases ϕ are immaterial, as they can be gauged away and have no bearing on the scattering process itself, which is parameterized by only one parameter θ.

This transmission coefficient of a free particle incident on a saddle point in the presence of a strong perpendicular magnetic field can be computed exactly as the potential is still harmonic (Fertig and Halperin, 1987). This determines the value of θ as a function of the energy offset from the saddle point, ϵ, in terms of the saddle point curvature $\gamma = \frac{4\epsilon}{l_c^2 \sqrt{V_{xx} V_{yy}}}$, where V_{xx} and V_{yy} are the second derivatives of the potential:

$$\sinh \theta = \exp(-\pi \gamma / 2) \,. \tag{8.19}$$

For energies well away from the saddle point, the leakage from one channel across the saddle is small, corresponding to the channels following the semiclassical trajectories, clockwise or counterclockwise depending on whether they are above or below the saddle point. As the energy is varied, the transmission interpolates continuously between these two limiting cases.

Chalker and Coddington simplified the nodes of the network to form a simple square lattice, that is, the locations of the saddle points are not endowed with any randomness. Where randomness does enter is via the different areas of the plaquettes whose corners are the saddle points forming the nodes of the network model. These give rise to different Aharonov–Bohm phases, and hence effectively random phases between outgoing and incoming links of the network. The hope is then that this random phase is now "enough" randomness to describe the system as a whole.

The tractability of the network model arises from the fact that the quantum mechanical problem has been reduced to one of multiplying scattering matrices.

These can be treated using an arsenal of methods developed for the study of transfer matrices, which are well suited for numerical computation.[2]

Thus, it was possible to obtain a value of $\nu_{CC} = 2.5 \pm 0.5$ for the critical exponent, which is distinct even within those sizeable error bars from the (semi)classical value $\nu_P = 4/3$. Later determinations remained in the range given by Chalker and Coddington, with smaller quoted uncertainties. However the ranges proposed in different studies are nonoverlapping, and it turns out that even after many years and much study, the critical properties of the plateau transition continue to be unsettled. So, for the moment, we know that the plateau transition is not the classical geometrical one, but we are not sure what its critical exponents are, and whether the Chalker–Coddington model is sufficient for capturing the actual experimental situation. The "obvious" remaining missing ingredient for the theory of the plateau transition are the interactions between the electrons.

8.4 Basic Ideas of Random Matrix Theory and the Tenfold Way

We would now like to think about how symmetry, localization, and topology are related more generally, starting with the two cases treated so far and concentrating on two spatial dimensions. We found Anderson localization of eigenstates in the real time-independent Schrödinger equation for a particle moving in a random potential. Adding a magnetic field led to more complex behavior, with the possibility of topological states (the integer quantum Hall effect) separated by extended states at isolated points in the set of energies. Clearly the addition of the magnetic field makes a considerable difference.

We saw earlier in this book that time-reversal-symmetric systems with spin-orbit coupling can also support a topological phase in 2D, the quantum spin Hall effect or 2D topological insulator. Numerical and analytical calculations predict that for random potentials in this class, the situation differs from the quantum Hall case: different insulating regions are separated, in general, not by critical behavior at a single value of the energy but by a metallic (i.e., delocalized) region over a nonzero interval of energies, with critical states only at the endpoints of the interval.

From the point of view of symmetry, these three cases (Anderson localization, quantum Hall effect, and $\mathbb{Z}_2$ topological insulator) correspond to the three classes of random matrices studied by Wigner and Dyson: orthogonal, unitary, and symplectic, respectively. They correspond to three different possibilities for

---

[2] Indeed, one of the motivations for the network model from the methodological side was that a computer was available at the time which was very good at doing such manipulations, and the network model was the nail which fit this hammer.

time-reversal symmetry. In the real Hamiltonians of simple Anderson localization, acting twice with time-reversal takes a state back to itself: $T^2 = 1$. In the symplectic case, appropriate for fermions with time-reversal symmetry, $T^2 = -1$ as discussed in Section 3.4 and Chapter 4. In the unitary case, there is no symmetry under time-reversal of either type.

So these three possibilities for time-reversal symmetry lead to three quite different behaviors in the same spatial dimension. Beyond time-reversal, there is one other physically relevant antiunitary symmetry[3] that systems can have even when they have no translational or rotational symmetries. This symmetry arises most often in the Bogoliubov–de Gennes description of quasiparticles in superconductors, which we review in a moment and will also be useful in the following chapter. We label the symmetry operation as P for particle-hole symmetry.[4] Finally, combining the antiunitary operations P and T leads to a chiral operation $C = PT$, which unlike the other two is unitary.[5]

Considering possible behavior under T and P leads to 10 symmetry classes, known as Altland–Zirnbauer classes (Altland and Zirnbauer, 1997) or the tenfold way.[6] Why ten, when there are three behaviors under T, and likewise three behaviors under P ($P^2 = +1, -1$, or 0)? There are nine possibilities arising because a system could have each possible kind of T or P symmetry; the tenth possibility arises because it is possible for a system to have symmetry under the product $C = PT$ without having symmetry under either operator separately. Remarkably, these ten classes support various types of topological behavior in a beautiful dimension-dependent way, which we return to after explaining where P comes from. The framework of our analysis is given by the Bogoliubov–de Gennes description of superconductivity, described in Box 8.2.

Box 8.2 Bogoliubov–de Gennes Formalism of Superconductivity

The Bogoliubov–de Gennes formalism is a way to describe quasiparticles in the mean-field description of superconductivity by Bardeen, Cooper, and Schrieffer, the famous BCS theory. There are many classes of superconducting states observed in experiment, which to date all seem to be characterized by a tendency of electrons to form (Cooper) pairs. Fortunately the description of a wide variety of superconductors *once they are*

---

[3] There are several different mathematical developments of the tenfold way, and it is possible to view the relevant symmetries as unitary rather than antiunitary; such subtleties are well beyond the scope of this introduction.

[4] The usage of that term varies somewhat in the community. For a specific exposition, see Zirnbauer (2020).

[5] Confusingly, some authors use C for what we have called P, and then $S = CP$ for the combined unitary symmetry.

[6] Presumably the name is a reference to the eightfold way of organizing hadrons in QCD, itself an homage to the Noble Eightfold Path of Buddhism.

superconducting seems to be captured well by a single formalism, whether the super-conductivity results from electron-phonon interactions, as in the first superconductors to be understood like Al or Nb, or something else. So in this book we will ignore the question of superconductivity's origin and instead focus on its consequences, which require a basic understanding of both the order parameter and excitations. A recently released graduate textbook that covers superconductivity with an eye to topological applications is Girvin and Yang (2019).

The basic meaning of mean-field theory here is that a four-fermion interaction term, with two annihilation and two creation operators, is reformulated by replacing two of the operators by a nonzero expectation value, that is, a number (if there is no spatiotem-poral variation) or, more generally, an order parameter field. This leaves two fermionic operators, that is, free fermions. Perhaps the greatest insight of the BCS theory is that the order parameter is the expectation value of an operator that does not conserve particle number: for example, in the BCS wavefunction for a spin-singlet, rotationally and translationally invariant superconductor,

$$|\Psi_{BCS}\rangle = \prod_k (u_k + v_k c_{k\uparrow}^\dagger c_{-k\downarrow}^\dagger)|0\rangle, \tag{8.20}$$

$$u_k{}^2 = \frac{1}{2}\left(1 + \frac{\epsilon_k - \mu}{E_k}\right), \quad v_k{}^2 = \frac{1}{2}\left(1 - \frac{\epsilon_k - \mu}{E_k}\right), \tag{8.21}$$

there is a coherence or off-diagonal long range order (ODLRO) between states of N and $N + 2$ electrons. Here ϵ_k is the energy of electrons before the pairing interaction is turned on, μ is chemical potential, $|0\rangle$ is the vacuum containing no electrons, and

$$E_k = \sqrt{(\epsilon_k - \mu)^2 + \Delta_k{}^2}. \tag{8.22}$$

We return to the physical significance of E_k in just a moment. Δ_k is related to the order parameter and measures the ODLRO in the state at momentum k. The reader may check that with $\Delta_k = 0$ the BCS wavefunction reduces to an ordinary filled Fermi sea, and that with nonzero Δ_k there is a nonzero expectation value in $|\Psi_{BCS}\rangle$ of the Cooper pair annihilation operator $c_{k\uparrow}c_{-k\downarrow}$.

The BCS wavefunction, or a similar superposition of different particle number states in a superfluid, provides a microscopic picture of why the order parameter field in these states has a phase and apparently breaks a U(1) symmetry, at least at the mean-field level. Hence the ODLRO concept explains how the order parameter phase is not the overall phase of a wavefunction, which is not meaningful in quantum mechanics, and that the symmetry breaking in the superconducting state is also not the U(1) "gauge symmetry" of electromagnetism, which can never really be broken since gauge free-dom is not truly a symmetry but rather an ambiguity. The fact that the order parameter wavefunction has this deep connection to the microscopic electrons suggests already that interesting things might happen to the excitations when the order parameter has nontrivial structure, such as a vortex.

What are the excitations above the BCS ground state like in a simple case? For this translation-invariant case, it is not hard to show that E_k is the energy above the ground state to create a Bogoliubov fermion excitation. Examples of the operators that create or annihilate these fermions are

$$\gamma_{p\uparrow}^{\dagger} = u_p c_{p\uparrow}^{\dagger} - v_p c_{-p\downarrow}, \quad \gamma_{-p\downarrow} = v_p c_{p\uparrow}^{\dagger} + u_p c_{-p\downarrow}. \tag{8.23}$$

Note that each of these adds momentum p and spin up. We present these operators in part to explain the difference between them and the Majorana fermion operators introduced in the following chapter. The Bogoliubov fermions are the usual complex fermions, with conventional commutation relations; their only novelty is that they are superpositions of electron and hole operators and hence may not carry definite charge, but they carry definite momentum and spin.[a]

Now let us incorporate disorder in order to understand what kind of fermionic excitations can exist and how the particle-hole symmetry P emerges. For simplicity we will continue to assume at some points unbroken SU(2) spin symmetry and spin-singlet pairing; an example of spin-triplet pairing of spin-polarized fermions is the Kitaev chain model discussed in Chapter 9. We use H_0 to denote the Hamiltonian including chemical potential for ordinary electrons without pairing, which we assume to be a free fermion model on discrete sites, with variation possible in both on-site and hopping terms. With i, j site labels and σ a spin index,

$$H_0 = \sum_{ij,\sigma} H_{ij}^0 c_{i\sigma}^{\dagger} c_{j\sigma}. \tag{8.24}$$

The physics of H_0 alone, incorporating hopping, disorder, and chemical potential, would generically be in the Anderson localization universality class with which we started the chapter, with doubly degenerate levels from spin.

Pairing is generated by an additional term

$$H_\Delta = -\sum_{ij} (\Delta_{ij} c_{i\uparrow}^{\dagger} c_{j\downarrow}^{\dagger} + \Delta_{ij}^* c_{j\downarrow} c_{i\uparrow}). \tag{8.25}$$

Again we allow for the pairing to be inhomogeneous, as is natural for a disordered system; this form would also allow us to include the order parameter vortices described in Section 8.5. Note that this form is consistent with Δ arising from expectation values of two c operators, as in the above translation-invariant case. We assume for now spin-singlet pairing with the orbitally symmetric pairing term $\Delta_{ij} = \Delta_{ji}$.

The full Hamiltonian whose eigenstates determine the quasiparticle excitations of this superconducting system is

$$H_{\text{BdG}} = H_0 + H_\Delta. \tag{8.26}$$

We would normally represent H_0 as two block-diagonal $N \times N$ matrices for N sites, one for spin-up and one for spin-down. H_Δ couples spin up and spin down, and a convenient way to write the Hamiltonian in $2N \times 2N$ matrix form to capture both pair-creation and pair-annihilation terms in H_{BdG} is

$$H_{\text{BdG}} = (c_{i\uparrow}^{\dagger} \quad c_{i\downarrow}) \begin{pmatrix} H_{ij}^0 & -\Delta_{ij} \\ -\Delta_{ij}^{\dagger} & -(H_{ij}^0)^* \end{pmatrix} \begin{pmatrix} c_{j\uparrow} \\ c_{j\downarrow}^{\dagger} \end{pmatrix}. \qquad (8.27)$$

The minus sign on H_{ij}^0 in the bottom right block compensates for the fermion operators being out of order, and $\Delta_{ij}^{\dagger}$ means the Hermitian conjugate with matrix elements $(\Delta_{ji})^*$. (For the singlet pairing we are currently considering, $\Delta^{\dagger} = \Delta^*$, while for triplet pairing $\Delta^{\dagger} = -\Delta^*$, and this is the origin of the difference in particle-hole symmetry operations mentioned below.) The notation means that the vector on the right of the matrix has components with j running from 1 to N twice, first for annihilation operators with spin up and then for creation operators with spin down, and similarly for the vector on the left. So just as in other quadratic fermion problems, solution of the eigenstates of the system reduces to finding eigenvectors of a certain matrix. The Bogoliubov–de Gennes form (8.27) is solved, often numerically, for a wide variety of problems in inhomogeneous superconductivity.

---

[a]   The Majorana fermion will be a more unusual object, a "real" fermion that is equal to its own Hermitian conjugate and hence carries no additive quantum numbers. It does have a definite fermion parity as it changes fermion number by an odd number, either $+1$ or -1.

The symmetry properties of the Bogoliubov–de Gennes form (8.27) turn out to be independent of the details of the inhomogeneities which may be present. Since it is these we are interested in for the classification, we turn to their analysis next. Suppose that we found an eigenstate, that is, a $2N$-component vector

$$(u_1, \ldots, u_N, v_1, \ldots, v_N) = (u, v) \qquad (8.28)$$

that when acted upon by the matrix $\mathbf{M}_{\text{BdG}}$ appearing in (8.27) just gets multiplied by an energy E:

$$\mathbf{M}_{\text{BdG}} \begin{pmatrix} u \\ v \end{pmatrix} = \begin{pmatrix} H^0 u - \Delta v \\ -\Delta^{\dagger} u - H^{*0} v \end{pmatrix} = E \begin{pmatrix} u \\ v \end{pmatrix}. \qquad (8.29)$$

Now consider the action of the same matrix on a different vector, and replace $\Delta^{\dagger}$ by Δ^* as we assume spin-singlet pairing:

$$\mathbf{M}_{\text{BdG}} \begin{pmatrix} -v^* \\ u^* \end{pmatrix} = \begin{pmatrix} H^0(-v^*) - \Delta u^* \\ -\Delta^*(-v^*) - H^{*0} u^* \end{pmatrix} = \begin{pmatrix} -(H^{*0}v + \Delta^* u)^* \\ -(H^0 u - \Delta v)^* \end{pmatrix} = -E \begin{pmatrix} -v^* \\ u^* \end{pmatrix}. \qquad (8.30)$$

The operation P taking (u, v) to $(-v^*, u^*)$, and $E \to -E$, is antiunitary because it involves complex conjugation, and applying it twice gives just a minus sign, that is,

$P^2 = -1$. In these respects it is exactly like the time-reversal operator on spin-half fermions from Section 3.3.

A useful exercise is to go back and check that, for a triplet superconductor, there is still an operator like P that takes E to $-E$, but now $P^2 = 1$. Note that a spin-triplet superconductor must have an orbitally antisymmetric wavefunction (e.g., the p-wave superconductor in Section 9.4, in order that the overall electronic wavefunction be antisymmetric under exchanging two electrons, which means $\Delta_{ij} = -\Delta_{ji}$ unlike above).

It may seem that the operator P is not exactly a symmetry since the energies have flipped. But note that emptying an occupied single-fermion level of energy $-E < 0$ can be viewed as a positive-energy excitation (a hole) of the system, of energy $E > 0$. This is why P is a symmetry, and called a particle-hole symmetry: the spectrum of the many-body system, including excitations above the ground state, is preserved. Note also that energy $E = 0$ is rather special: at any nonzero energy E, the action of P generates another level at energy $-E$, but at $E = 0$ there could in principle be a single unpaired level. Indeed the Majorana zero modes that are a main focus of Chapter 9 often appear as such zero-energy isolated solutions of the Bogoliubov–de Gennes equation.

We say that a metal or insulator is proximitized if it gains pairing terms such as those in Eq. (8.25) from proximity to a superconductor. This proximity effect greatly expands the family of accessible superconducting states by allowing the use of conventional s-wave superconductors to induce superconductivity in a wide variety of interesting materials that do not superconduct intrinsically. The Josephson effect in Box 8.3 is a simple example at the level of the order parameter of how superconductivity can penetrate through nonsuperconducting regions; indeed, a large part of the original controversy about whether the Josephson effect would be observable involved whether the coupling across the insulating barrier via electron tunneling would be appreciable or astronomically weak.

It is also possible for a particle-hole symmetry like P to arise in nonsuperconducting contexts, albeit usually with fine-tuning; a simple example is a one-dimensional tight-binding chain with nearest-neighbor hopping as discussed earlier in Box 4.1; even if the hoppings are random, the bipartite structure leads to a (now unitary) symmetry taking an eigenstate at energy $-E$ to one at energy $+E$. The random bipartite hopping problem will reappear in the context of disordered spin liquids below.

Now we are in a position to state how the symmetries T, P, and C control topological properties in independent-fermion systems. For example, in systems with none of these symmetries, we know that there is an integer quantum Hall effect in $d = 2$ with phases labeled by an integer. In the labeling of Lie algebras developed by Cartan, this is the statement that Class A in two dimensions has a $\mathbb{Z}$

Table 8.1 *Ten symmetry classes of free-fermion Hamiltonians in dimensions 1–4 and their topological possibilities.*

Cartan label	T	P	C	$d = 1$	2	3	4
A (unitary)	0	0	0	0	$\mathbb{Z}$	0	$\mathbb{Z}$
AIII	0	0	1	$\mathbb{Z}$	0	$\mathbb{Z}$	0
AI (orthogonal)	+1	0	0	0	0	0	$\mathbb{Z}$
BDI	+1	+1	0	$\mathbb{Z}$	0	0	0
D	0	+1	0	$\mathbb{Z}_2$	$\mathbb{Z}$	0	0
DIII	−1	+1	1	$\mathbb{Z}_2$	$\mathbb{Z}_2$	$\mathbb{Z}$	0
AII (symplectic)	−1	0	0	0	$\mathbb{Z}_2$	$\mathbb{Z}_2$	$\mathbb{Z}$
CII	−1	−1	1	$\mathbb{Z}$	0	$\mathbb{Z}_2$	$\mathbb{Z}_2$
C	0	−1	0	0	$\mathbb{Z}$	0	$\mathbb{Z}_2$
CI	+1	−1	1	0	0	$\mathbb{Z}$	0

Note: The possibility $+1$, -1, or 0 under T refers in the first two cases to having a T symmetry with eigenvalue $T^2 = +1$ or $T^2 = -1$, respectively, and to no symmetry in the third case, and likewise for P. Note that C is determined by the first two numbers, with the exception of distinguishing between classes A and AIII. The ordering of classes is chosen to show the regular precession of topological invariants with dimensionality, with period 2 for the first two complex classes and period 8 for the other real classes. The three Wigner–Dyson classes are A, AI, AII and are shown with their common names. Adapted from Ryu et al. (2010).

invariant. Similarly we know that with time-reversal symmetry $T^2 = -1$ in $d = 2$ or $d = 3$, which is Cartan's Class AII, there are $\mathbb{Z}_2$ invariants.

The remarkable fact is that in every dimension, three of the ten symmetry classes classes have $\mathbb{Z}$-valued topological invariants and two have $\mathbb{Z}_2$-valued topological invariants (Table 8.1). Furthermore, which class has which kind of invariant precesses regularly with increasing dimension. One set of eight real classes have a pattern that is periodic in dimension with dimension 8, and the two remaining complex classes alternate in dimension. The different cases in this decomposition $10 = 8 + 2$ depend on whether or not there is at least one antiunitary symmetry (P, T, or both), as antiunitary symmetries involve the complex conjugation operator, and the periodic behavior in dimensionality is known in topology as Bott periodicity.

The reader may wonder what general methods can be used to obtain the information in Table 8.1 without having to develop case-by-case explicit invariants as we have done for the IQHE (class A) and topological insulators (class AII). Several methods have been applied and agree on the answers: two examples close to the spirit of this book are to consider homotopy groups of the field theories used to describe disordered systems, or to think about possible stable gapless surfaces in one dimension lower (Ryu et al., 2010). A powerful and abstract approach that clarified the periodicity in dimensionality is to use methods from K-theory, which in heuristic terms generalizes the homotopy approach to situations where bands can be added or removed (Kitaev, 2009).

A question on which reasonable people still disagree is whether we have absolutely convincing evidence for the validity of the Bogoliubov–de Gennes description of quasiparticles beyond mean-field theory, in either numerics or experiment. The clearest evidence that something like the order parameter actually has experimental meaning is provided by the Josephson effect, discussed in Box 8.3. We have followed standard practice in this chapter in assuming that the quasiparticles found in the mean-field approach remain physically meaningful in the presence of corrections to the mean-field picture. One way to support this assumption is to relate some of the nontrivial topological states in the tenfold way (Table 8.1) to the presence of field theory anomalies (Ryu et al., 2012; Witten, 2016). Conversely, it is now understood via work on symmetry-protected topological order in interacting systems (Section 11.1) that, in some symmetry classes and dimensions, the invariants in Table 8.1 are not all stable; instead of an infinite, integer-valued set of possibilities there may be only a finite number (e.g., $\mathbb{Z}$ is replaced by $\mathbb{Z}$ mod 8). When crystal symmetries are enforced, there are many more than ten possible symmetry classes, and we will say a few words about this rapidly developing area in Section 11.3.

8.5 Vortices in Conventional Superconductors

We use the term defect to mean a discrete large perturbation to the system, possibly induced by an external impetus such as a magnetic flux. As our first step toward a theory of defects in topological phases, we first discuss how vortices work in a conventional superconductor, where they are topological defects of the order parameter introduced in the previous section.

Let us go back to the idea of Landau free energies from Chapter 2 and apply it to the superconducting state. The magnitude of the order parameter describes the fraction of electrons at a point that are superconducting, in a simple two fluid model where some electrons are superconducting and some are normal:

$$|\psi(\mathbf{r})|^2 = \frac{n_s(\mathbf{r})}{n}. \tag{8.31}$$

Here n is the total electron density, assumed constant in space. (In a more sophisticated theory, the interpretation of ψ is via the expectation value of a local two-fermion pairing operator like that described earlier in this chapter.)

First suppose that the wavefunction is also constant in space, and let $f(\psi, T)$ be the difference in free energy density between the superconducting and normal states if ψ is uniform. That is, for a system of volume V, the free energy difference is

$$\Delta F = F_N - F_S = V f(\psi, T). \tag{8.32}$$

Now, if the system is just below the superconducting transition T_c, then only a few electrons are superconducting, which means that we can expand f in a power series in ψ, since $|\psi|^2 \ll 1$:

$$f(\psi, T) \approx a(T)|\psi|^2 + \frac{1}{2}b(T)|\psi|^4. \tag{8.33}$$

We can make one more simplification. First, note that the free energy is minimized when

$$\frac{\partial f}{\partial |\psi|^2} = 0 \Rightarrow b(T)|\psi|^2 + a(T) = 0 \Rightarrow |\psi|^2 = -\frac{a(T)}{b(T)}. \tag{8.34}$$

At this magnitude, the free energy difference is

$$f(\psi, T) = -\frac{1}{2}\frac{a^2(T)}{b(T)}. \tag{8.35}$$

Recall that one of the defining features of a superconductor is the expulsion of magnetic field (the Meissner effect). The magnetic field reenters and drives the system normal once it is energetically favorable to do so; here we are assuming for simplicity that the superconductor is type I, so the magnetic field penetrates uniformly above the critical magnetic field H_c. The free energy difference per volume is therefore related to H_c:

$$f(\psi, T) = -\frac{1}{2}\frac{a^2(T)}{b(T)} = -\frac{H_c^2}{8\pi}. \tag{8.36}$$

We can obtain another relation involving a and b if we use the fact that the superfluid density goes as the inverse square of the penetration depth:

$$\frac{\lambda^2(0)}{\lambda^2(T)} = \frac{|\psi(T)|^2}{|\psi(0)|^2} = |\psi(T)|^2 = -\frac{a(T)}{b(T)}. \tag{8.37}$$

Here we have assumed that at zero temperature, all the electrons participate in the superconductivity. These two equations involving a and b can be used to reexpress everything in the Ginzburg–Landau equation in terms of the experimental quantities H_c and λ.

Now let us consider spatial variations in ψ. For slow variations, we can keep just the gradient-squared term, which leads to

$$\int \frac{n^*}{2m^*} \left| \frac{\hbar}{i} \nabla \psi(\mathbf{r}) + e^* \mathbf{A}(\mathbf{r}) \psi(\mathbf{r}) \right|^2 d\mathbf{r}. \tag{8.38}$$

Note that we have defined quantities n^*, m^*, e^* with the units of number density, mass, and charge. The BCS theory sketched above predicts charge $q = -e^* = -2e$ and $m^* = 2m$. We have also assumed that an external vector potential $\mathbf{A}$ enters in the same way as for a single particle.

Combining the constant and gradient terms, and the magnetic field energy, gives finally the free energy

$$F(\psi, T) = \int \frac{n^*}{2m^*} \left| \frac{\hbar}{i} \nabla \psi(\mathbf{r}) + e^* \mathbf{A}(\mathbf{r}) \psi(\mathbf{r}) \right|^2 d\mathbf{r}$$
$$+ \int \left[a(T)|\psi(\mathbf{r})|^2 + \frac{1}{2} b(T)|\psi(\mathbf{r})|^4 \right] d\mathbf{r} + \int \frac{H(\mathbf{r})^2}{8\pi} d\mathbf{r}. \tag{8.39}$$

Then the minimization of this functional over ψ gives the Ginzburg–Landau equation,[7]

$$\frac{\delta F}{\delta \psi(r)} = 0 \Rightarrow \frac{\hbar^2 n^*}{2m^*} \left[\nabla + \frac{ie^*}{\hbar} \mathbf{A}(\mathbf{r}) \right]^2 \psi(\mathbf{r}) + a(T)\psi(\mathbf{r}) + b(T)|\psi(\mathbf{r})|^2 \psi(\mathbf{r}) = 0.$$
$$\tag{8.40}$$

A large fraction of the long-distance, low-energy phenomenology of superconductors can be deduced from this equation.

The key difference between the super*conducting* and the super*fluid* vortices discussed in Chapter 2 comes from the nature of the current. Around a superfluid vortex, the phase winds and there is a current with a quantized vorticity, so the magnitude of vorticity falls off only inversely with radius. For the superconductor, the physical current must be gauge invariant and also involves the vector potential:

$$\mathbf{j} \propto \frac{\hbar e^*}{2im^*} \left[\psi^* \left(\nabla - \frac{ie^* \mathbf{A}}{\hbar} \right) \psi - \text{h.c.} \right]. \tag{8.41}$$

Vortices are created by applying a magnetic field to the superconductor. Far away from the center of a vortex, the screening currents described by (8.41) cancel the vector potential describing this magnetic field. Suppose the phase of ψ winds by 2π

---

[7] Ginzburg–Landau or Landau–Ginzburg? Both orders are found in the literature. Perhaps the majority view is to take Landau or Landau–Ginzburg to refer to the general concept of a free energy expansion in terms of the order parameter, and Ginzburg–Landau for its use in the specific case of a superconductor, for which Ginzburg shared the 2003 Nobel Prize.

over the circumference $2\pi r$ of the circle at radius r. The vector potential satisfies $|A|2\pi r = \Phi$, the magnetic flux enclosed. Then the total current can be zero if

$$|\nabla\phi| = \frac{1}{r} = \frac{|A|e^*}{\hbar} = \frac{\Phi e^*}{\hbar r}. \tag{8.42}$$

This is satisfied if $\Phi = \frac{h}{e^*} = h/2e$. More generally, the total flux, including screening from the supercurrent, through the total area of a superconducting vortex should be an integer multiple of the superconducting flux quantum $h/2e$. These currents fall away rapidly beyond the magnetic penetration depth. An easier and highly precise measurement of the same flux quantum is via the Josephson effect, which is another application of the gauge-invariant expression for the supercurrent (8.41).

Given the order parameter texture of a vortex, one can consider the quasiparticle states using the Bogoliubov–de Gennes approach from the preceding section. The result is that vortices can support trapped quasiparticle states, known as Caroli–Matricon–de Gennes (CMG) bound states, even for a simple s-wave superconductor. However, these states are ordinary Bogoliubov fermions, not the topological Majorana zero modes that will be a focus of Chapter 9, and finding experimental means to distinguish between ordinary and topological bound quasiparticle states is an important practical challenge.

Box 8.3 The Josephson Effect and Gauge Invariance

We have already discussed two examples of how topology leads to remarkably precise quantization in Chapters 3 and 5: the integer and fractional quantum Hall effects were discovered via experimentally observed Hall conductances quantized to a rational multiple of e^2/h. The quantum Hall effect was not actually the first example of such dramatic quantization in solids. We would like to explain how the simple Ginzburg–Landau theory of superconductivity in Eq. 8.39, with a U(1) order parameter coupled to an electromagnetic field, leads to the Josephson effects.

The main goal is to explain how the precise quantization of the AC Josephson effect is connected to the compactness of the U(1) gauge group of electromagnetism, via the way gauge invariance shows up in the Ginzburg–Landau formalism.[a] The relationship is very similar to that in the Laughlin flux insertion argument from Chapter 3: there, the key aspect of gauge invariance was that insertion of a flux quantum through a solenoid through a ring of electrons could be gauged away, in the sense that it did not modify the spectrum. Probably no one would believe the counterintuitive result that we shall find without a justification in terms of a microscopic theory like BCS. Even with a detailed argument in the BCS picture, the theoretical predictions of Josephson (1962) were quite controversial until their spectacular confirmation by experiments of Rowell (Anderson and Rowell, 1963), Shapiro (1963), and others.

Consider two identical superconducting regions separated by a potential barrier that allows Cooper pairs to pass through with some small probability. Suppose that one region has phase Φ_1 and the other has Φ_2. Just as the current density is related to the derivative of the phase in the Ginzburg–Landau equation, we expect the current through the barrier to be a function of the phase difference: $j(\Phi_1 - \Phi_2)$. The function j should be periodic with period 2π and also odd, and with a bit more insight we can find its exact form in this weak-coupling limit.

The key to understanding the Josephson effect is the requirement of gauge invariance under the nonrelativistic gauge transformation

$$V \to V - \frac{\partial \chi}{\partial t}, \quad A \to A + \nabla\chi, \quad \Phi \to \Phi + \frac{2e\chi}{\hbar}. \tag{8.43}$$

Here we have assumed that the order parameter in the Ginzburg–Landau equation describes Cooper pairs of charge $2e$. Keep in mind that the term flux quantum in discussions of superconductors is $h/2e$ rather than the single-electron flux quantum h/e used in the quantum Hall effect.

Let us start with the limit of an infinitely high tunnel barrier that separates the two superconductors completely. Then the condition on the Ginzburg–Landau fields ψ_1, ψ_2 at the two sides of the barrier, assuming the barrier is perpendicular to $\hat{\mathbf{x}}$, is

$$(\partial_x - \frac{ie^* A_x}{\hbar})\psi_1 = (\partial_x - \frac{ie^* A_x}{\hbar})\psi_2 = 0. \tag{8.44}$$

This is gauge-invariant and forces the supercurrent given below to equal zero. If the barrier is not infinitely high, then we expect that instead of 0, there will be some (to be determined) function of ψ. Expanding it in a power-series under the usual assumption in the Ginzburg–Landau equation that ψ is small, and replacing $e^* = 2e$, yields

$$(\partial_x - \frac{i2eA_x}{\hbar})\psi_1 = -\psi_2\lambda^{-1}, \quad (\partial_x - \frac{i2eA_x}{\hbar})\psi_2 = \psi_1\lambda^{-1}. \tag{8.45}$$

Here λ is some unknown length scale related to the barrier thickness and height. Now we can plug one of these equations into the expression for the supercurrent. At the left boundary, for ψ_1, one obtains

$$j = -\frac{i(2e)\hbar}{2m}\left(\psi_1^* \frac{\partial \psi_1}{\partial x} - \psi_1 \frac{\partial \psi_1^*}{\partial x}\right) - \frac{(2e)^2}{m} A_x \psi_1^* \psi_1. \tag{8.46}$$

If this form looks unfamiliar, note that the first part is just the ordinary particle current in quantum mechanics, and the second part is forced by gauge invariance. We are left with the simple form

$$j = \frac{i(2e)\hbar}{2m\lambda}(\psi_1^*\psi_2 - \psi_1\psi_2^*). \tag{8.47}$$

Now, for identical superconductors on both sides, the magnitude of ψ is the same, so that this predicts that $j \propto \sin(\Phi_1 - \Phi_2)$. So, even at no applied voltage, a supercurrent flows if there is a phase difference between the two superconductors.

What happens when a voltage is applied is quite amazing. The phase $\Phi = \Phi_1 - \Phi_2$ does not evolve at zero bias voltage: $\frac{\partial \Phi}{\partial t} = 0$. When a voltage is applied, we would like to find a gauge transformation that mostly eliminates the voltage but creates a phase difference, since we know how to handle phase differences. The system should be gauge invariant under transformations of the form

$$V \rightarrow V - \frac{\partial \chi}{\partial t}, \quad A \rightarrow A + \nabla \chi, \quad \Phi \rightarrow \Phi + \frac{2e\chi}{\hbar}. \tag{8.48}$$

The gauge transformation we need is a special case:

$$V \rightarrow V - \frac{\partial \chi(t)}{\partial t}, \quad \Phi \rightarrow \Phi + \frac{2e}{\hbar}\chi(t). \tag{8.49}$$

This leaves the vector potential constant since χ has no spatial dependence. Then a gauge-invariant relation between Φ and V is

$$\frac{\partial \Phi}{\partial t} + \frac{2e}{\hbar}V = 0. \tag{8.50}$$

At $V = 0$, this predicts constant phase difference, as expected. For nonzero V, we have

$$\Phi = \Phi(0) - \frac{2e}{\hbar}Vt \Rightarrow j = j_m \sin\left(\Phi(0) - \frac{2e}{\hbar}Vt\right). \tag{8.51}$$

So this simple theory predicts that, when a voltage is applied, there should be an oscillating current at frequency

$$\omega_J = 2|eV|/\hbar. \tag{8.52}$$

The Josephson frequency constant, which is just the inverse of the superconducting flux quantum, in useful units is $2e/\hbar = 483.5978484$ MHz/μV. Actually it has been measured to even more decimal places, comparable to the precision of the integer quantum Hall effect. The preciseness of its quantization (one part in 10^8 or more) is a consequence of its simplicity: it depends only on the assumptions of gauge invariance and quantum coherence. The sinusoidal variation of the current under a DC voltage can be taken as a direct measurement of the compactness of the U(1) gauge group of electromagnetism.

The Josephson effect may be one of the most useful discoveries in the physics of solids. The conventional Josephson effect described here is used to create superconducting quantum interference devices (SQUIDs), which are now widely used for precise magnetometry. Studies of SQUIDs have led directly to the development of several subtypes of conventional superconducting qubits, such as the flux and transmon qubits now in wide use. Another approach to quantum computation involves topological superconductors, which we discuss in more detail in the next chapter. These support exotic Josephson effects with 4π periodicity, that is, they only return to the original state after insertion of two superconducting flux quanta rather than one. The difference between states at zero and 2π flux is related to the existence of Majorana excitations.

---

[a] AC originally meant alternating current and has come to be used in physics contexts to mean finite-frequency. Similarly DC means direct current or zero frequency.

8.6 Flux and Crystalline Defects in Integer Topological Phases

Following standard practice, we use the term disorder to refer to possibly weak random fields distributed through an extended region, and defect to indicate an individual perturbation. Note that a single defect may nevertheless have dramatic effects over a large region: an edge dislocation in a three-dimensional crystal, for example, where a crystalline plane ends, is an extended object of the same type as the topological defects studied in Chapter 2.

We discuss some examples in this section of how even integer topological phases can show subtle behavior in response to various kinds of defects. That will serve as preparation for discussing defects in fractional phases and then, in the next chapter, in topological superconductors, which have both an order parameter and topological properties. Much of the magic of the superconducting vortex, giving rise to important differences between it and the relatively simple vortex of a neutral superfluid, originates in the coupling of the charged condensate to a magnetic field. We therefore start by thinking about how extra magnetic fields can modify the integer quantum Hall effect beyond the weak-field limit.

Consider an integer quantum Hall state, $\sigma_{xy} = \frac{ne^2}{h}$, with a large hole in it, and imagine that a solenoid of nonzero flux penetrates this hole. We expect that if the hole is large enough, we can start from the gapless, linearly dispersing electron state on an extended edge and quantize it to take account of the finite circumference of the hole. The flux in the solenoid modifies this quantization by the Aharonov–Bohm effect. As the flux in the hole is adiabatically changed, the spectrum of the edge is smoothly modified.

By the time the flux reaches one flux quantum Φ_0, the spectrum will be the same as at zero flux, but an integer number of electrons n will have been pumped to the inner edge via the continuous evolution of the spectrum with flux. However, we could guess from the solution of a particle on a ring (Box 3.1) that another value of flux might also be important if time-reversal symmetry is important. When a flux $\Phi_0/2$ (i.e., one half of a single-electron flux quantum; this is the flux through a vortex in a superconductor) pierces the hole, it does not break time-reversal symmetry; the phase change required in wavefunctions circling the flux is $e^{i\pi} = -1$, which is real.

So we might expect this π flux to have special properties in the spin-orbit-induced $\mathbb{Z}_2$ topological insulators, which depend on time-reversal symmetry. Indeed it does, and this gives a pumping interpretation of what the $\mathbb{Z}_2$ invariant discussed in Chapter 4 measures in two-dimensional topological insulators (Fu and Kane, 2006; Essin and Moore, 2007).

To start, think about two copies of the quantum Hall effect in opposite directions, say one for spin up and one for spin down. Suppose that at zero flux the quantization of edge states around a small hole leads to a gap above the ground state. One could view this as the highest occupied edge states around the hole consisting of a time-reversed pair states, one with spin-up going clockwise and one with spin-down going counterclockwise, say.

If a full flux quantum were inserted through the hole, an additional electron of one spin, say up, would be brought to the edge, and an electron of the other spin would be moved away from the edge. What happens as one half of a flux quantum is inserted through the hole? There will again be a degeneracy between two edge states because of time reversal (recall the discussion around Figure 4.3), but now one is occupied and one is not. Therefore the many-body excitation spectrum now has a zero-energy excitation, as one could move the electron from the occupied state to the unoccupied one.

Kramers's theorem, guaranteeing the existence of degenerate pairs of single-fermion states under time-reversal, shows that the above zero-energy excitation (i.e., a degeneracy of two ground states) remains present even when spin is not conserved. Hence the effect of adding the π flux is not to pump a definite charge, but rather to pump a twofold degeneracy, which we can think of as a spin-half degree of freedom with zero charge.

Similar excitations within the topological gap are created in three-dimensional $\mathbb{Z}_2$ phases. An important experiment to demonstrate the uniqueness of the surface metal of 3D strong topological insulators, which has been performed by several groups on multiple materials, is as follows. Consider a solid cylindrical wire that is large enough in cross section that a magnetic field *along the wire's axis* can be strong enough to have a few flux quanta enclosed by the wire's surface without being so large as to greatly modify the wire's bulk gap. The idea is that even though the field does break time-reversal symmetry and in principle removes the protection of the surface states, for a large enough wire, we could have the surface state respond primarily to the *flux* through the bulk, via the Aharonov–Bohm effect from Box 3.1, rather than the *field* at the surface.

We can then capture the wire's low-energy degrees of freedom by considering how the surface states, which would be a 2D metal in the thermodynamic limit, are modified by the flux. It turns out that the behavior of transport in metallic cylinders pierced by flux has a long history in mesoscopic physics. The novel feature when the metal cylinder is made using the surface of the topological insulator is similar in spirit to the creation of a zero-energy bound state in the two-dimensional case just treated.

Normally, this wire, if made long enough, would act as a one-dimensional system, and via either Ohm's law or localization physics we would expect its conductance to go to zero as its length increases. However, if there is one-half of a flux quantum piercing the wire, one finds that one helical mode remains conducting, that is, one left-moving and one right-moving mode survive, similar to the edge of the *two-dimensional* topological insulator (Bardarson and Moore, 2013). This unexpected prediction, that a wire with zero flux piercing it has significantly lower conductance than the e^2/h found in the wire pierced by an odd number of flux quanta, is strong confirmation via transport that a given material indeed hosts the unusual surface metal of a three-dimensional topological insulator.

This helical mode can be created by a crystalline defect, rather than an applied flux, in the weak topological insulator (TI) phase. Recall that the weak TI can be viewed in the simplest case as consisting of layers of the two-dimensional TI. Suppose that, because the crystal hosting the weak TI is imperfect, one of the crystal layers ends abruptly. This edge dislocation will carry a single helical edge. Even other kinds of dislocations that are harder to picture, such as a screw dislocation, can carry the same helical edge for topological reasons (Ran et al., 2009), and current carried by helical modes along dislocations may be an important contribution for bulk conduction in materials with nontrivial weak indices.

We now turn to interacting systems. Two general lessons that will recur are that it can be useful to think of the response of a system to a localized magnetic flux and that there is a direct relationship between extended edges, as at the boundary of a system, and the smaller edge of a vortex or other defect, where quantization of the spectrum appears.

8.7 Vortices in Quantum Hall States and Composite Fermions

Considering the response of a fractional quantum Hall state to an applied local flux turns out to lead to a useful construction allowing the Laughlin state at filling factor $\nu = 1/m$ to be generalized to many other experimentally observed fractions. We start from the trial wavefunction for a quasihole (5.21), which was obtained by multiplying the Laughlin wavefunction by a set of additional polynomial factors. For a quasihole at complex location $w = w_x + i w_y$ in the plane, the trial wavefunction is

$$\Psi_{\text{quasihole}}(z_1, \ldots, z_N) = \left[\prod_{i=1}^{N} (z_i - w) \right] \Psi_{1/m}(z_1, \ldots, z_N). \tag{8.53}$$

This added polynomial factor has a unit vorticity in the sense that the phase of the wavefunction moves through 2π as an electron orbits w (see also Section 5.1.4).

The quasihole wavefunction also describes a vortex in a physical sense similar to the superconducting case above. Suppose that we insert a magnetic flux quantum Φ_0 via a tiny solenoid passing through a point of the plane, as if we had a smaller inner circle in Figure 3.1b. To create a quasihole, we use a flux directed opposite to the uniform field used to create the state. Because of the Laughlin pumping relationship between flux insertion and electric fields from Chapter 3, and the fact that the Hall conductivity of the state is $\sigma_{xy} = \nu e^2/h$, we find that a fractional charge νe must have accumulated near the solenoid. So a vortex in a fractional quantum Hall state is not an entirely new object, as we implicitly discussed vortices when we discussed the quasihole in Chapter 5.

But now we return to the Laughlin wavefunction itself and let $m = 2p+1$ be odd, as is required if the wavefunction is for fermions. We can rewrite this as a product of a very large number of quasihole factors multiplying the *integer* quantum Hall state: the polynomial part of the wavefunction is

$$\Psi_{1/m} = \left[\prod_{i'<j'}^{N} (z_{i'} - z_{j'})^{2p} \right] \left[\prod_{i<j}^{N} (z_i - z_j) \right]. \tag{8.54}$$

Consider the form of this prefactor from the perspective of one electron, and hold that electron's position fixed. Now every other electron picks up $2p$ extra phase windings on circling that electron, which we view as $2p$ vortices attached to the electron.

This leads to a picture developed by Jain (2007) for how to view the Laughlin state and how to generalize it to the most clearly observed of the hierarchy states[8], which are now known as either the main sequence or Jain sequence. The idea is to view the Laughlin state $1/m$ as an integer quantum hall state of composite fermions, which are objects that combine an electron with the two flux quanta that serve to attach the vortices. Composite fermions are still fermions, because the Aharonov-Bohm phase of an electron on moving halfway around an even number of flux quanta is 2π. (Moving only halfway around the fluxes is appropriate to describe an exchange.) However, the composite fermions see effectively a reduced flux compared to the original electrons. The idea is that originally there were m flux quanta per electron, since the filling factor was $\nu = 1/m$. However, $2p$ of

---

[8] The hierarchy picture (Haldane, 1983a; Halperin, 1984) is that Abelian quantum Hall states can be obtained by forming Laughlin-like states out of the quasiparticles or quasiholes of a Laughlin state, and so on. While this leads to a variety of successful model wavefunctions, in principle any odd-denominator fraction can be obtained somewhere in the hierarchy, and the composite fermion picture presented here gives a more straightforward picture of why the main sequence states are particularly strong in experiment.

those flux were canceled by the fluxes attached to form composite fermions. The residual flux for a system of N electrons is

$$\Phi' = \frac{N\Phi_0}{\nu} - 2pN\Phi_0 = (2p + 1 - 2p)N\Phi_0 = N\Phi_0, \qquad (8.55)$$

which is exactly the right number of flux for the composite fermions to be in the $\nu = 1$ integer quantum Hall state.

We know how to construct other integer Hall states by filling higher Landau levels. What does the integer state $\nu = q$ become if we make it out of composite fermions rather than electrons? The residual flux should now be $N\Phi_0/q$ for N electrons. Then inverting (8.55) gives

$$\Phi' = \frac{N\Phi_0}{q} = \frac{N\Phi_0}{\nu} - 2pN\Phi_0 \Rightarrow \nu = \frac{q}{2pq + 1}. \qquad (8.56)$$

Hence, taking $p = 2$, the composite fermion theory predicts that there should be gapped fractional quantum Hall states at fillings $\nu = 1/3, 2/5, 3/7, \ldots$. The field theory description of these states is given by multicomponent generalizations of the Abelian Chern–Simons theories in Chapter 6. Consistent with that theory, the edges of these states are found to have q co-propagating chiral edge modes, like the $\nu = q$ integer Hall effect, but now with fractionalized excitations.

We could repeat the same process of forming composite fermions by flux attachment but instead attach flux parallel to the external field. This leads to fractions

$$\nu = \frac{q}{2pq - 1} \qquad (8.57)$$

and a series of states $\nu = 2/3, 3/5, 4/7, \ldots$. These are the particle-hole conjugates of the states above, in the sense that they are continuously connected to taking a filled Landau level and then adding positively charged holes into one of the states at $q/(2pq + 1)$. The wavefunctions constructed by the composite fermion procedure turn out to be rather accurate in comparison to small-system exact diagonalization, even though they have no adjustable parameter (Jain, 2007).

Note that this picture includes an oversimplification in one important respect. As we noted in the analysis of the fractional quantum Hall effect, its existence is owed to electron-electron interactions. In particular, in the case of the hardcore interaction going along with the Haldane pseudopotential $V_m = \delta_{m,1}$, there is only one energy scale available once the interaction has been projected to the lowest Landau level: the cyclotron energy has disappeared as an energy scale. In the integer quantum Hall effect, however, interactions are not needed, and it is the cyclotron energy which is the relevant energy scale. A theory which restores the role of interactions is available, but considerably more involved than the success of the Jain picture would lead one to hope (Halperin et al., 1993).

At any rate, these two lists of filling fractions are known as the main sequence. Both have an accumulation point at $\nu = 1/2$, corresponding to infinitely many filled Landau levels of composite fermions. But filling all Landau levels equally is as though the cyclotron energy has collapsed–the effective magnetic field remaining on the composite fermions is zero. The exotic composite fermion metal at $\nu = 1/2$ has been a topic of great interest for many years (Halperin et al., 1993), reinvigorated recently by a proposal to maintain explicit particle-hole symmetry at this point by Son (2015). The gapped Pfaffian state at $\nu = 5/2$ discussed in the next chapter can be viewed heuristically as an instability of the composite fermion metal toward a $p + ip$ superconductor of paired composite fermions.

8.8 Spin Liquids and Disorder

This section takes a quick detour into one of the messier aspects of materials physics, by discussing the effect of chemical and structural disorder on spin liquids. As advertised at the outset of this chapter, there is no shortage of different types of disorder. When synthesizing a chemical compound, one may end up with too few atoms of a given type, or too many; some may be in the wrong place, or the wrong ones. Misplaced or missing atoms may lead to structural distortions, which in turn can change nature (by altering local symmetries and crystal fields) and interactions (via exchange pathways) of nearby ions. Such problems are largely unavoidable – synthesis is not done at $T = 0$, and hence entropy will always favor a certain density of defects. Also, disorder can be hard to measure quantitatively, as microscopic information on, for example, exact correlations in the location of impurity atoms, is often not readily available. The need to engage with these items has led to much disparagement of the enterprise of condensed matter physics, famously encoded by the moniker "squalid state physics." Anyone who has pored over experimental data from an ill-characterized sample has experienced a certain level of this squalor, and we do not want to suppress this aspect of the field entirely in our book.

Condensed matter physics, however, has a habit of pulling itself, Münchhausen-like, out of the swamp by its own hair, and the next few examples illustrate the conversion of the nuisance of disorder into an asset, or even an interesting piece of new physics, at least in theory. To do so, however, requires engaging with some of the microscopic details of materials physics, which has a somewhat archaic flavor in the sense of involving lattice-scale, rather than long-wavelength, physics.

For cohesion of presentation, we select items for two families of models covered elsewhere in this book, namely, Coulomb spin liquids (as in spin ice on the pyrochlore lattice) and the honeycomb Kitaev spin liquid with its spectrum of gapless Majorana fermions coupled to an emergent $\mathbb{Z}_2$ gauge field.

The main theme we develop is how even simple disorder – we shall only consider dilution and distortions – can influence the system on many different levels. First, there are the immediate effects: dilution removes microscopic degrees of freedom; and distortions endow microscopic coupling constants with a disordered distribution. At the next level of complexity, disorder can also lower the (local) symmetry, and thence allow terms in the Hamiltonian which were symmetry-forbidden in the clean system. This happens in our first example (Section 8.8.1), where *static* distortions introduce quantum *dynamics* to spin ice.

Next, disorder can rearrange the low-energy spectrum, by nucleating excitations, that is, making them cheaper or moving them directly into the ground state. This is of particular interest in the case of fractionalized excitations, the existence of which can thus be probed in the limit of low temperatures already, even if their excitation energy in the clean system was high. These "excitations" can then exhibit collective behavior of their own. This we illustrate for the case of orphan spins in a classical Coulomb spin liquid (Section 8.8.2); there, their signatures are strong enough not only to be experimentally detected but one can even back out an estimate of their density, and hence the level of chemical disorder, from low-energy NMR experiments. This information would have been hard to obtain by other means.

Disorder in the Kitaev honeycomb model (Section 8.8.3) is similarly rich, and unusually tractable. As in the above case, vacancies can nucleate a gauge-charged defect by binding an emergent Ising flux. In the non-Abelian phase, each flux comes with a Majorana zero mode, which therefore generates a new low-energy set of degrees of freedom which can exhibit as yet poorly explored collective behavior.

Beyond this, their novel behavior comes from rearranging the spectral weight of the gapless emergent Majorana excitations with their Dirac spectrum. This turns out to be similar to what is observed in analogously disordered graphene. Here, the main attractions consist of the appearance of a random vector potential for slowly varying modulations of the bond strengths; for less gentle forms of disorder, this generates a more general disorder model known as random bipartite hopping. Both have rather remarkable low-energy spectra: the former exhibits a drifting exponent in the heat capacity, while the latter does its best to have a maximally divergent density of states at low energies. Both of these can be investigated with what is about the most basic probe available, namely, specific heat experiments.

8.8.1 Quantum Dynamics from Static Distortions in Spin Ice

Here we describe how distortions which are *static* can add quantum *dynamics* to an Ising model (Savary and Balents, 2017). The microscopic ingredient for this is an analysis of the crystal field scheme of the magnetic ion in the solid (Ashcroft and Mermin, 1976). In a nutshell, a rare earth ion like Ho^{3+} or Dy^{3+} in the spin ice crystal has electronic wavefunctions which differ from those in free space (but which, as the latter form a complete set, can be expressed as linear combinations of these). This is a result of the interactions of the electrons of an ion with the neighboring ions, which are themselves electrically charged. The determination of the crystal field scheme is a mature field in itself, and the output is a level scheme which describes at what energies and with what multiplicities which wavefunctions appear.

If the multiplicity of the lowest-lying state is twofold (i.e., if there are two states with the same energy), one labels one of the states as an Ising spin-up, and the other as an Ising spin-down. In this sense, the Ising degrees of freedom are typically actually *pseudo*-spins. Their coupling of which to, for example, external fields needs to be determined via the detour of the wavefunctions the Ising pseudo-spins encode, and the matrix elements within, and to excited states outside, the doublet.

For a lowest-lying state being a doublet, there are two fundamentally different options: either the ion in question has an even (like Ho^{3+}), or an odd (like Dy^{3+}), number of electrons. In the latter case, Kramers theorem guarantees that, in the presence of time-reversal symmetry, the degeneracy between the two states in the multiplet is exact (see Section 3.3). The alternative is not similarly protected, and this is the crux of the following idea.

In both cases, the crystal field levels are dominated by components with a large magnetic moments pointing along a particular high-symmetry axis, the [111]-direction (see Figure 5.13). The Ising pseudospin variable, S^z, encodes the direction the moment points in along the axis. The question then concerns the nature of the transverse degrees of freedom. Here, the (non-)Kramers corresponds to the transverse degrees of freedom being odd (even) under time-reversal symmetry, respectively. This can be seen most simply for an SU(2) algebra of the form $[S^x, S^y] = i S^z$. As the time-reversal symmetry includes complex conjugation, and given S^z is odd, the transverse components need to be either both odd, or both even.

The coupling to strain is fundamentally different between the two cases. Strain involves the displacement of electrically charged ions from their positions on the regular lattice. This changes electric fields, and hence is even under time-reversal symmetry. It is therefore impossible for strain to couple linearly to the transverse spin components in the Kramers case, as terms in the Hamiltonian have to be in the trivial representation for all symmetries, and thus cannot be odd. However, such a coupling exists in the non-Kramers case. It can be cast into the simple form of a transverse field,

$$H_{\text{spin}-\text{strain}} = -\sum_i \gamma_i S_i^x .$$

Here, both the strength, $\gamma_i \geq 0$, and the local direction of the x−component of spin depend on the strain configuration.

If the γ_i are smaller than the effective Ising exchange, Eq. 5.56, one can treat the transverse field perturbatively. The lowest-order off-diagonal term possible arises in sixth order in perturbation theory. This involves the γ_i on the bonds of a hexagon of the pyrochlore lattice, yielding the "ring-exchange" term around a hexagon, Eq. 6.18, with a coefficient

$$\Gamma = J \prod_\bigcirc (\gamma_i / J) \, . \tag{8.58}$$

This is the quantum dynamics in the manifold of spin ice states resulting from the static strain. The resulting many-body state depends on the probability distributions induced by the strain distribution. In the simplest case, where γ_i is distributed narrowly away from zero, this realizes a somewhat disordered but otherwise standard canonical quantum spin-ice Hamiltonian, equation 6.18. Static distortions can thus in principle turn a classical Coulomb phase into a quantum one.

8.8.2 Gauge-Charged Orphan Spins and Their Disorder Microscopy

The discovery of high-temperature superconductivity among many other things generated intense, and lasting, interest in the physics of unconventional magnets. A second experiment which stands out for the study of frustrated magnets specifically is an experiment on the magnetic insulator $SrCr_8Ga_4O_{19}$ (SCGO) (Obradors et al., 1988). This hosts Cr ions with a half-filled d-shell of electrons, resulting in nearly isotropic Heisenberg spins $S = 3/2$. The spins reside on well-decoupled two-dimensional slabs depicted in Figure 8.6. These are trilayers made up of a triangular layer sandwiched in between two kagome ones. The historical import of this experiment was twofold. First, it found that the characteristic energy scale of this magnet, encoded by the Curie-Weiss temperature Θ_{CW}, at which a mean-field theory would predict ordering, was more than two orders of magnitude higher than any discernible thermodynamic transition. Such a discrepancy is today taken as a defining feature of a frustrated magnetic material. Second, on account of the kagome layer, it set off a quest to understand the properties of kagome magnets, which for the case of the ground states of the $S = 1/2$ Heisenberg magnet continues to this day.

Here, we are interested in the effect of nonmagnetic vacancies in SCGO. It was found that it is not possible to synthesize samples with full coverage of the Cr-sites. Instead, nonmagnetic Ga ions replace these, with a density parameterized by x, although this quantity was not easy to establish experimentally for a given compound, all the more so since there are inequivalent sites for the Cr ion, each of which could come with distinct probabilities for dilution.

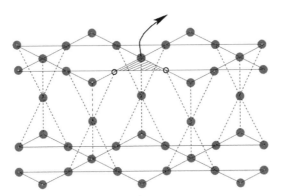

Fig. 8.6 The trilayer of $SrCr_8Ga_4O_{19}$ consists of a stack of kagome-triangular-kagome lattices. This can be cut out of the pyrochlore lattice in a [111] direction. Removing two spins in a triangle leaves behind a third, orphan, spin without interaction partners. This triangle, unlike all the other simplices (triangles/tetrahedra) of the lattice, cannot have vanishing total spin. Instead, at low temperatures, it acts as a paramagnetic moment of *half* the size of the microscopic spin moment.

In the intervening years, it has become clear that SCGO is "closer" to a pyrochlore than a kagome magnet, in that its Heisenberg model is in a classical U(1) spin liquid phase. This can be appreciated by taking up the idea of frustration leading to ground-state degeneracies mentioned in the context of the Ising magnet spin ice (Section 5.4). While for Ising spins, one can count the number of (discrete) spin configurations, for continuous Heisenberg spins, one considers the *dimensionality*, D, of the manifold of ground states. For a Heisenberg ferromagnet on any lattice, the only degrees of freedom in the ground state are the global spin rotations. By contrast, for the classical spin liquids, D can be *extensive*, that is, proportional to the system size. This can be estimated via a constraint counting argument pioneered in the context of the theory of elasticity by Maxwell (1864). The estimate is via the difference $D = F - K$ between the number of degrees of freedom of the spin system and the number of constraints imposed by the ground-state condition. The latter follows from the generalization of Eq. 5.56 from Ising spins σ to Heisenberg spins $\mathbf{S}$. In either case, each of the three components of the total spin of each simplex (i.e., each tetrahedron in the case of the pyrochlore lattice, and each triangle in the case of kagome and SCGO) needs to vanish: $\mathbf{L}_\alpha \equiv 0$. Now, a unit cell of the pyrochlore lattice contains four spins, each shared between two tetrahedra. Hence, $F = 4 \times 2$ and $K = 3 \times 2$, so that there are $D = 2$ degrees of freedom in the ground-state manifold per unit cell: a quarter of all available degrees of freedom remain unconstrained.

It is the fluctuations in this large but constrained ground-state space which underpin the Coulomb spin liquidity of pyrochlore and related magnets such as SCGO.

Indeed, the analogous counting for the SCGO slab, yields $F = 7 \times 2$ and $K = 4 \times 3$, as each unit cell hosts 7 spins, and contains four simplices (two triangles and two tetrahedra), so that $D = 2$ is again an extensive quantity. By contrast, for the kagome Heisenberg magnet, $F = K = 6$, so that $D = 0$ signals the absence of a spin liquid. (Instead, a very interesting phenomenon known as order by disorder occurs, where thermal fluctuations make the spins assume a coplanar arrangement.)

For Heisenberg spins, the genesis of the emergent gauge field upon coarse-graining is entirely analogous to the case of spin ice, with there now being three essentially independent flavors of gauge field, one for each spin component. The genesis of gauge charged defects, corresponding to $\mathbf{L}_\alpha \neq 0$, however, differs in detail from the magnetic monopoles in spin ice.

Upon dilution with one or several nonmagnetic ions, a simplex with two, three, or four occupied sites can still have these add up to $\mathbf{L}_\alpha = 0$. If only a single spin is left, this is obviously no longer possible, as $|\mathbf{L}_\alpha| = |\mathbf{S}_i| = 1$. This object, which has become known as an orphan spin (Schiffer and Daruka, 1997), acts like a free paramagnetic moment even at low temperatures. Note that this is not because it is a genuinely free spin, as it is *not* disconnected from the remainder of the system (Figure 8.6). Rather, like in the case of the edge spin $S = 1/2$ degree of freedom of the AKLT chain (Figure 5.7), it retains its interaction partners in one simplex, but not the other. And indeed, in analogy to this case, the effective free moment is fractionalized, and amounts to only half of that of the microscopic spin's magnetic moment, as can be established by a computation along the lines discussed in Box 5.3.

The simplex hosting the orphan spin violates the condition $\mathbf{L}_\alpha = 0$; as that condition underpins the emergent conservation law (see Section 5.4.2) underpinning the emergent gauge field, its violation implies that the orphan spin carries a gauge charge: it acts as a source/sink of the emergent gauge flux. In direct analogy with conventional electromagnetism, it therefore sets up a field decaying like $1/r^{d-1}$ with distance r, as the (emergent) field lines spread out over d-dimensional space following Gauss's law. Up to some coarse-graining and staggering factors, the magnetization density is given by these emergent field lines. This implies that spin correlations away from the orphan spin decay as $1/r$ for the $d = 2$-dimensional SCGO slab.

Again, as in the case of the Haldane chain, such spin correlations can be compared to NMR experiments (Limot et al., 2002), which measure the distribution of local fields. These can in turn be analytically and numerically modeled. Such a comparison with the level of dilution x as a fitting parameter can then be used to estimate the level of disorder in the sample. The value of dilution, x, thus obtained turned out to be in excess of the nominal value estimated after the growth of the sample. The origin of this discrepancy is unclear. The model may be incomplete, for example, there may be other sources of disorder reinforcing the NMR signal; or correlations between the locations of the vacancies enhancing the number of orphans; or it may reflect too optimistic an original estimate of the actual level disorder.

For completeness, we address the question why we should be allowed to model a quantum spin $S = 3/2$ with a classical Heisenberg model. The answer to this is primarily observational, and somewhat depressing, in nature: such modeling has in practice time and again turned out to work unreasonably well, even for $S = 1/2$, and the actual open questions is why, in many frustrated Heisenberg magnets, genuine quantum effects are so elusive.

8.8.3 Disorder in the Kitaev Honeycomb Spin Liquid

We next discuss both bond (distortion) and site (dilution) disorder in Kitaev's honeycomb model. Dilution has the tendency to be less gentle than distortions: it is strong (a spin disappears entirely rather than "a little bit" in a perturbative sense), and it is intrinsically short-wavelength: a vacancy is genuinely linked to the lattice scale. Distortions, by contrast, can be turned on perturbatively by taking the strength of the random stress that causes them to be small. Also, one can suppress short wavelength distortions explicitly. Technically, the resulting smooth strain configurations allow, for example, controlled gradient expansions; and physically, the properties of the distorted system may be perturbatively close to the undistorted one. In this account, we start with smooth and weak distortions, and then move on to the dilution problem.

The following discussions will be able to borrow results from studies of graphene (Vozmediano et al., 2010), as formally one obtains a fermionic problem on the honeycomb lattice. However, there are added features and distinctions, firstly due to Majorana nature of emergent degrees of freedom which translate nonlinearly into some observables, and also due to the existence of the emergent $\mathbb{Z}_2$ fluxes as their fractionalized companion degrees of freedom. These we will discuss toward the end of this chapter. In the meantime, it is sufficient to think of the hopping parameter t between sites in graphene as interchangeable with the exchange parameter J between spins in Kitaev's honeycomb model.

Dirac Fermions in an Artificial Gauge Field and Majorana Landau Levels

A spatially nonuniform hopping strength on the honeycomb lattice can give rise to an effective gauge field in the low-energy description near the Dirac points. This requires the presence of a strain field (say, a displacement $\mathbf{u}_i$ of lattice sites located in equilibrium at $\mathbf{r}_i$), with a minimal complexity. Obviously, a "hydrostatic" uniform compression of the lattice will simply change the overall strength of the hopping parameter t_{ij} uniformly. A spatially uniform compression, say along one axis, will lead to different values of the hopping parameters in the three different bond directions, t_α. This only leads to a displacement of the Dirac points away from $\pm\mathbf{K}$ in reciprocal space; indeed, once the change in the t_αs is large enough that the sum of two of them equals the third, the two Dirac points merge, and for an even stronger distortion, they are gapped out. (It is these considerations which lead to the phase diagram of Kitaev's honeycomb model depicted in Figure. 9.1.)

It is only when the strain field has nonvanishing second derivatives that an "artificial" gauge field emerges. This happens as follows. Let us assume the hopping matrix elements to depend on the strain field via $t_{ij} = t^0(1 + \alpha(\mathbf{r}_j - \mathbf{r}_i) \cdot (\mathbf{u}_j - \mathbf{u}_i))$, where the Grüneisen parameter α encodes the strength of the electron/spin-lattice coupling. In the continuum limit, one can carry out a gradient approximation which in the absence of strain yields the Dirac points at the zone corner wavevectors $\pm\mathbf{K}$ see (Section 2.5). When u varies only slowly, one can replace $\mathbf{u}_j - \mathbf{u}_i$ with the gradient of a continuum strain field $\mathbf{u}(\mathbf{r})$: $\mathbf{u}_j - \mathbf{u}_i \sim [(\mathbf{r}_j - \mathbf{r}_i) \cdot \nabla]\mathbf{u}(\mathbf{r})$. For nonvanishing second derivatives of the strain field, this can be manipulated to formally look like a (random) vector potential, that is, a term in the Hamiltonian minimally coupled like the vector potential corresponding to a standard magnetic field (Section 2.5.2). The resulting vector potential has the form

$$\mathbf{A} \sim \alpha \begin{pmatrix} u_{xx} - u_{yy} \\ -2u_{xy} \end{pmatrix} , \qquad (8.59)$$

where the subscripts denote the derivatives with respect to Cartesian coordinates. As an aside, we note that such engineering of "artificial" gauge fields has been very popular in the cold atomic community.

This case of a random vector potential for Dirac fermions has been analyzed in great detail by Ludwig et al. (1994). This has various fascinating properties, such as the existence of a nodeless delocalized state for any disorder realization, as well as the existence of multifractal scaling, which however take us beyond the scope of this treatment. Here, we content ourselves with extracting the results for what is arguably the most basic thermodynamic observable, namely, the heat capacity.

The heat capacity, $C(T) = \frac{dU}{dT}$, is a derivative of the internal energy $U = \sum_i E_i \exp(-\beta E_i)/Z$ of a system, where the sum on i runs over all many-body states of the system, and the Boltzmann factor depends on the (inverse) temperature $\beta = 1/T$, while $Z = \sum_i \exp(-\beta E_i)$ is the partition function. $C(T)$ therefore essentially encodes the many-body spectrum E_i.

For systems with simple (fermionic or bosonic) quasiparticle descriptions at low T, the internal energy takes on characteristic forms which can be used to diagnose those:

$$U = \int dE \, E \, \rho(E) \, n(E) . \qquad (8.60)$$

Here, $\rho(E)$ is the density of states of the quasiparticles (normalized to make U intensive), $n(E)$ is their occupancy, given by Bose-Einstein or Fermi-Dirac distributions $1/n(E) = \exp[\beta(E - \mu)] \mp 1$. The integral runs from $E = 0$ to $E = \infty$, effectively being cut off by $n(E)$ at high $\beta(E - \mu) \gg 1$. The elementary phase space density, $\rho(\mathbf{q}) \, d^d q \propto h^{-d} q^{d-1} \, dq$ depends on the dimension of space d; this is related to $\rho(E)$ via the dispersion of the quasiparticles, $E = \hbar\omega(q)$. Thus, for $\hbar\omega(q) = \Delta$, as in the Einstein models for phonons, the heat capacity vanishes exponentially, $C \sim \exp(-\beta\Delta)$ at low T, while it is linear in T for Fermi liquid, and quadratic for linearly dispersing magnons/phonons in $d = 2$, but not for quadratically dispersing magnons of a Heisenberg ferromagnet.

For the fermionic Dirac point corresponding to graphene and the Kitaev honeycomb model in $d = 2$ with the linearly vanishing density of states, $\rho(E) \propto E$, one also obtains a quadratic heat capacity, $C(T) \propto T^2$. This turns out not to be robust in the presence of the strain-induced random vector potential. Here, the density of states is filled in gently and continuously as the disorder strength increases. Its functional form remains algebraic, but its power law drifts *continuously*, even crossing over from a vanishing to a diverging density of states for $E \to 0$. This implies, in particular, that the characteristic exponent of the heat capacity in a Dirac system is not robustly pinned to its integer value, and therefore its absence need not be a reliable diagnostic of the absence of Dirac physics.

Uniform Synthetic Field

Beyond the disordered setting, one can also apply strain to produce a uniform effective magnetic field (Rachel et al., 2016). This, however, requires an increasing amount of strain as the system grows, as the field strength is given by the derivative of the vector potential, so that a uniform magnetic *field* implies unbounded growth of the vector potential with system size.

The reader may now wonder how strain, which as we emphasized in the context of spin ice preserves time-reversal symmetry, can give rise to a magnetic field, which does not. This is resolved by the magnetic field obtained in the gradient expansion pointing in *opposite* directions for the two Dirac points at $\pm\mathbf{K}$.

This has a further amusing consequence. It is a special feature of the wavefunctions of the central Landau level in graphene to reside on only one of the two sublattices of the honeycomb lattice (see Section 2.5.2). For the synthetic gauge field under consideration here, the wavefunctions of both Landau levels live on the same sublattice: the form of the strain removes the inversion symmetry underpinning their equivalence.

The natural follow-up question concerns the many-body state of the Majorana modes in this central Landau level when interactions are added. This poses the same question as that of the fractional quantum Hall effect of interacting electrons in a Landau level in Section 5.1. The main difference is the structure of allowed terms in the Hamiltonian, as well as the appearance of various types of symmetries. For instance, time-reversal symmetry translates into particle-hole symmetry of complex fermions obtained by pairing Majorana modes, so that the resulting many-body state effectively involves a half-filled Landau level. It appears that the resulting state is related to the physics of composite fermions in a half-filled Landau level discussed earlier in this chapter, but it is clear that much remains to be discovered here: the study of such Majorana many-body physics is still in its infancy (Rahmani and Franz, 2019).

Vacancies and Random Bipartite Hopping

We finally turn to the case of random dilution with nonmagnetic vacancies. As outlined above, this kind of disorder is less gentle than the weak and slowly varying bond disorder. Indeed, if one carries out the gradient expansion of the strain field **u** to higher order, further terms are generated corresponding not just to a random vector potential, but also to a random mass or a random scalar potential in the language of the relativistic Dirac equation.

The fact that random dilution immediately has a strong impact on the low-energy density of states can be seen quite directly from the structure of the hopping problem. The central ingredient here is that the honeycomb lattice is bipartite. As a result, the hopping problem on it yields a matrix of off-diagonal blocks when the sites are grouped separately into A- and B-sublattices:

$$\begin{pmatrix} 0 & h_{AB} \\ h_{AB}^{\dagger} & 0 \end{pmatrix}, \tag{8.61}$$

where the size of the rectangular $N_A \times N_B$ matrix h_{AB} is given by the number of sites on the sublattices, N_A and N_B. For a clean lattice, h_{AB} is therefore a square matrix. (This bipartite structure has played a role in previous sections, as it underpins the particle-hole symmetry of the spectrum of graphene.)

A vacancy, that is, deleting a single site from the honeycomb, thus necessarily leads to a zero-mode on the opposite sublattice. Indeed, for an imbalance $N_\delta = N_A - N_B$, there are guaranteed to be N_δ exact zero energy modes. This elementary fact of linear algebra follows from the observation that the matrix h_{AB} is rectangular, and even if it has maximal rank can only have as many nonzero eigenvalues as sites on the minority sublattice; the surplus of sites on the majority sublattice therefore must go along with zero eigenvalues.

For a random distribution of $N_{\mathrm{imp}} = N_{\mathrm{imp}}^A + N_{\mathrm{imp}}^B$ impurities, this imbalance scales as $\sqrt{N_{\mathrm{imp}}}$, so that the weight in the peak in the density of states at zero energy vanishes as $1/\sqrt{N_{\mathrm{imp}}}$ for a fixed vacancy concentration in the thermodynamic limit.

However, there are two other sources of zero modes. One is of vacancies surrounding a spin to disconnect it from the net entirely; this way of generating a zero mode is present on any lattice. The other corresponds to special vacancy constellation exactly localizing a mode inside a given region. Both of these are quite rare at low dilution, as they require a modest number of vacancies to assume correlated locations.

However, at low dilution probability $x \ll 1$, the vacancies will be well separated. If they were genuinely nonintercommunicating, there would be a peak in the density of states at zero energy given by N_{imp}, the total number of vacancies on both sublattices. At low dilution, there will therefore be two countervailing tendencies: the total weight of the peak decreases with dilution, $\sim x$. At the same time, its width also decreases, as the overlap between different vacancy modes, and hence their splitting, vanishes with increasing separation, the typical value of which scales as $1/\sqrt{x}$.

This set-up leads to a type of disorder problem known as a random bipartite hopping model, which has fascinated statistical physicists for many years. Unlike standard Anderson localization in $d = 2$, discussed earlier in this chapter, random bipartite hopping has a special energy, $E = 0$, around which the spectrum is symmetric. There turns out to be a diverging localization length as this special energy is approached . The resulting density of states has a strong divergence,

$$\rho(E) \sim \frac{1}{E} r_{GW}(E) \tag{8.62}$$

where, for some constant c, $r_{GW}(E) \sim \exp[-c |\ln E|^{2/3}]$ only just manages to cut off the divergence: in its absence, $\rho_E \propto E^{-1}$, would not even integrate to give a finite total number of states.

In fact the $\rho(E)$ is known in impressive detail (Motrunich et al., 2002). It even includes a crossover at a temperature which is unreasonably low from an experimental perspective. In practice, one would expect this intricate structure to translate into a slowly drifting exponent in the temperature dependence of the low-temperature heat capacity.

In the Kitaev honeycomb model, the density of states of the Majorana excitations is given by (half of) the expression of the electrons in graphene. However, in the former case, the fact that the Majoranas are derived from magnetic degrees of freedom suggests another thermodynamic quantity as a natural probe, namely, the magnetic susceptibility, which quantifies the size of the magnetization induced by an applied field, $M = \chi H$. A magnetic field applied to a Kitaev honeycomb magnet was touched upon in Eq. 7.34, where it was noted that its effect is most important at higher order in perturbation theory, as it gaps out the Dirac cone.

In fact, the linear susceptibility χ of the pure Kitaev honeycomb magnet vanishes. This results from the relation of the spin operator in terms of the fractionalized Majoranas: $\sigma_a = ib_a c$. By toggling a b-Majorana fermion, one of the u variables (see Figure 7.4) changes sign, and hence two $\mathbb{Z}_2$ fluxes are generated. The vanishing overlap between the flux variables in the states before and after action with the field make the lowest-order contribution of the field to the energy vanish.

In the presence of dilution, this is no longer the case. The b_a-Majorana belonging to the site whose neighbor in the a-direction is missing is now unpaired, and the magnetic field term $h_a \sigma_a = i h_a b_a c$ now amounts to adding a site, where the b-Majorana resides, to the hopping problem (Willans et al., 2010). This now allows for a stronger (lower-order in h_a) response to the field. The resulting form of the susceptibility then reflects the density of states, and would be expected to be of the form

$$\chi(T) \sim T^{1-\alpha(T)} , \qquad (8.63)$$

with $\lim_{T \to 0} \alpha(T) = 0$. Attractive as this is, we do caution that the vanishing of the linear susceptibility in the pure system is not stable to the introduction of generic perturbations, so that subtraction of a bulk term will be necessary in practice in experiments on a nonideal system.

There do in fact already exist experiments on a disordered candidate Kitaev spin liquid compound $H_3LiIr_2O_6$ (Kitagawa et al., 2018). These see an extended regime of $C \sim T^{3/2}$ at low T. This "reduced" exponent is consistent with such a picture; but possibly also with alternative explanations. A limitation in practice in such a setting is the capacity to probe the asymptotic low-energy behavior, which will inevitably be polluted by some other perturbations. Among these would be next-nearest-neighbor interactions, which destroy the bipartite structure of the hopping problem. These could hence cut off the low-energy divergence of the density of states even before its characteristic shape is visible. Nonetheless, the presence of stable intermediate-energy regimes tends to be a good indicator of underlying spin liquid physics.

Flux Binding

Besides rearranging the fermion density of states, the vacancy also changes the low-energy properties regarding the emergent $\mathbb{Z}_2$ fluxes. The ground state of the Kitaev honeycomb model is flux-free, that is to say, $W_p = 1$ (Eq. 7.32), for all elementary hexagonal plaquettes, ○. Diluted plaquettes, however, lower their energy by nucleating a flux. An easy way to see how this comes about is to consider the perimeter of a diluted plaquette, ⌂: this has length 12, rather than length 6 for the elementary honeycomb. A direct calculation of the zero point energy of a half-filled band involves summing up the energies of the occupied states, as sketched

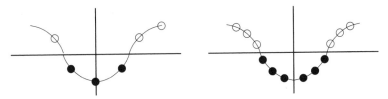

Fig. 8.7 An electron hopping on a ring of length L has allowed wavevector $k = 2\pi m/L$ in the absence of a flux, and $k = 2\pi(m + 1/2)/L$ in its presence, where m is an integer. The zero-point energy of a half-filled ring of length $L = 4n + 2$ is minimized if the ring is not pierced by a flux (left panel), while the presence of flux is preferable for $L = 4n + 2$ (right panel). For the Kitaev honeycomb model, an elementary plaquette $\bigcirc$ has $L = 6$, and hence no flux. By contrast, a nonmagnetic vacancy is located at the center of three merged plaquettes $\bigcirc$ with total circumference $L = 12$: the vacancy binds a flux.

in Figure 8.7. One finds that for loops of length $4n$, where n is an integer, the zero point energy is lower in the presence of a flux, while the converse is true for length $4n + 2$. Numerically, it has been confirmed that this is true not only for an isolated plaquette, but also for one embedded in the full honeycomb lattice in the Kitaev honeycomb model.

This flux bound to the plaquette implies that the above discussion was not strictly complete, as it made the connection to graphene without mentioning the presence of nucleated fluxes. However, both analytical and numerical observations suggest that the central qualitative properties of the random bipartite hopping problem are not affected by flux-binding.

Breaking time-reversal symmetry in the presence of the bound flux adds an intriguing twist to this story. This can be done by applying a field/three-spin term, see Eq. 7.34 and Eq. 7.35. The result is the appearance of a gap at the Dirac points. Through the equivalence of the gapped Kitaev honeycomb model with a p-wave superconductor, the flux is associated with a Majorana mode. This mode is localized at the vortex on the scale set by the inverse gap. Therefore its deviation from zero energy, on account of hybridization with other vacancy modes, is exponentially small in the distance, $\sim 1/\sqrt{x}$, to the other vacancies.

Hence, when the vacancies are well separated on the scale set by this localization length, these turn into a collection of Majorana zero modes. The cooperative behavior of these randomly distributed Majorana modes due to their residual interactions has thus far not been elucidated in detail.

As the gap increases with the size of the three-spin term, it turns out to be energetically favorable to revert to the zero-flux sector. This happens because in the absence of the flux, the Majorana mode is removed from the otherwise gapped region of the spectrum, and rejoins the other modes which have an energy of at least the size of the gap; thereby, the mode gains a zero-point energy of that order.

In the following chapter on topologically protected quantum computing, these Majorana modes will return and play a central role by virtue of their non-Abelian statistics. Therefore, being able to tune and manipulate the defects giving rise to the fluxes would turn the nuisance of disorder into an asset.

9

Topological Quantum Computation via Non-Abelian Statistics

Given the excitement generated by topological phases, with the concomitant expenditure of resources for their research, one might ask what they may actually be good for. One answer is as a precise and reproducible measurement of fundamental constants in the ratio e^2/h, and indeed the integer quantum Hall effect was used for this purpose in the 2018 redefinition of SI units; similarly the Josephson effect was used starting in 1990 to define a voltage standard via its precise measurement of the ratio e/h. The purpose of this chapter is to describe an even more ambitious application that depends not just on topological robustness but on fractionalization as introduced in Chapter 5. Of course, history cautions us against emphasizing potential applications when deciding what fundamental research to pursue. To give just one example, a medical MRI scan is a descendant of wartime radar research via studies of magnetic resonance physics in the postwar years: applications of fundamental research are typically apparent only a posteriori.

Nonetheless, there is one application of topological physics which is being pursued with a clear vision: topological quantum computing. Topological phases provide some unique advantages in the design of quantum computers, which hold the promise to be more powerful than any (classical) computer conceivable by extrapolation of present technology. This chapter is devoted to an explanation of the why, what and how of topological quantum computation. It is included not only because of the many aspects of topological physics it illustrates, as well as the potential technological importance of topological quantum computation, but also because it underlines one of the basic tenets of condensed matter research: new phenomena almost inevitably engender new applications, and perhaps we will one day consider a topological quantum computer to be as unremarkable as the hard disk drives for memory storage which appeared only a few years after the discovery of the GMR effect they utilize.

This chapter first reviews a couple of basic facts about computer science and computational complexity of classical computers, and discusses their scope for generalization to the quantum realm. It outlines the promise quantum computers hold, and mentions a couple of obstacles that will need to be removed for this promise to be realized. There are many books on quantum computing at various levels of sophistication, but few that discuss the topological approach in any detail; two exceptions with additional mathematics beyond our coverage are Wang (2010) and Pachos (2012). The key to the topological approach to quantum computing in this chapter will be that some topological phases support useful kinds of non-Abelian fractional statistics, which made an appearance as a theoretical possibility in Box 5.2.

We then present some aspects of the physical realization of non-Abelian statistics in a topological superconductor with a $p + ip$ order parameter, in both one and two dimensions. These aspects appear in other contexts, such as paired fractional quantum Hall states, which we also discuss. So far, topological superconductors have appeared in the tenfold way classification as superconductors whose Bogoliubov–de Gennes Hamiltonian for quasiparticles has topological properties such as surface states, similar to those of a topological insulator. The upshot of our discussion is that when those topological aspects are combined with topological defects in the order parameter, such as vortices, remarkable physics results. It may not be too surprising, given the existence of a topological edge state, that a vortex might have bound states, and indeed examples of this phenomenon were seen in Chapter 8. It is more surprising that the bound states generated through this process are the most experimentally accessible example of particles with non-Abelian statistics. An excellent reference for physicists who wish to delve deeper into non-Abelian topological phases is Nayak et al. (2008).

9.1 Quantum Computation: Universality and Complexity

The use of machines to assist humans with computational tasks goes back to the early civilizations. In the twentieth century, the invention of the microcomputer, and the exponential growth of its capacities following Moore's law, has been central to technological and social developments. Computer science, and in particular the field of complexity theory, has sprung up as the field of study that aims at understanding and quantifying the capabilities of computational devices.

The hope for the additional power of quantum computation springs from the fact that it acts not on classical bits, but on quantum wavefunctions, which are qualitatively richer than classical objects. This is embodied by the idea of quantum parallelism: if a classical computer is to obtain the value of a given integer-valued function, f, of an argument i consisting of a string of N bits, it needs to compute the function $f(i)$ for each of the 2^N values the input string can take. Similarly, a quantum computer can evaluate the function by taking a wavefunction $|i\rangle \otimes |1\rangle$

and returning $|i\rangle \otimes |f(i)\rangle$. Here the Hilbert space on which the quantum computer acts with unitary evolution is a product of dimension-2 Hilbert spaces or qubits, which one can think of as superpositions of the two states of a classical bit, and the Hilbert space is taken large enough to include the desired range of input and output integers.[1]

The linearity of quantum evolution permits the quantum computer to act, alternatively, on a wavefunction consisting of the equal amplitude superposition of all of the classical values, $|\phi\rangle = 2^{-N/2} \sum_{i=1}^{2^N} |i\rangle \otimes |1\rangle$, and obtain a wavefunction $|\psi\rangle = 2^{-N/2} \sum_{i=1}^{2^N} |i\rangle \otimes |f(i)\rangle$.

Computing the function like this, however, is not the same as tabulating it classically. In quantum mechanics, reading out the result of the computation leads to the collapse of the wavefunction, and tabulating $f(i)$ fully would thus still require at least 2^N operations, with an additional overhead due to the stochastic nature of the measurement process.

However, if what is required is not a list of the values $f(i)$ but some property of the ensemble of function values – such as a Fourier transform at a particular wavevector – the prospects for a speedup look more auspicious. Indeed, one problem in which the best classical computer is known to be outperformed by a quantum one is Grover's algorithm for an unstructured search. An example of an unstructured search is known as the inverse telephone number lookup: given a telephone number, how do you find out who it belongs to by consulting a telephone book? Classically, the best one can do is to go through the telephone book line by line, until one has located the number in question to find out who owns it. For 2^N numbers in the book (that is to say, binary telephone numbers with N digits), one will have to go through, on average 2^{N-1}, entries in the book. The time it takes to execute Grover's algorithm on a quantum computer, by contrast, has been shown to scale only as the square root of that number, $2^{N/2}$.

The unstructured search is simple enough that one can determine that under certain conditions these algorithms are optimal on classical and quantum computers, respectively. For larger classes of more complex algorithms, it is the field of complexity theory which investigates and classifies how efficient computers can solve problems. The usual definition of hardness – which may deviate considerably from what one experiences as difficult in everyday life when attempting to solve problems computationally – asks how the time to solve an instance of a problem class scales with its size. Here, the size is defined by a quantity such as the number of bits, N, needed to encode the entries in a phonebook for the example described in the previous paragraph.

---

[1] It may be useful to point out, in the context of the no-cloning theorem in Box 9.1, that different classical inputs i are represented in this construction by orthogonal states.

Problems whose run-time $T(N)$ scales no worse than polynomially, $T(N) \sim N^\alpha$ are known as easy; otherwise, they are called hard, in particular if the time scales exponentially, $T(N) \sim \exp(\alpha' N)$. The preference for *any* polynomial run-time, rather than each specific exponent α defining its own complexity class, allows this definition to be largely independent of the actual computer architecture used. This idea is embodied by the Church–Turing thesis, which essentially states that a computation based on any physical process cannot outperform a reference computing machine using the complexity defined above as a yardstick. It leads to the concept of universal computation, so that one can choose any (sufficiently powerful) computer architecture and determine its performance on a class of problems in order to determine the complexity of that class.

If the Church–Turing thesis were strictly applied, this could be the end of the story: there would be no hope to find a computer other than the universal Turing machine.[2] However, the hope formulated above is that a quantum computer could dramatically outdo a classical Turing machine, on account of its fundamentally different architecture allowing it to generate unitary time evolution of a full quantum wavefunction.

The speed-up provided by Grover's algorithm does not change the complexity class of the unstructured search between classical and quantum computers, because the relation between time taken on classical and quantum machines is just a square root. The hope is, nonetheless, that quantum computers may be able to provide an exponential speedup on some problems. The prime pedagogical example for this hope is embodied by the problem of prime number factorization. An algorithm due to Peter Shor (1994) allows a quantum computer to achieve factoring of large integers in polynomial time in the number of digits (i.e., polynomial in $\log N$ for integer N), which no known algorithm is known to achieve on a classical computer; indeed it was only shown a decade later that simply asking whether a number is prime *can* be done classically in polynomial time (Agrawal et al., 2004). The practical importance of this task lies in the widespread use of the hardness of factorizing prime numbers in public key encryption schemes. This would be vulnerable to breaking by a functioning quantum computer, although by the time that becomes available, encryption will presumably be available taking their existence into account.

It was hoped that, after Shor's algorithm was discovered, an avalanche of other provably fast quantum algorithms would be discovered for classically hard

---

[2] The Turing machine is an abstract model of a standard computer, free from the practical limitations of finite memory, errors, etc. Many different varieties of classical computer are known to be polynomially equivalent to a Turing machine, i.e., the time taken on the Turing machine to run some task scales as some power of the time taken on a different classical computer. The hope, based not on a proof but on our inability to date to find polynomial-time classical algorithms for problems such as factoring, is that quantum computers are not so equivalent.

problems. That has not yet happened,[3] and crucially there is not yet a quantum algorithm for an NP-complete classical problem. NP-completeness indicates classical problems which have a degree of universality in the sense that a polynomial-time algorithm for one such problem would imply a polynomial-time algorithm for the broad class. However, before one becomes unduly pessimistic, it is worth remembering that classical algorithms began to be discovered rapidly only after powerful classical computers became available, and also that many classical algorithms in wide use are faster in practical problems than mathematical analysis of the worst case would suggest. We hope that this history will repeat itself as we witness the advent of quantum computers that cannot be simulated on even the largest classical supercomputers.

9.2 Error Correction versus Fault-Tolerance

While considerable progress is happening on quantum computing hardware, there exists as yet no quantum computer that can scale to the size needed to surpass classical computation for tasks such as factoring. The main obstacle is to generate a coherent, tunable, and robust unitary time evolution of a wavefunction needed to carry out the quantum computation. The problem is already apparent on the scale of a quantum memory, which only requires robust encoding, rather than any actual manipulation, of quantum information. Consider a single qubit represented by the Pauli matrices, encoding information of a single degree of freedom corresponding to a two-state classical bit. Having initialized this qubit into a spinor $(\cos(\theta/2)\exp(i\phi/2), \sin(\theta/2)\exp(-i\phi/2))$, one would like to retrieve the information stored in it at a later time. However, in the meantime, this qubit is itself subject to quantum evolution. For instance, if it is physically represented by a quantum two-level system such as a spin in a solid state medium, it will be subject to unknown, and possibly uncontrollable and noisy, stray fields, which will affect its state in an unpredictable way, and hence degrade the information stored in it.

This can be counteracted by quantum error correction. Here, the quantum information is stored redundantly in a highly entangled state across several qubits. A quantum error correction algorithm can then try to spot the occurrence of an error and correct it. Indeed, one of the crucial advantages of quantum computation over other approaches that might also in principle be more powerful than the standard classical Turing machine, such as ideal analog computing with arbitrary-precision real numbers, is that quantum computing can retain its power even in the presence of a low level of error below some threshold rate. However, quantum error correction is much more subtle than classical error correction because of the no-cloning

---

[3] As quantum computing theorists have been heard to comment, Shor's algorithm was thought in 1994 to be the tip of the iceberg, but 25 years later looks more like the iceberg.

theorem (Box 9.1). Even though quantum information cannot be copied in the same way as classical information, it is possible to achieve redundancy without repetition, with the canonical example again due to Peter Shor (1995).

Box 9.1 The No-Cloning Theorem

Quantum information theory is by now a beautiful and rich topic, and we will see a more sophisticated application to topological phases in Section 11.2. Here we present a simple example of a nontrivial difference between quantum logic operations and classical logical operations, which is sufficient to make clear why error correction on quantum computers is much harder than on classical ones. Consider one of the simplest operations on a classical computer: given one bit stored in one memory location, copy, or clone, this bit into another memory location. In a classical computer this cloning is a simple way to generate redundancy and fault tolerance, as long as not all clones are subject to identical errors. We would like to understand whether such an operation is possible on quantum states.

More precisely, suppose that we have a unitary operator U such that, for two initial normalized states $|\psi\rangle$ and $|\phi\rangle$, its action is to clone them: for some initial normalized state $|s\rangle$, this means

$$U(|\psi\rangle \otimes |s\rangle) = |\psi\rangle \otimes |\psi\rangle, \quad U(|\phi\rangle \otimes |s\rangle) = |\phi\rangle \otimes |\phi\rangle. \qquad (9.1)$$

Here U could be in general some complicated unitary operator. We can derive one simple property, however, by taking the Hermitian conjugate of the first equation and then its inner product with the second. Since U is unitary and $\langle s|s\rangle = 1$,

$$\langle\psi|\phi\rangle = (\langle\psi|\phi\rangle)^2. \qquad (9.2)$$

Since the only solutions of $x^2 = x$ are $x = 0$ and $x = 1$, if a cloning operation is possible then either the initial states are identical or orthogonal.

Hence quantum states that are not orthogonal cannot be cloned by a unitary operation. We see that orthogonal states are more closely connected with the classical limit than are nonorthogonal states; if all of our qubits were in eigenstates with different eigenvalues, then they could be cloned like classical bits, but it is impossible to gain the power of quantum computation without working with superpositions.

A final note on entropy and classical computation is that erasure generates an energy $kT \log 2$, or an entropy $k \log 2$, per classical bit. Reversible computation, more analogous to the unitary information that is used above, is in principle possible without this entropy loss, but current computers do not bother to implement it. Conversely, there is an alternative approach to quantum computation based on performing measurements on the system and basing future operations on the outcomes of those measurements, so that the evolution is no longer strictly unitary. An accessible discussion of quantum information for physicists can be found in several textbooks, including Nielsen and Chuang (2010).

Such schemes come with considerable overhead – to start with, Shor's example spreads the information of a single qubit across a total of nine of them. In classical error correction, it is essentially sufficient to have logic elements that are more often right than wrong. In quantum computation, the overhead required to correct errors increases enormously rapidly as the error rate increases, and there is also (at least at the time of writing) a threshold error rate above which errors occur too rapidly to be corrected at all. The numerical value of the threshold depends on the particular quantum computing architecture, and considerable progress has been made in understanding quantum error correction through more general frameworks (Gaitan, 2013). To pick one example, an estimate for qubits with local interactions on a surface (Svore et al., 2007) is that the error rate needs to be below about 1 in 100,000 operations, and near this rate one requires thousands of physical qubits in order to synthesize one ideal, error-corrected "logical" qubit.

It would thus be nice to have access to a scheme which obviates such acrobatics and provides more directly usable qubits by being able to tolerate some level of noise-induced errors. Of course it seems likely that current noisy intermediate-scale quantum (NISQ) hardware, even without error correction, will still be useful, particularly for simulating problems in quantum materials and chemistry.[4] But current belief is that for typical computer science problems such as factoring, the potential of quantum computers will require a major step forward beyond all of the quantum computers we have now, with a dramatic reduction of error rates: a dramatic increase in the fault tolerance of hardware seems necessary for quantum error correction to be feasible. It turns out that the topological phases that are the focus of this book offer unique advantages for minimizing errors, which led to the proposal of topological quantum computing by Freedman and Kitaev.

9.2.1 Fault-Tolerant Topological Memories

It is here that one of the basic aspects of topological physics, their robustness, comes into play. While the state of a Josephson-junction qubit, or the spin of a trapped atom or a nitrogen-vacancy center in diamond, may show a surprisingly long coherence time, these two-level systems are all sensitive to local noise, and their sensitivity is found in practice to increase while gate operations are being performed, leading to fidelities (the faithfulness of a gate operation) measurably less than unity. It would certainly be nice to identify degrees of freedom which cannot be thus influenced by local noise. Such degrees of freedom can be constructed from the excitations of topological phases.

---

[4] Indeed, this was the original vision for where quantum computers were needed, as articulated independently by Yuri Manin and Richard Feynman in the early 1980s.

One instance, discussed in Section 5.3.2, is the winding parity of an Ising topological phase, which distinguishes nonlocally between two topological sectors. Crucially, any local operation – for example, measuring dimer density – cannot distinguish between whether the parity is even or odd. (In fact, for a system with finite width L, this statement is true only to within exponential accuracy in L.) For this reason, local noise also cannot distinguish between the two winding parities, and therefore not influence their time-evolution differentially.

It is this feature which underpins fault tolerance: even in the presence of noise, the qubit evolves in a predictable fashion without degradation of the information stored in it. The winding parity therefore qualifies as a fault-tolerant topological memory, at least at zero temperature.

9.3 Nonlocal Operations for Quantum Computing

It was in this context that Kitaev (2003) introduced the toric code, discussed in Section 6.3. Fault tolerance need not be restricted to memory elements but can also be used to effect quantum computation itself. Here, we return to the notion of universal computation, that is, the quantum version of the Church–Turing thesis. Rather than studying what is the best architecture *in principle*, it is sufficient to consider any (sufficiently powerful) scheme for building a quantum computer and study its properties. *In practice*, of course, the architecture of choice will depend on all the details this approach does not consider.

9.3.1 Universal Gates

For maximal simplicity, one identifies a set of components sufficient for building a universal computer. In the case of a classical computer, this set of components is small. With a classical computer evaluating a sequence of Boolean logic clauses, one appeals to the fact that an Boolean logic clause can be decomposed into a sequence of Boolean NAND operations each acting only on a pair of bits. Here, NAND stands for the operation which returns 1 (true) unless both input bits were themselves 1.

For quantum computers, things are somewhat analogous, with the following differences. First, the evolution of a quantum wavefunction is unitary, and hence logic gates of the classical NAND type are forbidden as they are obviously irreversible already on the level of having less information in the output than in the input (one bit versus two). Instead, the basic building blocks are represented by unitary matrices. It turns out that, again, it is sufficient to consider matrices (gates) acting on only a pair of qubits.

In addition, the space of unitary evolutions is continuous, and cannot thus be parameterized by a finite discrete set of operations, which are only countable in number. Instead, by considering a discrete set of gates, one can *approximate* any member of the continuous family of unitary operations arbitrarily well by executing a sequence of operations from the discrete set. This is the content of the Solovay–Kitaev theorem (Kitaev, 1997). However, one needs a sufficiently powerful or universal discrete gate set in order for generic unitary evolution to be well approximated.

Different choices for this discrete set are possible (Wong, 2018). One of them consists of only a triplet of operations. These contain two gates only acting on a single qubit, namely, the phase gate P and the $\pi/8$ gate $R_{\frac{\pi}{8}}$; and one gate acting on a pair of qubits, the controlled-NOT gate, which flips one provided the other takes a specific value:

$$P = \begin{pmatrix} 1 & 0 \\ 0 & i \end{pmatrix} \;\; ; \;\; R_{\frac{\pi}{8}} = \begin{pmatrix} \cos\frac{\pi}{8} & -\sin\frac{\pi}{8} \\ \sin\frac{\pi}{8} & \cos\frac{\pi}{8} \end{pmatrix} \; ;$$

$$CNOT = \begin{pmatrix} 1 & 0 & 0 & 0 \\ 0 & 1 & 0 & 0 \\ 0 & 0 & 0 & 1 \\ 0 & 0 & 1 & 0 \end{pmatrix}. \tag{9.3}$$

With these, one can construct a universal quantum computer in much the same way as a general Boolean circuit consisting of NAND gates provides the basis for a general classical computer.

As a note of caution, there are theorems showing that some sets of quantum gates are polynomially equivalent to universal *classical* computation, and hence provide no quantum speedup in the complexity theoretic sense. For instance, the apparently innocuous replacement of $R_{\frac{\pi}{8}}$ by the Hadamard gate $H = R_{\frac{\pi}{4}} = \frac{1}{\sqrt{2}}\begin{pmatrix} +1 & -1 \\ +1 & +1 \end{pmatrix}$ is equivalent to a set of gates known as Clifford gates, which are universal for classical but not quantum computation by the Gottesman–Knill theorem.

9.3.2 Nonlocal Operations via Non-Abelian Anyons

With the possibility of constructing a universal quantum computer via a small set of gates, the next question is how best to implement these operations. For this one needs to choose a set of degrees of freedom, the qubits, to encode the information to be processed; as well as a physical process to effect the gate operation. The following describes how these choices are made and implemented for the topological quantum computation architecture.

The starting point is that excitations of a topological phase can have internal degrees of freedom which can act as qubits. This is in itself not a big deal –

conventional excitations of conventional systems, such as a hole in a Fermi liquid, can also have internal degrees of freedom, such as a spin quantum number.

The basic ingredient provided by the topological phase is that the information may be stored in this internal degree of freedom nonlocally, and thus protected against degradation from local noise, just like the quantum memory described above. And the central idea of topological quantum computation is that this information may nonetheless be processed with physically realizable operations, which however now need to be topological in nature. If the unitary operations are generated by braiding quasiparticles (i.e., moving them around each other) while they remain spatially separated from each other, then the fault tolerance of a quantum memory will be retained even while nontrivial operations are performed. We now explore the current approaches to this goal in more detail.

Non-local encoding of quantum information

The primary physical example of particles allowing nonlocal storage of quantum information are the non-Abelian anyons introduced in Box 5.2. The simplest members of this class of particles are Majorana zero modes. The search for these has been an intense endeavor, and their status is described further down. Their importance was introduced by Kitaev in the simple setting of the Kitaev or Kitaev–Majorana chain (Kitaev, 2001; Kitaev and Laumann, 2009).

Box 9.2 What Is a Majorana Fermion or Zero Mode?

The study of Majorana fermions is interdisciplinary, at least in the sense that the term "Majorana" has different meanings in different communities. The original idea came soon after Dirac's discovery of his relativistic equation for an electron. The electron in three dimensions is described by a four-component spinor field, but there are two ways that the four-component Dirac equation can decouple into two independent two-component equations. We saw condensed matter realizations of one such decoupling for massless fermions, found by Weyl, in Chapter 7. Another decoupling was found by Majorana, and the search for its realizations is an active topic in multiple areas of physics.

In high-energy physics, a Majorana fermion, or Majorana particle, is an elementary particle that is at the same time its own antiparticle. This is clearly not the case with almost all the elementary fermionic particles we know: positively charged particles like the proton or the positron cannot be their own antiparticles, as those need to carry negative electric charges. The only remaining candidate is the neutrino.

This possibility is under active current investigation in high-energy experiments, where one looks for violation of lepton number conservation. This would most simply show up in neutrinoless double β-decay. In the normal (single) β-decay, a neutron decays into a proton, emitting an electron and an antineutrino in the process: the electron carries electric charge opposite to the proton's, while electron and antineutrino have zero net lepton number. In double β-decay, one thus ends up with two protons, two electrons and two antineutrinos. If the (anti-)neutrino is its own antiparticle, two antineutrinos can therefore pair annihilate, sharing their energy and momentum between the other decay products, thereby violating lepton number conservation. Having zero mass is not a prerequisite for this – indeed, given the existence of neutrino oscillations, it is no longer believed that the neutrino mass is zero, although previously neutrinos were speculated to be Weyl fermions.

We do not dwell on this high-energy aspect of Majorana fermions further. In the following, we concentrate on two guises this term takes in many-body physics. First, there is a much more pedestrian meaning to Majorana fermion operators as a real fermion operator, colloquially known as half (or square root of) a complex fermion. We record some basic properties of such operators first and then start the discussion of realizations as emergent particles.

Given a standard spinless fermion created by $c^\dagger$ and annihilated by c, we define two Majorana operators γ_1, γ_2 by

$$\gamma_1 = c + c^\dagger; \quad \gamma_2 = i(c^\dagger - c). \tag{9.4}$$

The Majorana particles described by γ_1, γ_2 are their own antiparticles in the sense that $\gamma_1^\dagger = \gamma_1, \gamma_2^\dagger = \gamma_2$. Thus "creation" and "annihilation" are the same operation, although this terminology is somewhat misleading as there no longer is a number operator to go with each Majorana fermion.

They have a fermion-like anticommutation relation

$$\{\gamma_j, \gamma_k\} = 2\delta_{jk}. \tag{9.5}$$

This implies that $i\gamma_1\gamma_2$ is Hermitian and indeed related simply to the number operator of the original complex fermion (try it!). We also see from the definition that $\gamma_1^2 = \gamma_2^2 = 1$.

Their second appearance is that, as discussed in some length in the remainder of this chapter, a Majorana (zero) mode is a fractionalized degree of freedom which can be robust to any form of local noise by virtue of its nonlocal nature. Equation 9.4 is obviously more of a mathematical construct, in the sense that what variables to use to describe a physical problem is to large degree a question of personal preference and convenience. At the same time, the physical connection between the two roles is that an (approximate) Majorana zero mode is most simply expressed in terms of Majorana fermion operators. This expresses the converse of Weinberg's third law of progress in theoretical physics: "You may use any degrees of freedom you like to describe a physical system, but if you use the wrong ones, you'll be sorry."

An example of why the use of Majorana fermion operators is physically justified comes from counting states. Since one complex fermion gives two Majorana fermions, the dimension of the Hilbert space of a system with an even number of Majorana fermions is not 2^N but $2^{N/2}$. Indeed, in the states described below such as the $p + ip$ superconductor or the Moore–Read quantum Hall state, one finds that the ground-state degeneracy of the system with an even number N of vortices is not 2^N, as if we had conventional bound zero-energy states in each vortex, but $2^{N/2}$. For the superconductor, the Majorana fermions represent the Hilbert space obtained from the zero-energy solutions of the Bogoliubov–de Gennes equation in vortices.

To summarize, the decoupling of a complex fermion into two real fermions is always possible mathematically, and may become relevant physically for massive particles such as the neutrino. These real fermion operators also are appropriate to label degenerate ground states in certain topological states of matter in two spatial dimensions, where the ground state remains degenerate even once quasiparticle locations are specified. The exchange statistics from moving these particles around each other will turn out not to be fermionic but to be non-Abelian, that is, determined by non-Abelian representations of the braid group (Eq. 5.20). The mathematical object that includes both the braid group and operations such as fusion (which determines the quantum numbers of the joint state of two quasiparticles) is known as a modular tensor category (Nayak et al., 2008). So it seems fair to say that the Majorana zero mode of condensed matter is considerably different from the Majorana fermion of high-energy physics, and arguably much more subtle.

Braiding Supplies Unitary Gates

The exchange of anyons and its relation to their relative statistics was already discussed in Section 5.1.4 for abelian anyons in the form of Laughlin quasiparticles. The phase picked up by the many-body wavefunction reflects the statistical angles of the anyons.

The effect of particle exchange for non-Abelian quasiparticles is more complex. This is captured by an analysis in terms of the braid group describing particle exchange (see Box 5.2). Generally, braiding quasiparticles is described by matrices acting on the space of degenerate states of the anyons. For Abelian anyons, whose braid group has one-dimensional irreducible representations, these matrices can be reduced to simple scalars encoding the phase picked up by the exchange.

By contrast, for non-Abelian quasiparticles, the matrix structure is unavoidable. Topological quantum computation can use the transformations encoded by these matrices as fundamental building blocks for the quantum gates. When a degeneracy arises due to the Majorana zero modes, braiding can lead to nontrivial unitary transformations within this multiplet of degenerate states. If the concomitant matrix structure is so rich as to represent a universal set of gates able for quantum computation, then the system of non-Abelian anyons can form the basis for a topologically protected quantum computer. We will see how Majoranas provide a non-Abelian representation of the braid group in Section 9.5, but first it seems worthwhile to describe simpler physical systems where Majoranas can be seen to appear.

9.4 Majoranas in One Dimension: The Kitaev Chain

The simplest instance of non-Abelian anyons is provided by Majorana zero modes. These are by now known to occur in a number of different settings in theory. The experimental search for them has been intense and things look promising on several fronts. As in the case of fractionalization described in Section 5.2.3, they appear most transparently in a microscopic model in one dimension, where this discussion starts with the case of the Kitaev chain, before working its way up to two dimensions, where a less microscopic effective field-theoretic description becomes more useful.

The starting point is a one-dimensional open chain modeling a spin-polarized p-wave superconductor (Kitaev and Laumann, 2009):

$$H_{KM} = \sum_{j=1}^{N-1} \left[-t \left(a_j^\dagger a_{j+1} + a_{j+1}^\dagger a_j \right) + \Delta \left(a_j a_{j+1} - a_j^\dagger a_{j+1}^\dagger \right) \right]$$

$$- \mu \sum_{j=1}^{N} \left(a_j^\dagger a_j - 1/2 \right) . \tag{9.6}$$

Here, following Kitaev's notation, the a_i operators are annihilation operators of fermions, where the spin index is suppressed as the particles are supposed to be fully spin polarized, so that only one spin direction appears. t is the strength of the hopping matrix element, giving rise to a kinetic energy $E_k(q) = -2t \cos(q)$ as a function of momentum q, with the lattice constant set equal to unity. The p-wave BCS gap is denoted by Δ, and μ is the chemical potential.

A transformation of the usual (complex) fermions into Majorana fermions is always possible (Box 9.2). This proceeds by decomposing the complex operators into a pair of "real" ones, which we label by c:

$$c_{2j-1} = a_j + a_j^\dagger \; ; \; i c_{2j} = a_j - a_j^\dagger . \tag{9.7}$$

Their anticommutation properties are also simple:

$$\{c_j, c_k\} = 2\delta_{jk}. \tag{9.8}$$

Rewriting H_{KM} yields the simple form

$$H_{KM}^{\text{maj}} = \frac{i}{2}\left[\frac{\mu}{2}\sum_{j=1}^{N} c_{2j-1}c_{2j} + \Delta \sum_{j=1}^{N-1} c_{2j}c_{2j+1}\right], \qquad (9.9)$$

where we have set $\Delta = t$ for the purpose of the present analysis.

H_{KM}^{maj} is still Hermitian. It can be diagonalized using an orthogonal $2N$-dimensional square matrix. Its eigenvectors come in pairs with eigenvalues $\pm\epsilon_\alpha$.

Now, the mode we are interested in has $\epsilon_0 = 0$; this eigenvalue is special as it is the only one for which the accompanying energy $-\epsilon_0$ is degenerate. It corresponds to the eigenvectors

$$v_0^L = \left(1, 0, \mu/(2\Delta), 0, (\mu/(2\Delta))^2, 0, (\mu/(2\Delta))^3, \dots\right), \qquad (9.10)$$
$$v_0^R = \left(\dots, (\mu/(2\Delta))^3, 0, (\mu/(2\Delta))^2, 0, (\mu/(2\Delta)), 0, 1\right).$$

In the limit $N \to \infty$ this expression is only normalizable for $\mu/(2\Delta) < 1$; if it is, the mode is localized near the edge. This condition therefore describes the range of the topological phase of the chain, directly diagnosed here by the presence of an edge mode.

Of course, any linear combination of these two modes is an eigenstate at zero energy also. However, for any finite-size system, one should expect a mixing and splitting of these two, just like the bonding-antibonding splitting between the levels of two initially well-separated particles. Here, the wavefunctions at the ends of the chains have localization length $\xi_{LR} = 1/\ln(2\Delta/\mu)$. Their splitting $\delta_{B/AB}$ is related to their overlap, which therefore decays exponentially with system size N:

$$\ln \delta_{B/AB} \sim -N\xi_{LR}. \qquad (9.11)$$

There are several interrelated ways to see what is special about this "excitation" at zero energy, in particular its intrinsic robustness for use in topological quantum computing. Note that in the above derivation of $\delta_{B/AB}$, it was necessary to combine the two edge modes in order to change their energy. But, if one makes the chain infinitely long, the two modes cannot communicate with one another.

If one now applies any local perturbation at one end which keeps the bulk gap intact, the two modes still cannot communicate; as one of the chain ends does not see the perturbation, and therefore remains unchanged, so does the other – this is the robustness to local noise. In practice, the effect of noise is thus exponentially suppressed by the same matrix elements responsible for the smallness of $\delta_{B/AB}$.

One can also look at the spectrum of the Hamiltonian directly. Still considering an isolated edge mode at zero energy, one can ask why it cannot simply move to a finite energy $\epsilon_0 \to \epsilon_{0'} \neq 0$. However, the condition that eigenvalues come in pairs, $\pm\epsilon$, would require there to appear a second mode at $-\epsilon_{0'} \neq \epsilon_{0'}$. However, there is no second such mode available: the remainder of the local spectral weight is located above the gap, and hence cannot perturbatively move to $\epsilon_{0'}$. To achieve this, a partner is needed – which is the one out of reach at the other, well-separated, edge. We will return to the topological protection of the Majorana edge modes in the context of symmetry protected topological order in Chapter 11.

In the case of the superconductor, there is a separate reason for the robustness of the zero mode. To see how this comes about, note that the topological degeneracy can be traced back to the parity, $P_a = (-1)^{\sum_j a_j^\dagger a_j}$, of the complex a fermions in Eq. 9.6. [In Majorana language, this operator can be accordingly rewritten, $P = \prod_{j=1}^{N} (-i c_{2j-1} c_{2j})$.] This operator commutes with the Hamiltonian; therefore, if it acts nontrivially on a state, it will generate a second, perfectly degenerate one. In an isolated fermionic system, only operators respecting the fermionic parity are believed to be allowed, that is to say, any *physical* perturbation will commute with P_a. The degeneracy between the two parity sectors in the superconductor can therefore not be lifted as no operators are available, even in principle, which can generate the off-diagonal terms which lead to a mixing, and hence lifting of the degeneracy, of the states.

Another way to look at this one-dimensional model is discussed in Box 9.3, using the nonlocal Jordan-Wigner transformation between spins and ordinary complex fermions in one spatial dimension. It turns out that the Kitaev chain can be transformed into a very well-studied model, the transverse-field Ising model. (This is an indication that in one dimension, particles of different statistics may not actually be very different from each other, consistent in spirit with the bosonization approach to interacting fermion systems mentioned in Section 6.6.1.) One reason that the transverse-field Ising model is well studied is that its zero-temperature critical point is the most widely used pedagogical model of a quantum critical point (Sachdev, 1999); indeed, the universal features of this quantum critical point are essentially the same as those of the finite-temperature Ising critical point in two spatial dimensions. The Kitaev chain thus furnishes a simple, exactly solvable example of a surprisingly broad range of important physics.

Box 9.3 The Jordan–Wigner Transformation and Statistics in 1D

The Jordan-Wigner transformation relates the algebra of spin operators to that of fermions. It can be used to turn a spin chain into a fermionic one, in particular relating the Kitaev chain to a transverse-field Ising model. It proceeds as follows.

The local Hilbert space is two-dimensional. For spins, the two conventional basis states are a spin pointing up, $|\uparrow\rangle$, or down, $|\downarrow\rangle$, in some preferred (typically, z-, but frequently also x-) direction. For particles, the conventional choice is to have the site occupied or empty. Notice that an Ising symmetry operation, exchanging $|\uparrow\rangle$ and $|\downarrow\rangle$, amounts to a particle-hole symmetry operation, exchanging occupied and empty sites.

One might be tempted to then identify the complex fermion operators exchanging empty and occupied sites, a, $a^\dagger$, with the spin raising and lowering operators, σ^+, σ^-. However, the algebra of Pauli matrices does not produce the fermionic anticommutation relations in such a simple way. Rather, to achieve this, it is necessary to attach a string to the spin operators, so that the transformation becomes a nonlocal one:

$$a_j = \left(\prod_{k=1}^{j-1} \sigma_k^z\right) \sigma_j^+ \, , \ a_j^\dagger = \left(\prod_{k=1}^{j-1} \sigma_k^z\right) \sigma_j^- \, , \tag{9.12}$$

with the string starting at a boundary site labeled $k = 1$. This identification does indeed reproduce the fermionic anticommutation relation $\{a_j, a_j^\dagger\} = 1$.

The conversion of the Kitaev chain (Eq. 9.6) with $\Delta = \Delta^* = t$, is then equivalent to a transverse field Ising magnet

$$H_{\text{TFIM}} = -J \sum_{j=1}^{N-1} \sigma_j^x \sigma_{j+1}^x - h \sum_{j=1}^{N} \sigma_j^z \, , \tag{9.13}$$

with $t = J$ and $\mu = -2h$.[a]

Several additional applications of this transformation are discussed in Auerbach (1994). For example, if we remove the transverse field in Eq. 9.13 and replace it with an interaction between y spin components $\sigma_j^y \sigma_{j+1}^y$ of equal strength to the $\sigma_j^x \sigma_{j+1}^x$ term, we obtain the so-called XX spin chain, which under the Jordan–Wigner transformation becomes a problem of free complex fermions.

It appears that in one dimension, the statistics we assign to the particles in a system is quite flexible. A deep reason for this is that in one dimension particles cannot move around each other to accomplish an exchange without passing through each other. The effect of an exchange, namely, a phase shift, is the same as the effect of an elastic collision. So the interaction-induced phase shift can be used to mimic the effect of particle statistics. In two dimensions the distinction between interaction-induced scattering and exchange statistics is clearer since the former is not just a phase, and it is a highly nontrivial challenge to find microscopic interactions that generate a specified type of exchange statistics. In two dimensions we can use flux attachment in principle to generate exchange statistics from bosons (Section 6.6.1), and we saw in the preceding chapter how adding flux to a microscopic electron system in a fractional quantum Hall state can localize fractional quasiparticles. Applying that idea to more complicated states will now lead to localized Majoranas in two-dimensional systems.

---

[a] To remain in keeping with the conventional presentation of the Kitaev chain, the transverse field acts in the z-direction here, while the Ising symmetry corresponds to a flip $\sigma^x \leftrightarrow -\sigma^x$. By contrast, in most of this book, we use the more frequent convention of the transverse field acting on the x-components, and the exchange between z-components.

9.5 Majoranas in Two Dimensions

Moving on to two dimensions, there are several physical realizations of such Majorana zero modes, including in superconductors, quantum Hall states, and frustrated magnets. Ultimately, these can all be pictured using the physics of vortices in p-wave superconductors.[5] Crucially, these vortices, unlike in the case of an s-wave superconductor, are endowed with an internal degree of freedom of Majorana nature. As a result, a pair of vortices exhibits one (complex) in-gap state. Its properties are analogous to those of the edge states of the chain discussed in the previous section. The nature of vortices in a p-wave system has been worked out in detail in Read and Green (2000), but we start here from a different and historically essential perspective, that of fractional quantum Hall states.

9.5.1 Quantum Hall Physics at $\nu = 5/2$

The history of non-Abelian statistics in two dimensions involves a beautiful connection between seemingly different areas of physics, which is well worthy of a brief diversion. A great success of theoretical physics in the 1980s was the realization that conformal field theories in two spacetime dimensions (roughly, conformal invariance combines scale and rotational invariance) are incredibly constrained (DiFrancesco et al., 1997). Their symmetries are not just the finite-dimensional global conformal group, but instead an infinite-dimensional group, essentially because, using a complex coordinate $z = x + iy$ for the plane, local conformal invariance is a property of the map $z \to f(z)$ for any meromorphic function f.

Conformal field theory (CFT) is by now a major subject in its own right, with impact across physics and mathematics. It is far from the least surprising of CFT's uses in physics that the conformal blocks appearing in multipoint correlation functions of CFTs, which are gapless theories in 2 spacetime dimensions, turn out to give a recipe for wavefunctions showing nontrivial braiding for states with an energy gap in $2 + 1$ spacetime dimensions. In addition to their braiding, operator product expansions, which describe the behavior of multipoint correlators in a field theory when two points become close to each other, turn out to give the fusion rules that describe how quasiparticles merge into either triviality or new quasiparticles when brought close to each other in the topological phase.

This observation by Moore and Seiberg (1989), made via the connection to Chern–Simons theories mentioned in a moment, was rapidly followed by the

---

[5] The main example is the spin-polarized $p + ip$ superconductor, which breaks time-reversal symmetry. However, there are other p-wave superconductors that do not break time reversal but still support Majorana modes, so we sometimes use the broader term. However, a gapless p-wave superconductor would not be topological in the sense we describe here.

seminal proposal that a wavefunction constructed in this way might describe a fractional quantum Hall state at an *even-denominator* fraction (Moore and Read, 1991). Remarkably, the first even-denominator quantum Hall plateau was observed at approximately the same time in experiment (Willett et al., 1987), at filling factor $\nu = 5/2$. We present this Moore–Read or Pfaffian wavefunction in the quantum Hall context and then its connection to $p + ip$ superconductivity. Recall that all of the quantum Hall wavefunctions constructed in Section 8.7 occur at odd-denominator fractions, as do the basic Laughlin states at $\nu = 1/m$.

Hence the observation of a plateau at $\nu = 5/2$ in very clean samples and low temperatures came as a shock. For our purposes, this state should be thought of as a partially filled Landau level with $\nu = 1/2$ on top of an inert $\nu = 2$ integer quantum Hall state; being in a higher Landau level modifies the Haldane pseudopotentials and affects the energetics of which state turns out to be the ground state, but focusing on $\nu = 1/2$ is sufficient to describe the key aspects. The wavefunction of the Moore–Read state, in a notation analogous to that of Eq. 5.3, is given by

$$\Psi_{5/2} = \prod_{i<j}(z_i - z_j)^2 \, \mathrm{Pf}\left[\frac{1}{z_i - z_j}\right] ; \tag{9.14}$$

this polynomial factor multiplies the "inert" part of the wavefunction related to the filled lower Landau level. Here, the Pfaffian PfM – a less familiar and also harder-to-evaluate property of a matrix than its determinant, to which it squares – of a skew-symmetric matrix M of (even) size p, with entries $M_{ij} = \frac{1}{z_i - z_j}$ is given by

$$\mathrm{Pf}(M) = \sum_{\pi_\alpha \in P} \mathrm{sgn}(\pi_\alpha) m_{i_1, j_1} m_{i_2, j_2} m_{i_3, j_3} \cdots , \tag{9.15}$$

where the sum runs over all pairings of the numbers $1 \ldots p$ such that $i_k < j_k$ and $i_j < i_k$ for $j < k$, with $\mathrm{sgn}(\pi_\alpha)$ the sign of the permutation mapping $1 \ldots p$ to $i_1, j_1, i_2, j_2, i_3, j_3, \ldots$.

This state can be thought of as a p-wave superconductor of spin-polarized composite fermions. This proceeds in several steps. First, one lets the filled lowest Landau level account for $\nu = 2$, that is, as a straightforward integer quantum Hall effect accounting for 2 out of the 5/2 flux quanta per electron, and this part of the system is inert. One is then left with a half-filled first excited ($n = 1$) Landau level.

Attaching two flux quanta to each particle then keeps the their quantum statistics intact but (on average) cancels the external field – one now has composite fermions which are spin polarized in zero field. Having these composite fermions undergo a phase transition into a p-wave superconductor then produces the Moore–Read state, along with the phenomenology of the p-wave superconductor. Indeed, the BCS wavefunction (Eq. 8.21) for an order parameter with $p + ip$ symmetry in two dimensions can be transformed from momentum space to real space and has a Pfaffian form (Read and Green, 2000).

More precisely, there are separate, weak-coupling and strong-coupling, phases of the spinless two-dimensional $p + ip$ superconductor, unlike the more familiar s-wave superconductor that occurs in many bulk metals. A typical superconductor, such as aluminum or niobium, has interparticle spacing much less than the Cooper pair size, given by the coherence length $\xi \approx \hbar v_F / \Delta$, where v_F is the Fermi velocity and Δ is the superconducting gap. In other words, if we transform to a real-space form of the BCS wavefunction, we find that the pair wavefunction has a relatively large length scale, with overlapping pairs: the total number of electrons within the size of the sphere, $n\xi^3$, is much greater than two. However, there is no gap closing or phase transition as we bind the pairs more closely, until the state can be viewed as a condensate of nonoverlapping bosonic pairs; this process is often referred to as the BEC-BCS crossover.

However, in the two-dimensional $p + ip$ superconductor, there is a gap closing and phase transition between two phases. The strong-coupling phase does not have zero-energy bound states in vortices, or the connection to the fractional quantum Hall state given above. It is the weak-coupling phase that has Majoranas in the vortex core. These can be found by solving numerically the Bogoliubov–de Gennes equation, but a reason why they have to exist follows from analyzing the Aharonov–Bohm effect of flux $h/2e$ on the Majorana edge state of this superconductor, similarly to what was found for topological insulators in Chapter 8. The critical point between the weak-coupling and strong-coupling phase has a gapless Dirac spectrum of quasiparticles at the mean-field level (Read and Green, 2000). The discussion of the standard vortex in the spinless superconductor can be adapted to a half-quantum vortex in a spinful triplet superconductor, around which the phase rotates by π but the pairing direction vector $\mathbf{d}$ also rotates by π.

A model form for the bound Majorana fermion operator at a vortex core is (Stern et al., 2004)

$$\gamma = \frac{1}{\sqrt{2}} \int d\mathbf{r} \left[f(\mathbf{r}) e^{-i\Omega/2} \psi(\mathbf{r}) + f^*(\mathbf{r}) e^{i\Omega/2} \psi^\dagger(\mathbf{r}) \right]. \qquad (9.16)$$

Here $f(\mathbf{r})$ falls off at sufficiently large distance r from the vortex core and ψ is the continuum electron annihilation operator. The robust existence of this zero-energy state, whatever the details of the superconducting properties near the vortex, is similar in spirit to the zero-energy bound state of the Jackiw–Rebbi model from Box 6.2. Two key differences between this and a conventional bound state are that γ is real (i.e., its own Hermitian conjugate), as expected for a Majorana, and has a $1/2$ factor in its dependence on the superconducting phase Ω along some reference line, say the $\theta = 0$ line from the vortex core.

When there are multiple vortices at locations $\mathbf{R}_i$, well separated in space, each has a bound Majorana of this form but the phase factor replacing Ω depends on the position of all the vortices, as the simplest model for the superconducting phase satisfying the winding conditions around each vortex is

$$\Delta(\mathbf{r}) = |\Delta(\mathbf{r})|e^{i\theta_i + i\Omega_i}, \quad \theta_i = \arg(\mathbf{r} - \mathbf{R}_i), \quad \Omega_i = \sum_{j \neq i} \arg(\mathbf{R}_i - \mathbf{R}_j). \quad (9.17)$$

Crucially, the γ_i are not uniquely defined given the positions of the vortices, but can be modified by an adiabatic braiding process. For example, on moving one vortex entirely around another, far away from other vortices, the spatial separation vector $\mathbf{R}_1 - \mathbf{R}_2$ passes through 2π, and hence the superconducting phase passes through 2π. From the Majorana wavefunction in Eq. 9.16, this process takes $\gamma \to -\gamma$ for the two Majoranas in these vortices.

Now we can finally give an example of what it means to say that braiding these Majoranas in the vortex cores of the weak-coupling $p + ip$ superconductor induces a unitary operation on the degenerate ground states. Consider the simplest case of four vortices. There are then four ground states since each vortex has one bound Majorana and two Majoranas make a degeneracy of two. Now moving the vortices around each other will generate some Abelian phase, including energy-dependent factors in general (Section 2.1) which we ignore; this corresponds to focusing on projective representations of the braid group. The operation of braiding also conserves the fermion parity, which means that the four-dimensional ground-state Hilbert space separates into two two-dimensional sectors of even and odd particle number.

Now the effect of moving the vortices around each other can be deduced from the microscopic physics of the superconducting wavefunction. We consider the spin-polarized $p + ip$ case here; an instructive paper with further details for the half-quantum-vortex case in a triplet superconductor is Ivanov (2001). Moving one vortex around another essentially amounts to two exchanges performed in the same sense, so a check on our result is that repeating it twice should give the two minus signs on the two vortex Majoranas found above. The action of exchanging two vortices i and $i + 1$ generates the following transformation on the Majorana operators:

$$\gamma_i \to \gamma_{i+1}, \tag{9.18}$$

$$\gamma_{i+1} \to -\gamma_i, \tag{9.19}$$

$$\gamma_j \to \gamma_j \quad \text{if } j \notin \{i, i + 1\}. \tag{9.20}$$

This specifies the action of one of the braid group generators $P_{i,i+1}$ (see the text around Eq. 5.20 in Box 5.2 for the definition of these generators) on Majorana operators. Indeed we see that performing this operation twice would give the two minus signs found above. The last step is to show that this action on operators is of the form $\gamma \rightarrow U^\dagger \gamma U$, that is, equivalent to a unitary action $U(P_{i,i+1})$ on states in the Hilbert space, which is intuitively like going from the Heisenberg to the Schrödinger representation. The resulting unitary action is

$$U(P_{i,i+1}) = \exp\left(\frac{\pi}{4}\gamma_{i+1}\gamma_i\right). \tag{9.21}$$

$U(P_{i,i+1})$ is unitary because the product of two Majoranas is anti-Hermitian, not Hermitian. The same representation had previously been derived for quasiholes in the Moore–Read Pfaffian state (Nayak and Wilczek, 1996), so the above calculation demonstrates the close relationship between the non-Abelian statistics of Majoranas in superconductor vortices and those believed to exist in the fractional quantum Hall effect at $\nu = 5/2$. It is easiest to understand Eq. 9.21 by repeating the braiding of these two vortices, which gives

$$U_2 \equiv U(P_{i,i+1})^2 = \exp\left(\frac{\pi}{2}\gamma_{i+1}\gamma_i\right) = \cos(\pi/2) + \sin(\pi/2)\gamma_{i+1}\gamma_i = \gamma_{i+1}\gamma_i. \tag{9.22}$$

We can verify that U_2 has the desired effect on the two Majorana operators:

$$U_2^\dagger \gamma_i U_2 = \gamma_i \gamma_{i+1} \gamma_i \gamma_{i+1} \gamma_i = -\gamma_i, \tag{9.23}$$

$$U_2^\dagger \gamma_{i+1} U_2 = \gamma_i \gamma_{i+1} \gamma_{i+1} \gamma_{i+1} \gamma_i = -\gamma_{i+1}. \tag{9.24}$$

So the single exchange will be given by a square root of U_2 such as in Eq. 9.21. Note that what we have written down as a braid group representation generates the physically correct action on Majoranas, but so would any version with different Abelian phase factors, since those drop out of the unitary transformation of operators.

Now that we have the representation of the braid group generators, we could in principle build up much more complicated unitary actions simply by concatenation. The braiding process in the Moore–Read wavefunction with four quasiholes was computed explicitly by numerical means and shown to generate the predicted unitary operation on the space of ground states (Tserkovnyak and Simon, 2003).

We make two more remarks about the fractional quantum Hall realization of Majoranas and then turn to spin systems. The first is on energetics. A parent Hamiltonian does exist to make the Moore–Read state the exact ground state, but it consists of three-body rather than two-body interactions, so its direct physical relevance is unclear. However, numerical simulations suggest that the Moore–Read state and its particle-hole conjugate are competitive in energy for physical Hamiltonians with two-body interactions with the composite fermion metal and other gapless states at half-integer filling fractions. Small differences, such as whether one is in the lowest Landau level at $\nu = 1/2$ or has the different wavefunctions and pseudopotentials that arise at $\nu = 5/2$, pick out which state is actually realized.

The second remark is about the edges of non-Abelian quantum Hall states, which is another example of the deep connection to conformal field theory. As mentioned above, the $p + ip$ superconductor has a propagating Majorana edge mode, as does the $\nu = 5/2$ state, along with additional Abelian edge modes. The Abelian modes are described by the chiral Luttinger liquid theory from Chapter 6, which resulted from connecting the Abelian Chern–Simons mode in the bulk to a chiral boson at the boundary, which may be the simplest conformal field theory. Non-Abelian Chern–Simons theories also exist, and their edges are described by more complicated conformal field theories (mostly, chiral versions of the Wess–Zumino–Witten models that made a brief appearance in Box 4.2); the single propagating Majorana is the same conformal field theory that describes (the holomorphic sector of) the two-dimensional Ising critical point. Further examples of how bulk, edge, and vortices are connected from the powerful field-theoretic perspective are given in Nayak et al. (2008). For examples of how superconducting devices might reveal non-Abelian statistics, see Beenakker (2013).

9.5.2 The Non-Abelian Phase of the Kitaev Honeycomb Model

The p-wave superconductor also has an incarnation in the field of topological spin liquids. This again starts by a purely formal rewriting of a Hamiltonian in terms of Majorana fermion degrees of freedom, and then finding that these in fact provide a natural description of the emergent, fractionalized low-energy degrees of freedom. The model in which this happens was proposed, again, by Kitaev, and goes under the name of the Kitaev honeycomb model, introduced in Section 7.4, also discussed in the context of disorder and topology (Section 8.8.3). Its exact solution, to which we have appealed repeatedly, is presented in Box 9.4.

Box 9.4   Solution and Phase Diagram of the Kitaev Honeycomb Model

Kitaev's original solution of the eponymous honeycomb proceeds via a mapping to Majorana fermions. Indeed, the bond-directional Ising interactions encoded in the honeycomb model, Eq. 7.31, can be thought of as decomposing the lattice into one-dimensional chains with alternating Ising couplings of the x- and y- spin components. The interchain Ising couplings of the z- components do not interfere with the strings of the Jordan-Wigner transformation (Box 9.3) in the way that a Heisenberg interaction would, so that the solvability is preserved (Feng et al., 2007). Here, we follow the original derivation of Kitaev, see the Les Houches lecture notes (Kitaev and Laumann, 2009), which cover Kitaev's seminal contributions to the field, as in the Kitaev chain of the previous subsection, as well as the toric code in Section 6.3.

In this box, we present the exact solution, and use it to identify the phases of the Kitaev honeycomb model. These include the gapless spin liquid with the characteristic Dirac spectrum; the anisotropic gapped phase; as well as the non-Abelian gapped phase in the absence of time-reversal symmetry. For details of conventions and variable definitions, it is useful to refer back to Figure 7.4.

The crux of the exact solution is to reduce the spin Hamiltonian to a quadratic form, which can then be diagonalized in a standard way. To achieve this, the three Pauli operators are represented by four Majorana fermion species: c_i, known as the matter fermions, and the bond fermions b_i^a with $a = x, y, z$ on every lattice site j:

$$\sigma_i^a = i c_i b_i^a. \tag{9.25}$$

The bond fermions turn out to be nondynamical. They are subsumed into bond operators $u_{ij} = i b_i^a b_j^a$ with i, j labeling nearest-neighbor sites at the ends of bond a. In terms of these, the Hamiltonian of the Kitaev honeycomb model reads

$$H = \sum_{\langle ij \rangle_a} i J_a u_{ij} c_i c_j + i K \sum_{\langle ij \rangle_a, \langle jk \rangle_b} u_{ij} u_{jk} c_i c_k. \tag{9.26}$$

The first term corresponds to the bond-directional nearest-neighbor exchange, while the second term derives from the three-spin interaction introduced perturbatively in Eq. 7.35.

The bond operators are static, with eigenvalues $u_{ij} = \pm 1$. Thus the Hilbert space in which H acts can be decomposed into gauge $|F\rangle$ and matter $|M\rangle$ sectors. Replacing the bond operators by their eigenvalues yields the quadratic matter Majorana fermion Hamiltonian. The reason the three-spin term does not spoil solvability is that (part of) it yields a next-nearest-neighbor hopping for the matter fermions, that is, the resulting term in the Hamiltonian is still quadratic.

The introduction of four Majorana species has doubled the size of the Hilbert space on each site. This implies a redundancy which is in the local $\mathbb{Z}_2$ gauge structure of the problem. Namely, the physical properties (including the spectrum) depend on the configurations $\Phi_\bigcirc$ of $\mathbb{Z}_2$ fluxes W_p (Eq. 7.32) on the plaquettes of the lattice, rather than configuration of bond variables u_{ij} individually. The flux on each hexagon is given by a product of bond variables $W_p = \prod_{\langle ij \rangle \in \bigcirc} u_{ij}$. The physical eigenstates $|\Psi_{\text{phys}}\rangle = P|\Psi\rangle$ are obtained using a projector to the physical subspace. This projector is proportional to $\left[1 + (-1)^{N_\chi} (-1)^{N_f} \right]$, with $N_{\chi/f}$ denote bond/matter fermion number operators, see below. This in particular implies that physical states have fixed overall fermion parity. As there exist a range of observables which can be computed in unprojected states, we do not dwell further on subtleties involving the projection operation, but do caution that the projection may not be entirely innocuous, especially in small systems (Pedrocchi et al., 2011); and that specifically the fermion parity conservation can have qualitative signatures in the dynamical response of the spin liquid.

For a given configuration of bond variables $\{u_{ij}\}$ the Hamiltonian assumes the form

$$H = \frac{i}{2} \begin{pmatrix} c_A & c_B \end{pmatrix} \begin{pmatrix} F & M \\ -M^T & -D \end{pmatrix} \begin{pmatrix} c_A \\ c_B \end{pmatrix}. \tag{9.27}$$

Here, $M_{ij} = u_{\langle ij \rangle_a} J_a$, and c_A/c_B denote for the N-component vectors c_{Ar}/c_{Br} for a lattice with N unit cells. The next-nearest-neighbor matrices F_{ij} and D_{ij} connect sites on the same lattice and thus vanish in the absence of the three-spin term, $K = 0$.

Since the complex diagonalization of a quadratic complex fermion Hamiltonian is relatively straightforward (and much practiced), it is preferable to combine the Majorana fermions into pairs to make up complex fermions. This analogously leads to two complex fermion species for bond and matter fermions

$$\chi^{\dagger}_{\langle ij \rangle_a} = \frac{1}{2}(b_i^a - ib_j^a) \, , \, f_{\mathbf{r}} = \frac{1}{2}(c_{Ar} + ic_{Br}) \, , \tag{9.28}$$

where A, B label the sublattice of the honeycomb lattice. The u_{ij} variables are given by

$$u_{ij} = 2\chi^{\dagger}_{\langle ij \rangle_a} \chi_{\langle ij \rangle_a} - 1. \tag{9.29}$$

The quadratic Hamiltonian in terms of the complex fermions, written as vectors $c_A = f^{\dagger} + f, c_B = i(f^{\dagger} - f)$, takes a Bogoliubov–de Gennes form (Box 8.2), including anomalous terms $f^{\dagger} f^{\dagger}$:

$$H = \frac{1}{2} \begin{pmatrix} f^{\dagger} & f \end{pmatrix} \begin{pmatrix} h & \Delta \\ \Delta^{\dagger} & -h^T \end{pmatrix} \begin{pmatrix} f \\ f^{\dagger} \end{pmatrix}, \tag{9.30}$$

where $h = (M + M^T) + i(F - D)$ and $\Delta = (M^T - M) + i(F + D)$. This can be diagonalized via a transformation T (Blaizot and Ripka, 1986)

$$T \begin{pmatrix} h & \Delta \\ \Delta^{\dagger} & -h^T \end{pmatrix} T^{\dagger} = \begin{pmatrix} E & 0 \\ 0 & -E \end{pmatrix}, \tag{9.31}$$

yielding

$$H = \sum_{n>0} E_n a_n^{\dagger} a_n - \frac{1}{2} \sum_{n>0} E_n, \tag{9.32}$$

where the eigenvalues $E_n \geq 0$, $n = 1 \ldots N$, depend on the flux configuration, $E_n \equiv E_n(\Phi_{\bigcirc})$.

The ground state of the matter fermion Hamiltonian, Eq. (9.32), is the one annihilated by all $a_i = X_{ik}^* f_k + Y_{ik}^* f_k^{\dagger}$. The ground-state energy is therefore $E_{gs} = -\frac{1}{2} \sum_n E_n$.

While the bond variables are static, they have of course not gone away: to determine the global ground state of the spin system, it is necessary to find the flux sector with the minimal ground-state energies $E_{gs}(\Phi_{\bigcirc})$. A convenient theorem by Lieb (1994) states that the ground state in a translationally invariant honeycomb lattice is flux-free, that is, it corresponds to the familiar graphene hopping problem. Thanks to the translational invariance, a Fourier transform $f_{\mathbf{r}} = \frac{1}{\sqrt{N}} \sum_{\mathbf{q} \in \text{BZ}} e^{-i\mathbf{q}\mathbf{r}} f_{\mathbf{q}}$ yields

$$H_0 = \sum_{\mathbf{q} \in \text{BZ}} \begin{pmatrix} f_{\mathbf{q}}^{\dagger} & f_{-\mathbf{q}} \end{pmatrix} \begin{pmatrix} \xi_{\mathbf{q}} & -\Delta_{\mathbf{q}} \\ -\Delta_{\mathbf{q}}^* & -\xi_{\mathbf{q}} \end{pmatrix} \begin{pmatrix} f_{\mathbf{q}} \\ f_{-\mathbf{q}}^{\dagger} \end{pmatrix}. \tag{9.33}$$

In this representation H_0 is equivalent to a BCS Hamiltonian describing a supercon-ductor with a momentum-dependent gap $\Delta_{\mathbf{q}} = -i\,\text{Im}\,s_{\mathbf{q}} - \kappa_{\mathbf{q}}$ (complex for $K \neq 0$), whose quasiparticle dispersion is $\xi_{\mathbf{q}} = \text{Re}\,s_{\mathbf{q}}$, where $s_{\mathbf{q}} = \sum_{i=0,1,2} J_{a_i} e^{i\mathbf{q}\cdot\mathbf{n}_i}$, and $\kappa_{\mathbf{q}} = 4K\,(\sin\mathbf{q}\cdot\mathbf{n}_1 - \sin\mathbf{q}\cdot\mathbf{n}_2 + \sin\mathbf{q}\cdot\mathbf{n}_3)$. Here $a_0 = z, a_1 = x, a_2 = y$, and $\mathbf{n}_0 = (0,0)$, $\mathbf{n}_1 = (1/2, \sqrt{3}/2)$, $\mathbf{n}_2 = (-1/2, \sqrt{3}/2)$, as in Figure 2.1.

Denoting $\Delta_{\mathbf{q}} = |\Delta_{\mathbf{q}}| e^{i\phi_{\mathbf{q}}}$, and under the Bogoliubov transformation

$$\begin{pmatrix} f_{\mathbf{q}} \\ f_{-\mathbf{q}}^{\dagger} \end{pmatrix} = \begin{pmatrix} \cos\theta_{\mathbf{q}} & e^{i\phi_{\mathbf{q}}} \sin\theta_{\mathbf{q}} \\ -e^{-i\phi_{\mathbf{q}}} \sin\theta_{\mathbf{q}} & \cos\theta_{\mathbf{q}} \end{pmatrix} \begin{pmatrix} a_{\mathbf{q}} \\ a_{-\mathbf{q}}^{\dagger} \end{pmatrix}, \tag{9.34}$$

with $\theta_{\mathbf{q}}$ fixed via $\tan 2\theta_{\mathbf{q}} = |\Delta_{\mathbf{q}}|/\xi_{\mathbf{q}}$ the Hamiltonian is diagonalized as

$$H_0 = \sum_{\mathbf{q}} E_{\mathbf{q}}(a_{\mathbf{q}}^{\dagger} a_{\mathbf{q}} - 1/2), \tag{9.35}$$

whose spectrum is given by

$$E_{\mathbf{q}} = 2\sqrt{\xi_{\mathbf{q}}^2 + |\Delta_{\mathbf{q}}|^2}. \tag{9.36}$$

For $K = 0$ the spectrum $E_{\mathbf{q}} = 2|s_{\mathbf{q}}|$ of matter fermions is gapless if $|J_z| < |J_x| + |J_y|$ (and permutations). At the isotropic point $J_x = J_y = J_z$ there are two Dirac cones positioned at $\mathbf{K} = \pm(2\pi/3, -2\pi/3)$ with a linear energy spectrum $E(\mathbf{q}) \propto |\mathbf{q}|$ at small energies, as in graphene (Section 2.5) (Figure 9.1). In the presence of exchange anisotropy the Dirac cones move in the Brillouin zone, and merge at the transition line

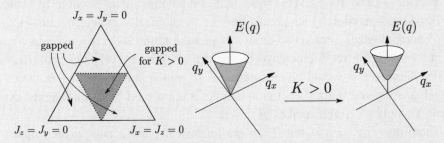

Fig. 9.1 (left) Phase diagram of Kitaev model as function of couplings anisotropy and strength of three-spin term, K_3. (middle) The Dirac cone familiar from graphene acquires a gap (right) in the presence of the three-spin term K_3. From Knolle et al. (2015). Reprinted with permission by Springer International Publishing.

(between the gapped and gapless QSLs), so that for $|J_z| > |J_x| + |J_y|$ the spectrum is gapped. The phase diagram of the Kitaev model through the cut in the parameter space defined by $J_x + J_y + J_z = 1$ is shown in Figure 9.1.

The Dirac cones of the gapless phase (shown in grey) acquire a gap for nonzero K_3 (see Figure 9.1, middle panel). The spectrum remains gapless only along the dashed lines in Figure 9.1, right panel, with quadratic band touching at zero energy. The outer triangles of the phase diagram correspond to gapped Abelian QSLs whose fermionic bands are characterized by a zero Chern number. The formerly gapless phase around the isotropic point possesses non-Abelian excitations like those of the $p + ip$ superconductor discussed above.

The upshot (Eq. 9.33) is that the spin model takes on the form of a p-wave superconductor. Its complex order parameter can be obtained by expanding the gap function $\Delta_{\mathbf{q}}$ near the Dirac points. With this as a starting point, one can then investigate the properties of vortices along the same lines as was done for the Pfaffian quantum Hall state in the previous section: the vortices carry Majorana zero modes, and therefore, the Kitaev honeycomb model with the three-spin term is a non-Abelian spin liquid. The question whether and how such Majorana degrees of freedom in a magnetic insulator can be manipulated, for example, for the purposes of implementing braiding operations, is an open one. We will return to the experimental search for this phase at the end of this chapter.

9.6 Universal Computation and the Read–Rezayi States

The Majorana systems lack one desirable feature outlined above – the matrices which represent their braiding operations are not in fact rich enough in themselves to provide a basis for universal quantum computation. It is possible to achieve universal computation by supplementing the topologically protected gate set with two nontopological gates, a one-qubit $\pi/8$ phase rotation and a two-qubit measurement, and the error requirements on these extra gates are not very stringent (Bravyi, 2006). However, it would be advantageous in principle to make use of states in which any desired unitary operation can be implemented, with sufficiently small error, via only protected operations.

There do exist such states. They are intrinsically more complicated than the Majorana-supporting states that have been the focus on this chapter, because unlike the Majorana states they do not have a representation in terms of quadratic Hamiltonians of Bogoliubov–de Gennes type. Experimentally, there may exist one quantum Hall state with non-Abelian excitations which does not suffer this shortcoming. This state is known as the Read–Rezayi (parafermion) state (Read and Rezayi,

1999), which has been proposed for filling fraction $\nu = 12/5$, but it is yet unclear how to create a robust plateau at that fraction or whether the Read–Rezayi state is energetically favorable over an Abelian hierarchy state also possible at that fraction.

9.7 Experimental Implementations of Majorana Modes

In two dimensions, multiple vortices can in principle be created concurrently, thereby allowing storage and manipulation of an increasing amount of quantum information. The most efficient use of resources is then determined by considerations such as maximizing the total amount of information capacity versus its robustness. For instance the full Hilbert space for $2n$ vortices has size 2^n, but typical states in that Hilbert space are so highly entangled that processing information becomes very delicate. It can therefore be advantageous to restrict the space under consideration to subspaces exhibiting simpler structures. If the qubit is encoded in the operators $ic_{2j-1}c_{2j} = \pm 1$ then it may be advantageous to insist that four consecutive Majorana modes always have even parity, $i^2 c_{2j-3}c_{2j-2}c_{2j-1}c_{2j} = 1$, so that the total fermion parity of the system does not imply entanglement between all the individual Majorana modes.

Much effort has been expended on advancing experimental realizations of the above states, mostly focusing on the case of Majorana modes which are within closest reach. In fact, each of these cases has seen encouraging, yet at present not fully conclusive, progress. One of the most active in recent years has been the search for the Majorana zero modes in quantum wires. This was partly enabled by the realization that a semiconductor wire with sufficiently strong spin-orbit coupling, and in a magnetic field, can be driven into the topological phase by the proximity effect from a conventional *s*-wave superconductor (Oreg et al., 2010; Sau et al., 2010): when there is only one sheet of the Fermi surface, similar to what was shown in Figure 3.9, the induced superconductivity in the wire is effectively *p*-wave.

The search for Majoranas in quantum wires was also encouraged by a theoretical suggestion that braiding operations would in fact be possible for such a system, even though particle exchange seems rather unnatural for a wire geometry (Alicea et al., 2011). The idea is to construct *junctions* between wires, so that one zero mode can be moved aside for the other to move past (Figure 9.2). For this, it is required to arrange for a domain wall between a topological and nontopological superconductor. Such a domain wall will host a Majorana zero mode, and if it can be constructed to be tunably mobile, so will be the location of the Majorana zero mode.

While a direct observation of quasiparticle braiding is at present beyond reach, it has been argued that a fingerprint of the Majorana zero mode should be the

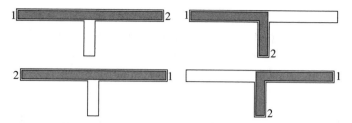

Fig. 9.2 A T-junction to allow braiding of Majorana zero modes located at the endpoints. By changing gates, the endpoints (labeled 1,2) of the superconducting wire (shaded) are interchanged by the sequence of operations to be executed in clockwise sequence.

possibility of tunneling into the system at zero bias, that is, voltage difference between the wire and the external electron reservoir. This is a form of Andreev reflection: the superconductor accepts the electron, and reflects a hole into the external reservoir; it can do this by using the condensate of Cooper pairs as a reservoir of charge pairs. Such a zero-bias signal has indeed been observed, but questions remain regarding the uniqueness of that interpretation, and also regarding a possible inconsistency with the size of the finite-size splitting (such as $\delta_{B/AB}$ mentioned above) which for the experimental geometries in question is not negligible.

Moving on to quantum Hall physics at $\nu = 5/2$, here experiment is similarly encouraging. Among over 100 plateaux determined in the Hall conductance in quantum Hall samples, the one at $\nu = 5/2$ has long stood out as being the only one with an even denominator. Much effort has gone into trying to pin down its nature, and the Moore–Read state has been a prominent contender for a long time. Direct tunneling of an electron into the Moore–Read state is not as simple as in the case of the superconducting wires: the quasiparticles in the Moore–Read state have charge $e/4$, so that four of them are involved when an electron is transferred from the external reservoir.

The exceptionally large quantum coherence lengths in quantum Hall systems offer a separate avenue for demonstrating the existence of Majorana excitations. Indeed, perhaps the strongest experimental evidence in favor of the Moore–Read state comes from an interferometer experiment. The basic idea (see Figure 9.3) is to interfere quasiparticles traveling along the edges of a sample which have taken different paths (Bonderson et al., 2006). These alternative paths arise as constrictions induce nonzero tunneling between the edges in a controlled way.

The relative phase between different paths then depends not only on the flux threading the area enclosed by the edges and the two constrictions, that is, the standard Aharonov–Bohm phase. In addition, the phase is sensitive to the number

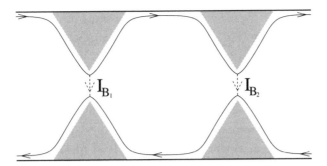

Fig. 9.3 Sketch of quantum Hall interferometer. Current is injected at the source (top left) and extracted at the drain (bottom left). A quasiparticle can tunnel between top and bottom edge at two constrictions, B_1 and B_2. The quasiparticle having taken the longer path will have braided with any quasiparticle present in the area enclosed by the two paths and the two edges. The number of such quasiparticles can be changed by tuning the external magnetic field or an electrostatic potential. The interference signal between the two paths thus provides information about the nature of the braiding of the quasiparticles.

of vortices present in that area of the system: the longer path, labeled by I_{B_2}, effectively includes a braiding operation involving each vortex located in that area.

While it is hard to have detailed knowledge about all the quasiparticles in the bulk, some features are nonetheless robust. A particularly striking such feature of the interference pattern is that it is sensitive to the parity of the number of vortices in the enclosed region. Indeed, when the gate voltage which determines the charge, and hence the number of vortices, in the enclosed region, is varied, one finds an alternation of signals with and without interference; these are approximately equal in width, signaling the successive addition of charge.

The origin of the sensitivity of the interference signal to the enclosed fermion parity can be seen when comparing the paths of a vortex on the edge. Let us label the Majorana mode on the edge by γ_0, and the ones in the bulk by γ_i, with $i = 1 \ldots n_b$. As the edge Majorana encircles a bulk one, it effects unitary transformations of the type discussed around Eqs. 9.18–9.24. When the number of bulk vortices, n_b, is even, the resulting unitary includes the factor $\gamma_0^{n_b} = 1$, so that it does not depend on the state of the edge mode at all. By contrast, for odd n_b, the unitary does depend on γ_0 of the edge vortex, so that the signal from successively passing vortices does not add coherently, thereby removing the interference signal.

Perhaps a more physical line of argument is the following. If there is an odd number of vortices in the bulk, there must be a further one outside, say to the right of the paths. Now, consider the state of a pair formed by this outside vortex and the last unpaired one inside. A quasiparticle tunneling through the left constriction does not change that state whereas the one tunneling through the right constriction does.

The two resulting states for the particles to have passed through the different constrictions are hence orthogonal, so that no interference between them takes place.

Finally, the situation in Kitaev's honeycomb model is even harder to determine directly – while non-Abelian defects may very well be present in a system, they are hard to move around in a controlled fashion. Unlike in the superconductors, where they were mobile by dint of being tied to a domain wall, or by being a charged excitation in a system of mobile electrons, Kitaev's honeycomb model describes an insulating spin system where the only conserved quantity that is reasonably straightforwardly transported is energy. However, phonons are also involved in heat transport, and disentangling the contribution of a possibly topological spin system is a hard task.

One promising experiment has been done on the quasi–two-dimensional compound $\alpha-\mathrm{RuCl}_3$. As mentioned in Section 7.4, first principles calculations and quantum chemical arguments suggest this to be a system in which the Kitaev honeycomb model is approximately realized. In an applied magnetic field, experiments on the thermal Hall transport coefficient, κ_{xy}, were carried out. This quantity measures the ratio of the energy current to a temperature gradient applied in a perpendicular direction.

For a gapped bulk, the heat current for infinitesimal temperature changes is carried by the gapless edge. Like in the case of the quantized electrical conductivity of the quantum Hall effect, one also finds a thermal conductivity in terms of only fundamental constants. However, while there is a quantum of charge, e, there is no such quantum of heat, and therefore, it turns out that it is κ_{xy}/T which depends only on the fundamental constants k_B and $\hbar$. A precision measurement of the appropriate combination, $k_B^2/\hbar$, would therefore require an equally precise measurement of T, which renders this quantity less useful from a metrological perspective.

A simple model for determining the prefactor linking $\kappa_{xy}/T \propto k_B^2/\hbar$ is to consider a fermionic edge channel with linear dispersion given by $\omega(q) = vq$, where $\omega(q)$ is the energy of an excitation with wavevector q, and v is the quasiparticle velocity along the edge. The temperature dependent energy flux, $\Phi_\varepsilon(T) = v\varepsilon(T)$ is then given by the dependence of the energy density ε on T:

$$\Phi_\varepsilon(T) = v \int_0^\infty \frac{E\,dE}{hv} \frac{1}{1+\exp[E/(k_BT)]} = \frac{k_B^2}{\hbar} \frac{\pi T^2}{24} . \tag{9.37}$$

The integral runs over positive energies only, as appropriate for a Majorana mode, and $1/(hv)$ is the density of states in $d = 1$. This gives

$$\frac{\kappa_{xy}}{T} = \frac{1}{T} \frac{d\Phi_\varepsilon(T)}{dT} = \frac{1}{2} \frac{\pi k_B^2}{6\hbar} . \tag{9.38}$$

Note that the nonuniversal velocity v has dropped out of the final expression. The reason for separating out the factor of $1/2$ is to indicate that the Majorana result is half of what would have been obtained for a complex Fermion edge state.

The value obtained in one experiment on $RuCl_3$ in a magnetic field is indeed close to this value (Kasahara et al., 2018). Attempts to replicate those experimental results are under way, with the main uncertainty on the theory side being the question what is the appropriate Hamiltonian to model the magnetic compound. At any rate, this striking result could imply that $RuCl_3$ in a field harbors non-Abelian quasiparticles.

While all of these experiments are very encouraging, at the time of writing this book, much work remains to be done by theorists and experimentalists alike to advance toward an unambiguous identification of non-Abelian anyons, and their utilization for topological quantum computing.

10

Topology out of Equilibrium

When we say we consider the ground states of closed quantum systems, or their equilibrium finite-temperature properties, this implies considerable restrictions on the kind of behavior we can expect to encounter. For instance, the ground state at zero temperature, being an eigenstate, is stationary under the evolution generated by the system Hamiltonian, itself taken to be time-independent; while at nonzero temperature, considerations of entropy maximization proscribe certain types of ordering, as exemplified by the impossibility, encoded in the Mermin–Wagner theorem, of breaking continuous symmetries in $d = 2$.

While there is, more or less, one type of thermal equilibrium, many nonequilibrium settings are possible. For example, by considering an initial state which is not an eigenstate of the system Hamiltonian, one obtains a time dependent problem even for a time-independent Hamiltonian; this has become to be known as a quantum quench. If the Hamiltonian itself is time dependent, one refers to a driven system.

For the finite-temperature case, the implicit assumption is that a concept of temperature is defined in the first place, and that the system thermalizes, that is, reaches an effectively time-independent (as far as local observables are concerned) steady state, the properties of which depend only on a small number of parameters, such as the energy density of the initial state. However, there are cases where these assumptions are not satisfied; these include glasses and localized (see Chapter 8) systems. In particular many-body localization, introduced below, furnishes a generic route to nonthermalization.

In the following, we cover material which touches on each of the above items. This chapter will necessitate a fair amount of background material, which we provide as we go along. This includes sections on thermalization and the lack thereof; the description of periodically driven (Floquet) systems, as well as Floquet engineering; and the possibility of defining phase structure out of equilibrium via the notion of eigenstate order.

The nonequilibrium behavior of matter is such a broad, complex and rich field – as the reader can verify by taking a quick look out of the window – that it is not yet possible to write anything approaching as comprehensive an account as can be done in the equilibrium setting: at present, we quite simply even lack a similarly systematic framework. This statement holds true both for topological aspects and more broadly. However, a focus on periodic driving creates a useful structure for nonequilibrium systems loosely analogous to how assuming perfect crystals simplifies equilibrium topology via Bloch's theorem. Nonetheless, despite the formal similarity between Floquet and Bloch structures, there are numerous differences in how they apply in realistic physical systems, as we start to see in the following section.

10.1 Time-Dependent and Time-Periodic (Floquet) Hamiltonians

The aim of this section is to follow an avenue in which a relatively gentle deviation from the equilibrium setting has proven still to be tractable while yielding qualitatively new phenomena. We first provide a compendium of simple facts about unitary time evolution generated by time dependent Hamiltonians. This will serve as a springboard for the special case of when this time dependence is periodic. Such systems are known as Floquet systems, and we will find that they host a number of interesting new topological phenomena which are beyond the reach of static Hamiltonians.

The formal solution of the time-dependent Schrödinger equation

$$i\hbar\frac{d|\psi\rangle}{dt} = H(t)|\psi\rangle \tag{10.1}$$

is given by the unitary time evolution operator

$$\mathcal{U}(t, t_0) = \mathcal{T}\exp\left[-\frac{i}{\hbar}\int_{t_0}^{t}dt\, H(t)\right]. \tag{10.2}$$

This can in turn be used to implicitly define an *effective Hamiltonian*,

$$\exp\left[-\frac{i}{\hbar}(t - t_0)H_{\text{eff}}\right] := \mathcal{U}(t, t_0). \tag{10.3}$$

At this stage, this is simply a formal definition, and we will in particular have to return to the issue of the non–single-valuedness of the logarithm. However, the usefulness of this is that it indicates the possibility of transferring a lot of the intuition we have from regular Hamiltonians. In particular, it is guaranteed that there is a complete set of orthonormal states for the Hilbert space of the system, ensuring that any state $|\psi(t)\rangle$ can be written as a linear combination of eigenstates of the time-evolution operator.

However, at this stage, it is then natural to pose the question what this setting can possibly achieve that cannot also be achieved by, say, a time-independent Hamiltonian H_{eff}.

The answer is that a Hamiltonian thus defined does not inherit all physically important properties from the instantaneous Hamiltonians $H(t)$ on which it depends. Perhaps most importantly, H_{eff} need not be, and will in general not be, local, even if $H(t)$ is local for all times t. (Loosely speaking, local means involving only combinations of operators which are nearby in real space, like exchange interactions between spins at most a few lattice spacings apart.) Since several properties of many-body systems which we take for granted actually depend on the locality of Hamiltonians, systems described by a nonlocal H_{eff} can thus exhibit unexpected (and novel) behavior. The time crystal described in detail below is a case in point.

10.2 Floquet Basics

In the case of periodically driven systems, $H(t+T) \equiv H(t)$, the effective Hamiltonian is generally referred to as the Floquet Hamiltonian, H_F, and $\mathcal{U}_F = \mathcal{U}(0, T) = \exp\left[-\frac{i}{\hbar} T H_{\text{F}}\right]$ is the Floquet unitary.[1]

10.2.1 Discrete Time Translation Symmetry

Floquet systems have considerable additional structure compared to the general time-dependent case. While both have discarded the invariance of the static case with respect to infinitesimal time translations, the Floquet problem retains a symmetry with respect to *discrete* time translations. This leads to energy conservation, which follows from Noether's theorem, to be replaced by quasi-energy conservation: quasi-energies are only defined modulo $\hbar\Omega = 2\pi\hbar/T$.

This is completely analogous to the vestigial conservation of crystal momentum in a periodic potential, as opposed to momentum conservation in the continuum. There, physically distinct momenta of a Bravais lattice are restricted to a Brioullin zone. For the case of a chain of lattice constant a, the allowed crystal momenta thus lie in the interval $[-\pi/a, \pi/a)$, with momenta differing by a reciprocal lattice vector $2\pi/a$ being equivalent. For a periodically driven system, quasienergies are similarly restricted to lie in a Floquet Brioullin zone ranging from $-\hbar\pi/T$ to $\hbar\pi/T$. The multi-valuedness of the logarithm in the above definition of the effective Hamiltonian is related to the choice of Brioullin zone, with choices differing by addition of a multiple of the "reciprocal lattice vector" $2\hbar\pi/T$.

---

[1] One is free to fix the "gauge choice" of the time of the "beginning" of the period, t_0, a point which will not be important in what follows.

10.2.2 Floquet Ensembles

In a nonequilibrium setting, the familiar constraints imposed by equilibrium thermodynamics need to be rethought: concepts like thermodynamic potentials or temperature need no longer be useful or even exist. In Box. 10.1, we give a brief summary of the issues involved in order to make this treatment self-contained, but which fails to do full justice to the rich and interesting field of nonequilibrium quantum dynamics.

Let us first address the issue of temperature, whose existence is related to energy conservation, in the same way that a chemical potential is defined only in the presence of particle number conservation. By its very nature, this item has been abandoned in the Floquet setting, as quanta of energy $2\pi\hbar/T$ can be added or subtracted, and hence there simply is no concept of temperature. Staying within the framework of thermodynamics, what we now need to do is maximize entropy without this constraint – which effectively means giving each state the same weight. This is also known colloquially as an "infinite temperature ensemble," as the concomitant Boltzmann factors $\exp[-E/(k_B T)]$ also become state-independent when one sets $T = \infty$.

This turns out to be the generic setting for Floquet systems. It is known simply as Floquet-ETH, in analogy to the static systems obeying eigenstate thermalization. In the Floquet case, there is obviously nothing to talk about in terms of nontrivial correlations in the longtime limit.

However, energy conservation need not be the only constraint present in a system. There can also be particle number conservation, global symmetries like a U(1) spin symmetry or, in an integrable system, any number of other constants of motion. The most straightforward way of obtaining an integrable system in the Floquet setting is to consider a set of free fermions, subject to a periodic drive, but which remain noninteracting. This setting allows importing ideas from the corresponding static setting, in particular the notion of a generalized Gibbs ensemble (see Section 10.2.3). This programme can be carried out entirely analogously for the Floquet case, where it has been christened the (Floquet-) periodic Gibbs ensemble, Floquet-PGE.

A more involved way of avoiding equilibration to an infinite temperature Floquet-ETH ensemble involves avoiding the process of equilibration altogether. This can be achieved by adding disorder to the system, so that the system ceases to be ergodic. It turns out that this can be arranged for in a way which is generic, that is to say which – unlike the integrable case mentioned above – is stable to any small perturbation of the Hamiltonian. The underlying phenomenon is known as many-body localization (Nandkishore and Huse, 2015). We will discuss this option after covering the integrable cases in the next sections.

Box 10.1 Phase Structure in and out of Equilibrium

The aim of this section is to explain how one can generalize the notions of phases, and transitions between them, beyond the familiar setting of equilibrium thermodynamics. To do this, we first need to illuminate the connection between quantum many-body physics and equilibrium thermodynamics, where the notion of eigenstate thermalization plays a central role. We then explain how the phenomenon of many-body localization (MBL) presents an alternative to thermalization. And finally, how MBL allows for the identification of nonequilibrium phases in a crisp way.

10.2.3 Equilibration and Thermalization – and Absence Thereof

Basic thermodynamics is built on the twin concepts of equilibration and thermalization. The former states that a system, left to its own devices, will eventually reach a time-independent steady state. The latter implies that this steady state is determined by only a small number of parameters – such as conserved quantities like energy or particle density (or temperature/chemical potential, depending on the choice of ensemble).

Eigenstate Thermalization

This is a far cry from the microscopic picture of quantum mechanics embodied by the Schrödinger equation, which is a fully microscopic theory in which a general wave-function is determined by the amplitudes of the different basis states it contains. For a lattice system of N spins-1/2, there are exponentially many, 2^N, of these. This is unimaginably far more information than is encoded in the thermodynamic description. Also, it is in fact impossible to prepare an exact generic eigenstate at a finite energy density above the ground state, since the adjacent levels are only an energy of order $O(2^{-N})$ away, so that Heisenberg's uncertainty principle states it would take a time $O(2^N)$ to prepare them. For a macroscopic N of a thermodynamically large sample, this time is beyond our lifetimes, if not that of the universe.

The resolution is provided by the eigenstate thermalization hypothesis (ETH), which essentially states that local observables and correlators in generic eigenstates take on the values characteristic of thermodynamic equilibrium at the energy/particle density of the eigenstate under consideration. Then, it is no longer necessary to know all the basis state amplitudes; nor indeed is it necessary to have an exact eigenstate: a combination of quantum states with the same energy/particle densities will do.

Generalized Gibbs Ensemble

It is possible to increase the number of conserved quantities yet further. There may for example also be symmetries leading to the conservation of spin or momentum density. Indeed, in integrable systems, there may be an extensive number of conserved quantities.[a] Each of these then enforces its own constraint. This leads to what is known as a generalized Gibbs ensemble (GGE): each conserved quantity leads to a "Lagrange multiplier" in the way that energy/particle density lead to the notions of temperature and chemical potential. These integrable systems are fine-tuned, in the sense that generic perturbations typically destroy the supernumerary conservation laws, and collapse the GGE to a standard thermodynamic Gibbs ensemble. The counterpart of the generalized Gibbs ensemble in Floquet systems with additional conservation laws is known as Floquet periodic Gibbs ensemble (PGE).

10.2.4 Eigenstate Thermalization, Phase Transitions, and order

The notion of phases and phase transitions then transfers neatly from thermodynamics to eigenstates. Consider a setting with a disordered high-temperature state and a low-temperature ordered one, such as in a transverse field Ising ferromagnet in dimension $d \geq 2$ (Eq. 10.4 with couplings independent of j). In that case, generic eigenstates at high energy density do not exhibit long-range ferromagnetic order, while those at low energy do. The critical energy corresponds to the energy density of the system at the temperature of the thermodynamic phase transition.

This allows to make a connection to the quantum quenches mentioned in the introduction. Consider evolving an initial state, $|\psi_0\rangle$, with such a Hamiltonian (generally not an eigenstate thereof). Whether expectation values at long times will exhibit long-range order or not then depends only on the energy density of the initial state, that is, the expectation value $\langle\psi_0|H|\psi_0\rangle$.

10.2.5 Many-Body Localization

We next present MBL as an alternative to thermalization in a generic many-body system; that is to say, unlike Anderson localization (discussed in Box 8.1), MBL does not require any specific fine-tuning but is stable to arbitrary perturbations. Put differently, MBL is the fully interacting version of localization, while Anderson localization describes a noninteracting single-particle phenomenon.

Having said this, there is a cartoon limit of many-body localization which is extremely transparent. We illustrate this using a disordered transverse field Ising model in $d = 1$:

$$- H_{\text{TFIM}} = \sum_{j=1}^{N} J_j \sigma_j^z \sigma_{j+1}^z + \sum_{j=1}^{N} \Gamma_j \sigma_j^x \,, \tag{10.4}$$

where exchange couplings J_j and fields h_j are random variables, and we use periodic boundary conditions $\sigma_1 = \sigma_{N+1}$.

The case with $J_j \equiv 0$ is very simple–each spin aligns with its local field Γ_j, and the eigenstates are the "classical" configurations of the spins along the field axis. In this case, states nearby in energy clearly need not look similar: if one finds a large set of sites λ with $\sum_{j\in\lambda} \Gamma_i \approx 0$, one can flip this entire set and end up with a very different, but near-degenerate, state: eigenstate thermalization is manifestly violated.

The amazing feature of MBL, despite all its subtleties, is that this picture is a good starting point for the description of the generic situation. This idea is captured by the idea of so-called l-bits, which states that a *local* change of variables turns an MBL-Hamiltonian into an entirely classical-looking *local* one. That is to say, one can define a set of l-bits, τ_j^z, to diagonalize the Hamiltonian as follows:

$$\tau_j^z = \sum_{k,\alpha} A_{j,k}^\alpha \sigma_k^\alpha + \sum_{kl,\alpha\beta} A_{j,kl}^{\alpha\beta} \sigma_k^\alpha \sigma_l^\beta + \sum_{klm,\alpha\beta\gamma} A_{j,klm}^{\alpha\beta\gamma} \sigma_k^\alpha \sigma_l^\beta \sigma_m^\gamma + \dots \quad (10.5)$$

$$H = \sum_j h_j \tau_j^z + \sum_{jk} h_{jk} \tau_j^z \tau_k^z + \sum_{jkl} h_{jkl} \tau_j^z \tau_k^z \tau_l^z + \dots \quad (10.6)$$

Crucially, the coefficients A and h in these expressions vanish rapidly unless their indices refer to sites nearby to the reference site j. The above cartoon essentially consists of keeping only the first term in each line (and more formally corresponds to a noninteracting limit in which the change of basis between physical and l-bits is linear).

10.2.6 Eigenstate (or Eigenspectrum) Order

The Hamiltonian in Eq. 10.6 being local and classical (i.e., consisting of fully mutually commuting terms), the spectrum does not exhibit level repulsion like in random matrix theory. Therefore, unlike the case of a system obeying eigenstate thermalization, the level statistics will be Poissonian.

There can, in addition, be further correlations between energy levels, which signal the existence of different phases even in this nonequilibrium setting. This possibility of defining crisply separated phases outside of thermodynamic equilibrium is all the more remarkable for the twin facts that (i) it can exist in settings where the related disorder-free Hamiltonian obeying ETH does not support nontrivial order and (ii) new types of order appear, without a counterpart in equilibrium systems. The latter is the subject of Section 10.6, on time crystals, and the former the subject of the following discussion.

The above cartoon argument could alternatively been made setting not the $J_j \equiv 0$, but the $\Gamma_j \equiv 0$. Not much would have changed. On the surface, the natural basis choice would have been the classical one for the exchange, and more deeply, the Ising symmetry would then result in *pairs* of degenerate states, related by a global Ising spin flip.

It turns out that this feature also persists beyond the cartoon limit: the full many-body states remain paired into quasidegenerate doublets. Such a degeneracy in turn implies that in the Hamiltonian, Eq. 10.6, the "Ising-odd" terms, namely, those with an odd number of τ^zs must vanish: $h_j = h_{jkl} = 0$, while $h_{jk} \neq 0$.

Note that a spectral degeneracy has already appeared in Section 5.1.4, where a three-fold ground-state degeneracy even in the absence of a local order parameter acted as a diagnostic of topological order in the fractional quantum Hall effect. The present notion of eigenspectrum order works to diagnose both topological and local types of order. It is in fact used as a standard diagnostic for the breaking of discrete symmetries in exact diagonalization studies of finite-size lattice systems. In this setting, the ground states are Schrödinger cat states. For the transverse field Ising ferromagnets, say, these are not the oppositely magnetized ordered "up" and "down" states, but rather their symmetric and antisymmetric combinations,

$$|\pm\rangle = \frac{1}{\sqrt{2}} \left[|\text{up}\rangle \pm |\text{down}\rangle \right] . \tag{10.7}$$

This mixing results from the possibility of sweeping a domain wall across the system to connect the two states; as a state with a domain wall has a nonzero activation energy, this requires a virtual process of $O(N)$ steps, so that the resulting splitting is exponentially small in N, and vanishes in the thermodynamic limit $N \to \infty$.

Conventionally, this degeneracy is present in the ground state only. Excited states, by contrast, tend to form bands. For instance, in the Ising chain, all states with a single domain wall are degenerate as far as the exchange is concerned, while the transverse field allows the domain wall to move by flipping a spin adjacent to it:

$$\langle \uparrow\uparrow\uparrow\downarrow\downarrow | (-\Gamma\sigma^x) | \uparrow\uparrow\downarrow\downarrow\downarrow \rangle = -\Gamma . \tag{10.8}$$

This thus gives rise to a hopping problem with dispersion $-2\Gamma \cos(k a)$ with $k = 2\pi n/N$ so that so that the resulting splitting is only algebraically small in N.

However, for nonuniform values of J_j, the disorder in J_j localizes the domain walls. This happens because the energy of a domain wall depends on its location: the process depicted in Eq. 10.8 no longer connects states at the same energy. Therefore, even excited states containing such domain walls need not mix to yield the momentum eigenstates labeled by k that appear in the dispersive band for the clean system.

As a result, eigenspectrum order indicating the breaking of an Ising symmetry can be present even at finite energy densities above the ground state, where ergodic systems in low dimension cannot support such order. Figure 10.3 illustrates the notion of eigenspectrum order for the case of periodically driven Ising chains.

The use of eigenspectrum order at finite energy densities as an ordering diagnostic is limited by the fact that the energy level spacing itself is exponentially small, $O(2^{-N})$, so that an exponentially nearby second cat state can get lost in the sea of other nearby states. However, both cat states exhibit the same long-range order in their spin correlations. In disordered magnets, such order is not measured by a straightforward Fourier component of the magnetization (such as the one at $k = 0$ for a ferromagnet and at $k = \pi$ for an antiferromagnet), but by an Edwards–Anderson order parameter, Eq. 10.18.

---

[a] The precise definition of what constitutes a relevant conserved quantity is not entirely settled. Indeed, the modulus of the projection, $|a_i|^2$, of an arbitrary wavefunction, $|\psi(t)\rangle = \sum_i a_i(t)|\phi_i\rangle$ on any given eigenstate, $|\phi_i\rangle$, of the time evolution operator, is time-independent by the very definition of an eigenstate. The number of such constants of motion equals the size of the Hilbert space, and is hence exponentially large in system size for, say, a spin system on a lattice. However, a projection onto an individual eigenstate, $|\phi_i\rangle\langle\phi_i|$, is in general a highly nonlocal operator and thus not relevant for consideration as an observable in the conventional sense.

10.3 Floquet Topological Insulators

We start off with a discussion which is quite analogous to the study of topological band structures presented earlier (see, e.g., Section 3.2). We are therefore interested in the properties of the spectrum of the single particle states of a noninteracting Floquet Hamiltonian. As this treats single-particle states as essentially independent, it manifestly falls under the heading of noninteracting "integrable" systems.

The central result is the following: by subjecting a nontopological static band structure to a periodic modulation with zero average, Oka and Aoki (2009) showed that one can obtain a topological Floquet band structure, the first instance of what is now known as a Floquet topological insulator (Lindner et al., 2011).

10.3.1 Floquet Engineering

This is an instance of an application of a general set of ideas which go under the heading of Floquet engineering (Oka and Kitamura, 2019; Rudner and Lindner, 2020). This appeals to a separation of timescales between a fast driving and a much longer timescale on which the response of the system is probed. Historically, the phenomenon of dynamic localization was a first application of this kind of Floquet engineering (Dunlap and Kenkre, 1986). This was concerned with a particle hopping on a one-dimensional lattice, subject to a uniform electric field applied at frequency Ω. In the static case, $\Omega = 0$, localization arises via Bloch oscillations. For sinusoidal driving of the potential difference between adjacent sites, $V(t) = V \sin(\Omega t)$, the particle also executes a micromotion but for fine-tuned values of V/Ω, the roots of the Bessel function $J_0(V/\Omega) = 0$, it unfailingly returns periodically to its original location: when observed on a timescale much larger than the driving period, the particle appears localized, no matter how strong its bare hopping matrix element.

The basic ingredient for analyzing such high-frequency Floquet engineering is a perturbative expansion which is controlled in the smallness of the driving period, $T \sim 1/\Omega$. This so-called Magnus expansion for the Floquet Hamiltonian, $H_F = \sum_i H_i$, consists of a sequence of nested commutators of depth i, the first two terms of which are simple:

$$H_0 = \frac{1}{T} \int_0^T dt \, H(t) \tag{10.9}$$

$$H_1 = \frac{1}{2} \left(\frac{1}{T} \right)^2 \int_0^T dt \int_0^t dt_1 \, [H(t), H(t_1)] . \tag{10.10}$$

H_0 is just the average Hamiltonian, and H_1 encodes to what degree the instantaneous Hamiltonians at different times fail to commute.

The generation of a topological band structure was first demonstrated for the case of graphene subjected to a time-varying field (Oka and Aoki, 2009). The simplest way to derive this result is to consider the following cartoon of a circularly polarized light-field (Rudner and Lindner, 2020): the drive period T is subdivided into four portions of equal duration, during which the field successively points in the x-, y-, $-x$-, and $-y$-directions, that is, it rotates by $90°$ at each step. This can be encoded by a vector potential $\mathbf{A}_0^{(n)}$, $n = 1 \ldots 4$, which points along these directions during the corresponding parts of the drive. The Floquet unitary over a full period for the mode at momentum $\mathbf{k}$ is then given by

$$\mathcal{U}_F(\mathbf{k}) = \mathcal{U}^{(4)}(\mathbf{k})\mathcal{U}^{(3)}(\mathbf{k})\mathcal{U}^{(2)}(\mathbf{k})\mathcal{U}^{(1)}(\mathbf{k}), \tag{10.11}$$

$$\mathcal{U}^{(n)}(\mathbf{k}) = \exp\left(\frac{i H^{(n)}(\mathbf{k})T}{4\hbar}\right), \tag{10.12}$$

$$H^{(n)} = v_F\left(\hbar\mathbf{k} - e\mathbf{A}_0^{(n)}\right) \cdot \boldsymbol{\sigma}, \tag{10.13}$$

where the Pauli matrix refers to the graphene sublattice as in Section 2.5.

At the Dirac point, the system is gapless in the absence of driving. As the fast drive is switched on, this ceases to be the case, as can be seen by expanding the above equation to second order in the small parameter $\kappa = ev_F|\mathbf{A}_0|T/(4\hbar)$ to obtain a mass term:

$$\mathcal{U}_F(\mathbf{k} = 0) = \exp(-i\kappa\sigma_y)\exp(-i\kappa\sigma_x)\exp(i\kappa\sigma_y)\exp(i\kappa\sigma_y) \tag{10.14}$$

$$\approx \mathbf{1} + \kappa^2[\sigma_x, \sigma_y] \approx \exp(2i\kappa^2\sigma_z). \tag{10.15}$$

This amounts to a Floquet engineered Hamiltonian

$$H_F(\mathbf{k} = 0) = \frac{2\kappa^2\hbar}{T}\sigma_z. \tag{10.16}$$

The field-induced mass term/gap thus scales with the intensity of the periodic electric field, and vanishes with the driving period: $\kappa^2/T \sim A_0^2 T$. The sign of the mass term is set by the sign of the polarization (clockwise or counterclockwise) of the oscillating electric field. Under time-reversal, these are interchanged, and hence the circularly polarized electric field breaks time-reversal invariance, thus removing the symmetry which protects the gapless Dirac points. In this sense, the appearance of a gap in this setting is unavoidable.

So far, so good. However, there are two flies in the ointment. First, in condensed matter physics, effective Hamiltonians tend to be effective ones in the sense of describing the low-energy physics. This limits the scope for having a simple high-frequency expansion, as there will be higher-energy degrees of freedom to which one can potentially couple, which are omitted in the above treatment.

This is related to the question of single-particle versus many-body physics: the spectrum of a many-body system is unbounded above, although the matrix elements to the very high energy states may be very small. The fundamental difference between static and Floquet systems in this language is that the nonineracting band picture can be a good starting point even for the interacting many-body system, as the band gap between the highest filled and the lowest empty band can allow for an adiabatic switching on of the interactions. In Floquet systems, where all many-body states are crowded into the Floquet Brioullin zone of size $2\pi\hbar/T$, such a gap is generally not available.

In an actual Floquet experiment, it is hence a question of detail what level of heating one can live with. In practice, this will also depend on the observable in question. (For a general discussion of "shaking" in many-body physics/optical lattices, see Eckardt, 2017.)

Second, and relatedly, fixing the initial state is another challenge. Ideally in the static situation, in order to measure the Chern number of a band, one puts the chemical potential in the gap above that band, so that all the states in the band in question are filled, and empty in the bands above; and then one measures the relevant transport coefficient.

This cannot be done in Floquet systems, for essentially the same reason as described above: there is strictly no concept of high and low energy; put differently, bands do not naturally fill up one after the other as a chemical potential is increased. One thus needs to devise a separate part of the experimental protocol about how to prepare an appropriate initial state.

Nonetheless, the situation is far from hopeless. For one thing, heating may be so slow to allow for a perthermal regime which persists on a timescale parametrically large in drive parameters (Abanin et al., 2015; Kuwahara et al., 2016). On this timescale, the system does not maximize entropy, and interesting phenomena may be observable starting from an appropriate, for example, thermal, initial state. Thus, in a system driven periodically for a finite rather than an infinite amount of time, the energy spectrum is in practice not perfectly periodic but one can still see a significant number of Floquet replicas, as in the angle-resolved photoemission spectroscopy of an optically pumped topological insulator surface (Wang et al., 2013).

10.4 Anomalous Floquet–Anderson Insulator

The Floquet topological insulator demonstrates how one can change the topology of a band structure using periodic driving. The resulting graphene Floquet band-structure discussed above is essentially identical to what could have been achieved – at least theoretically – by directly adding a *static* mass term $\propto \sigma^z$. The question which naturally poses itself is whether the Floquet setting allows for topological band structures without a static counterpart. In the following, we present two such instances, both of which turn out to have rather simple rationalizations.

The first appears already for a two-dimensional band structure consisting of only two bands in which disorder can localize every single particle state, but which nonetheless exhibits stable chiral edge states; this is known as the anomalous Floquet–Anderson insulator (AFAI) (Titum et al., 2016). The second is a band structure with a topologically protected Majorana zero mode away from (quasi-)energy 0, which is known as the π-Majorana fermion. In both cases, the fact that the quasi-energy is periodic, that is, that it "lives on a circle," is the enabling new ingredient.

To see how the AFAI comes about, it is useful to recall once more Laughlin's flux insertion argument linking the existence of the gapless chiral edge mode to a bulk property, the existence of a delocalized state, and to see how its strictures can be avoided in a Floquet system. To recapitulate this argument for the static case, consider an annulus (a circular disk with a circular hole at its origin). Upon adiabatic insertion of a *unit* flux, the initial and final Hamiltonians, H_i and H_f, are the same, up to a gauge transformation which removes the flux: a unit flux cannot yield a nontrivial Aharanov-Bohm phase.

This does, however, not mean that the state of the system remains unchanged: it is possible that the final state, $|\psi_f\rangle$, is distinct from the initial state, $|\psi_i\rangle$. This is the case for the topologically protected chiral edge state, which increases in energy ("an excitation is added" to one edge upon flux insertion). A little bit more specifically, if there is a chiral edge state connecting the lower to the upper band, flux insertion amounts to an uphill spectral flow in energy. To keep the overall spectrum unchanged, there must therefore also be a downhill flow ("an excitation is removed" from the opposite edge). If all the bulk states are localized, there is no way of transporting the charge from one edge to the other, and hence a bulk delocalized state must exist in this setting.

What changes in the case of the AFAI is that the overall spectrum need not be unchanged: due to the periodicity of the quasienergy, one only needs to demand that the sum of quasienergies remains unchanged *modulo* $2\pi/T$. That means that the spectral flow of the edge state can wrap around the periodic quasienergy direction while the bulk states all remain fully localized throughout.

A process leading to such a Floquet Hamiltonian is readily sketched. As in the case of the Floquet topological insulator, we subdivide the drive period T into several (in this case, five) segments, during which the Hamiltonian H_n is constant.

As a brief aside, we note that this kind of construction has a number of desirable features. First, the piecewise constant segments can often straightforwardly be chosen to be easily visualized; by contrast, a continuously time-dependent Hamiltonian will generally have intermediate forms during its smooth development which may be rather more complex. In the study of time crystals below, we will encounter the case of alternating Hamiltonians both of which are "classical," in the sense of consisting of manifestly commuting operators, even though the Hamiltonians do not commute with each other. Also, this structure is much more easily simulated: multiplication of a handful of unitaries is much simpler than integration over a continuous family. Nonetheless, these formulations are not just equivalent: a discontinuous drive profile contains more higher Fourier components than a smooth one, so that the equilibration properties of the different drives may differ considerably.

In practice, one thus devises a simple model drive the desired properties of which are present on a certain level of intuitive obviousness. This then needs to be supplemented by a detailed study of the robustness of the desired phenomenon. This is typically done numerically, and we encourage the reader to follow up on this important, but not very pedagogically instructive, aspect in the original literature.

Returning to the AFAI drive protocol, this consists of one piece where disorder generates Anderson localization (Figure 10.1a) and a second in which a four-step hopping Hamiltonian provides for a motion with chiral edge states. The latter is constructed using the fact that the sites of the square lattice can be subdivided into

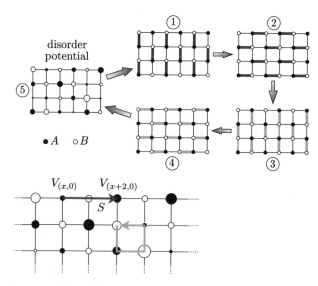

Fig. 10.1 An anomalous Floquet–Anderson insulator can be obtained via the quinary drive sketched in the top panel. The crux is the sequence of hoppings, labeled 1–4, which leads to a particle simply moving around a plaquette in the course of the period, as indicated in the bottom panel. (The disorder term, 5, in the drive is responsible for localizing the states in the bulk.) At an edge, however, the motion is systematically disrupted in favor of a clockwise displacement. From Titum et al. (2016). Reprinted with permission by *Physical Review*.

two sublattices, A and B, so that bonds only join different sublattices; and that the bonds can be subdivided into four groups, those pointing north/east/south/west from sublattice A to sublattice B. A particle in the bulk hopping along the bonds in the driving sequence executes a clockwise motion around a plaquette. However, at the boundary, the particles on one sublattice have to skip the hop corresponding to the missing bond at the boundary, thereby missing a sublattice change. The hopping sequence is constructed so that then continuing its motion from the other sublattice just moves it along the edge as shown in Figure 10.1. This skipping motion leads to the desired delocalized edge state.

It is possible to define a bulk topological invariant which goes along with the edge states discussed here. It is the time coordinate within a period which furnishes the extra dimension entering the definition of the invariant compared to the static case; the original paper on bulk-edge correspondence in Floquet systems by Rudner et al. (2013) contains a detailed account of this construction. A more recent review of the universe of Floquet drives is by Harper et al. (2020).

10.5 Driven Kitaev Chain and π-Majorana Fermions

In the anomalous Floquet–Anderson insulator, the periodicity in quasienergy of the Floquet Brioullin zone played an important role, as it allowed for the existence of edge states winding around that periodic direction.

This periodicity underpins the existence of another type of protected feature of the band structure, the so-called π-Majorana fermion. This can be accounted for straightforwardly following the line of argument in Section 9.4 for the robustness of the Majorana zero mode, which to preclude confusion we will refer to as 0-Majorana in the following.

The crucial item of the argument concerned the symmetry ensuring that states come in pairs at energies $\pm\epsilon$, so that an isolated state in a gap would be pinned at $\epsilon = 0$. This generalizes to the quasienergy in an obvious way, so that the (quasienergy-)0-Majorana fermion directly corresponds to the (energy-)0 Majorana mode in the undriven case.

Now, the quasienergy being defined modulo $2\pi/T$ means that there is another location which is special: $\pi/T = -\pi/T$ modulo $2\pi/T$. If there is a gap at π/T, an isolated mode at this quasienergy will also be pinned at that energy, and thus analogously topologically protected. Such a mode is known as the π-Majorana. Crucially, this has no undriven counterpart: the periodicity of the quasienergy cannot vanish continuously, as the continuous time-translation symmetry is either present or not. The π-Majorana is therefore present only in the nonequilibrium setting.

These two modes can exist entirely independently of each other, so that one can have four combinations of Majoranas: trivial (i.e., none), 0, π, and 0π. These can be realized straightforwardly by constructing a binary drive along the lines described above: one combines two Hamiltonians to construct a driven version of the Kitaev chain discussed in Section 9.4. As explained there (see also Box 9.3), this can alternatively be viewed as a (driven) Ising chain, and we adapt that picture for ease of visualization, as the binary drive is physically particularly transparent there.

The first member of the binary drive is an Ising exchange, while the second is a transverse field. The unitary over a Floquet cycle then take the form $\mathcal{U} = \mathcal{U}_\Gamma \mathcal{U}_J$, with

$$\mathcal{U}_J = \exp\left[-i \sum_j J_j \sigma_j^z \sigma_{j+1}^z\right] , \ \mathcal{U}_\Gamma = \exp\left[-i \sum_j \Gamma_j \sigma_j^x\right] , \quad (10.17)$$

where we will be considering both open and periodic boundary conditions in what follows. Note that we have committed an abuse of notation by suppressing the explicit role of the drive period T in Eq. 10.17, by identifying $J_j T/2$ and $\Gamma_j T/2$ with J_j and Γ_j, respectively. Like this, these two variables still encode the relative strengths of the exchange and the field; and they also signal more directly a periodicity in their strength in this setting: there is no high-field limit as such. Rather, shifting each coupling constant by, say, 2π has no effect, so that the phase diagram in the space of couplings is itself also periodic.

Since this is a disordered problem, the distributions of J_j and Γ_j need to be specified. The result does not depend on these distributions in detail, but what does matter are the mean values $\bar{J}$ and $\bar{\Gamma}$; and for concreteness, it may be useful to think of a box-like probability distribution of width of a small fraction of 2π. Figure 10.2 shows the phase diagram for these parameters.

Let us now consider the four possible situations, with none, one (0 or π), or both (0 and π)-Majoranas in turn. The conceptually most straightforward cases are those

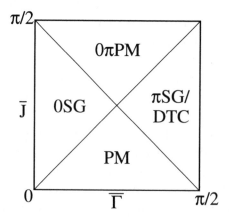

Fig. 10.2 Phase diagram of the Ising spin chain subject to the binary drive, Eq. 10.17 (Khemani et al., 2016).

which are analogues of the static cases, namely, those without the π-Majorana. These can be considered by setting one of the couplings to vanish entirely, so that the resulting problem is the static one discussed in Box 10.1 on phase structure.

The eigenstates of the paramagnet then are just the "classical" ones diagonal in the preferred basis of $\mathcal{U}_\Gamma$ (see Section 10.2.5). Switching the noncommuting exchange term, $\mathcal{U}_J$, back on then leads to a dressing of these states but, like in a paramagnet, all correlators remain short-ranged.

10.5.1 0 and π Spin Glasses

Considering first the case of vanishing fields $\Gamma_j \equiv 0$, we end up with a disordered classical magnet, as in Section 10.2.6. As the couplings are disordered, the order parameter diagnosing the difference to a paramagnet is not a simple (anti)ferromagnetic one in the form of a magnetization at a given wavevector. Instead, it is an Edwards–Anderson spin glass order parameter, which is designed to distinguish whether correlations are large (but of random sign), or instead are small in modulus. This is done by simply squaring the correlator before averaging over space:

$$q_{\text{EA}} = \frac{1}{N^2} \sum_{i,j=1}^{N} \langle \sigma_i^z \sigma_j^z \rangle^2 \,. \tag{10.18}$$

Notably, the Edwards–Anderson order parameter, q_{EA}, can be nonzero no matter which eigenstate it is evaluated in for the fully many-body localized system we consider: this is an instance of a type of order in $d = 1$ which cannot be present in a thermal setting.

The spin glass state with nonvanishing q_{EA} is called the 0-spin glass, 0SG, to distinguish it from the πSG (Khemani et al., 2016), which we discuss next.

The case of a π-Majorana fermion is not reducible to a simple undriven system, as explained above. The cartoon picture of this state is a combination of $\mathcal{U}_J$ as in the 0SG, but with an intervening $\mathcal{U}_\Gamma$, with $\overline{\Gamma}$ chosen so that it effects a global spin flip $\mathcal{P}_\pi$ interchanging the states with the opposite Ising polarizations: $| \uparrow\downarrow\downarrow\uparrow\downarrow\rangle \leftrightarrow | \downarrow\uparrow\uparrow\downarrow\uparrow\rangle$. The overall unitary thus reads

$$\mathcal{U}_{\pi\text{SG}} = \mathcal{U}_J \, \mathcal{P}_\pi \,. \tag{10.19}$$

It is easy to see that the two finite-size eigenstates, the Schrödinger cats of Section 10.2.6, remain eigenstates of this $\mathcal{U}_{\pi\text{SG}}$ but – crucially – the antisymmetric one picks up an overall minus sign due to the spin flip:

$$\mathcal{U}_{\pi\text{SG}}|\psi_\pm\rangle := \mathcal{U}_{\pi\text{SG}} \frac{1}{\sqrt{2}} [| \uparrow\downarrow\downarrow\uparrow\downarrow\rangle \pm | \downarrow\uparrow\uparrow\downarrow\uparrow\rangle]$$

$$= \pm \exp(i\phi) \frac{1}{\sqrt{2}} [| \uparrow\downarrow\downarrow\uparrow\downarrow\rangle \pm | \downarrow\uparrow\uparrow\downarrow\uparrow\rangle] = \pm \exp(i\phi)|\psi_\pm\rangle, \tag{10.20}$$

where the angle ϕ is the same for both states. As advertised, the minus sign corresponds to an extra phase $-1 = \exp(i\pi)$ picked up in the time evolution over the course of the period. Therefore, the two states are now separated by a quasienergy π. This amounts to them being located a maximal distance from each other in quasienergy (see Figure 10.3). Note that the two states $|\psi_\pm\rangle$ are locally indistinguishable, yet they are nondegenerate; this reflects the fact that the unitary $\mathcal{U}_{\pi SG}$ *cannot* be written in terms of a local, static Floquet Hamiltonian. This is a reflection of the fact that the π-SG is a genuine new *Floquet* phase.

This quasienergy difference in the single-particle picture is "supplied" by the occupancy of the π-Majorana mode: the two states in the doublets for the 0SG and those for the πSG differ by the occupancies of the 0- and π-Majorana modes, respectively.

10.5.2 The 0π Paramagnet as a Symmetry-Protected Topological Phase

The situation for the joint presence of both 0 and π-Majorana modes then naturally gives a quartet of states, consisting of two quasidegenerate pairs a distance π apart. Starting from any given state, one can cycle between these by toggling occupancies of the Majorana modes at $0, \pi$, and then again at 0. The phase characterized by this eigenstate order is called the 0π paramagnet. Unlike the prior two phases, it has no spin-glass order; but unlike the trivial paramagnet, it is a symmetry-protected topological state (see Chapter 11). Like in the case of the AKLT chain (Section 5.2.4), this means that its behavior with periodic boundary conditions is trivial, but for open boundary conditions, topologically protected edge modes appear.

This can be seen by analyzing the unitary underlying the 0π Floquet paramagnet (0πPM), again by considering the simple special case $J_i \equiv \pi/2$, where

$$\mathcal{U}_{0\pi PM} = \mathcal{U}_\Gamma \prod_j \exp[i\frac{\pi}{2}\sigma_j^z \sigma_{j+1}^z] = \mathcal{U}_\Gamma \begin{cases} (-i)^N \text{ periodic b.c.} \\ (-i)^{N-1}\sigma_1^z \sigma_N^z \text{ open b.c.} \end{cases} \quad (10.21)$$

The exchange terms hence contributes a state-independent global phase in the case of periodic boundary conditions; the physical behavior of the periodic chain is hence essentially that of the trivial paramagnet.

For an open chain, however, the unitary depends on the state of the spins at the endpoints of the chain. The resulting unitary is thence that of the paramagnet, multiplied by $\sigma_1^z \sigma_N^z$. Regarding all the spins in the interior of the chain, $\mathcal{U}_J$ still has no influence, and all the action takes place at the surface. The surface spins are subject to their local fields, Γ_1 and Γ_N, and the "exchange" term $\sigma_1 \sigma_N$ from $\mathcal{U}_J$. The state of the full system thence factorizes into a product state of the interior and the edge spins, and one can essentially ignore the interior part. The four eigenstates of the edge spins are then labeled by two Ising variables. First, the product of the

PM	0SG	πSG	0πPM
		$\|\uparrow\downarrow\downarrow\uparrow\downarrow\rangle$ $\|\uparrow\downarrow\downarrow\uparrow\downarrow\rangle$	$\|\rightarrow\rightarrow\leftarrow\rightarrow\rightarrow\rangle$ $\|\rightarrow\rightarrow\leftarrow\rightarrow\leftarrow\rangle$
$\|\rightarrow\rightarrow\leftarrow\rightarrow\rightarrow\rangle$	$\|\uparrow\downarrow\downarrow\uparrow\downarrow\rangle$ $\|\downarrow\uparrow\uparrow\downarrow\uparrow\rangle$	$+\|\downarrow\uparrow\uparrow\downarrow\uparrow\rangle$ $-\|\downarrow\uparrow\uparrow\downarrow\uparrow\rangle$	$\pm\|\leftarrow\rightarrow\leftarrow\rightarrow\leftarrow\rangle$ $\pm\|\leftarrow\rightarrow\leftarrow\rightarrow\rightarrow\rangle$
$\langle\sigma^z\rangle=0$ $\langle\sigma^x\rangle\neq 0$	$\langle\sigma^z\rangle\neq 0$	subh. $\langle\sigma^z\rangle\neq 0$ discrete time crystal	$\langle\sigma^z\rangle=0$, $\langle\sigma^x\rangle\neq 0$ subh. $\langle\sigma^x_{\mathrm{edge}}\rangle\neq 0$

Fig. 10.3 Observables corresponding to the phases of the Ising spin chain subject to the binary drive, Eq. 10.17. The quasienergy "axis" is compactified into a circle of unit radius, as it is periodic with 2π. The eigenstates are distributed on this circle randomly, in pairs either exponentially close, or exponentially close to being separated by π; or in quartets consisting of two pairs separated by π. In the cartoon of the states, $\sigma^x=+1$ ($\sigma^z=+1$) corresponds to an arrow pointing right (up).

spins expressed in the basis along the field direction; and second, the parity which encodes whether the edge spins are in an even or an odd superposition with their Ising-reversed copies. One of these labels the states within the degenerate doublet, the other the doublets separated by quasienergy π, as depicted in Figure 10.3.

Putting all of this together yields an inert bulk, while the edge spins do exhibit *period-doubling*, that is, their dynamics has a component at a frequency equal to half of the drive's. This can most easily be seen by explicitly time-evolving the spin operators at the edge with the unitary $\mathcal{U}$ (Eq. 10.21).

Such period doubling is a most remarkable phenomenon, and it is its presence in the π spin glass which is responsible for it being called a discrete time crystal, as we discuss in more detail next.

10.5.3 Temporal Correlations and the Floquet Discrete Time Crystal

The temporal correlations of the driven Ising chains turn out to contain the fundamental novelty of these Floquet phases. Having already established long-range spatial (spin-glass) order in the πSG, we now ask about its correlations in time. To set the stage, recall that equilibration in the conventional sense implied the presence of a time-independent steady state. Here, time-independence is not a natural option as the Hamiltonian itself changes over the course of a period, and the concept of equilibration is replaced by that of synchronization: one now observes the correlations stroboscopically, say at the beginning of each drive period, and asks if the sequence thus obtained is time-independent. We will see that it is not.

Macroscopically distinct Schrödinger cat states cannot easily be stabilized in experiment. It is therefore very difficult to study the longtime behavior of these eigenstates of $\mathcal{U}_{\pi SG}$. Instead, a natural starting state is the simple state

$$| \uparrow\downarrow\downarrow\uparrow\downarrow\rangle = \frac{1}{\sqrt{2}}\left[|\psi_+\rangle + |\psi_-\rangle\right].$$

After one period, this becomes (omitting the global phase $\exp(i\epsilon)$ due to the quasienergy of $|\psi_+\rangle$):

$$\frac{1}{\sqrt{2}}\left[|\psi_+\rangle - |\psi_-\rangle\right] = | \downarrow\uparrow\uparrow\downarrow\uparrow\rangle.$$

Another period later, the state is then back to the original $| \uparrow\downarrow\downarrow\uparrow\downarrow\rangle$.

Therefore, there is temporal symmetry breaking as well: the periodicity of the correlations is double the period of the drive. In analogy to (space) crystals, which have a lowered space-translational symmetry, this system is hence called a time crystal. Since it is the discrete time translational invariance of the Floquet drive which has been broken, rather than the continuous spatial translation symmetry of free space, the more precise moniker is Floquet discrete time crystal (DTC).

10.6 Many-Body Floquet Discrete Time Crystal

At this stage, the reader may feel somewhat underwhelmed. After centuries of unsuccessfully searching for a system spontaneously breaking time-translation symmetry – for the longest time, these went under the name perpetuum mobile – we have found a magnet which, essentially, takes two half-rotations in order to execute a full one. And indeed, as such, this is barely front-page news. What is remarkable is the stability of this phenomenon to all sorts of perturbations. Indeed, time-translational symmetry breaking is a conceptually subtle business, and we refer the reader for details to the review Khemani et al. (2019), on which parts of this chapter are based, for details.

The stability of the above cartoon pictures to tuning the drive parameters away from the simplest values 0 and $\pi/2$ follows by the same arguments as those presented for the Kiteav Majorana chain. While, for example, the edge states of the 0πPM will leak into the bulk, they will remain localized there. And, more importantly, the quasienergy difference π is pinned to this value, so that the period doubled response will remain, even if the Ising spin flip is not a perfect inversion, $\mathcal{P}_\pi$, but offset by a small but finite angle, $\mathcal{P}_{\pi-\epsilon}$. This is in stark contrast to

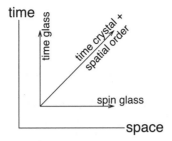

Fig. 10.4 Sketch of the spatiotemporal nature of correlations in the Floquet DTC. At long distances and long times, a novel combined spatiotemporal from of order is observed. Moving along the time- (or the space-) axes only leads to "local" glassy correlations. In particular, on-site correlations at long times exhibit oscillations at many frequencies, characteristic of the local disordered environment. Equal-time correlations at large distance evidence a standard spin glass.

noninteracting spins, where an offset of the flip by an angle ϵ leads to a continuous drift in the response frequency (see Figure 10.5).

Absolute Stability

These insights, however, pertain to the single-particle picture made possible thanks to the integrability of the driven Ising chain, which it inherits from the static transverse field Ising model. What happens if generic (small) terms are added to the Hamiltonian which spoil the integrability? This question is important, as the definition of a phase requires stability to perturbations–it is phase transitions, not phases, which require fine-tuning microscopic parameters.

The answer is that the πSG persists, and indeed, that it is more stable than even conventional static spin glass phases (Else et al., 2016; von Keyserlingk et al., 2016). A flavor for why this may be the case can already be obtained by asking what is the consequence of applying an infinitesimal Ising symmetry-breaking field in the direction of the ordered moments. In the case of the 0SG, this immediately breaks the exponentially small quasidegeneracy, as a state $|\uparrow\downarrow\downarrow\uparrow\downarrow\rangle$ will in general have a nonvanishing magnetization. By contrast, the above cat states $|\psi_\pm\rangle = \frac{1}{\sqrt{2}}[|\uparrow\downarrow\downarrow\uparrow\downarrow\rangle \pm |\downarrow\uparrow\uparrow\downarrow\uparrow\rangle]$ have a nonvanishing splitting of π/T, that is, of $O(N^0)$; they will therefore not mix appreciably when an infinitesimal field is applied. This result can be reexpressed in a real-time picture. Formally, instead of inverting the spins by $\mathcal{U}_\Gamma$, one can choose to enter a moving reference frame which sees an inversion of the "Ising-odd" applied field instead, while leaving the "Ising-even" exchange invariant. This means that the field points alternatingly up or down. It hence averages out, and with the field, the Ising symmetry breaking disappears.

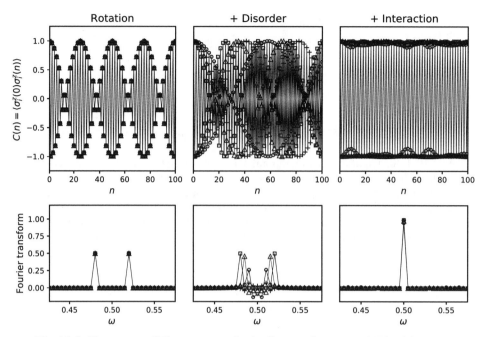

Fig. 10.5 Signatures of Floquet many-body discrete time crystal. The drive consists of an imperfect half-rotation (left panel), with added disorder in the field (middle panel). Finally, upon addition of a disordered exchange, the spins exhibit robust period doubling. The top panels show the corresponding stroboscopic real-time evolution of the temporal on-site correlators of the σ^z. The bottom panels show the same information in Fourier space, where the locking into period doubling manifests itself in a peak locked at $\omega/\Omega = 1/2$. Figure courtesy of Matteo Ippoliti.

The reader who is uncomfortable with such wordy explanations will need to delve somewhat more deeply into the physics of time-translation symmetry-breaking, MBL, and Floquet unitaries (see, e.g., Section 5 of Khemani et al. (2019)). Put concisely, the "topological" formulation of the difference between 0 and π SG lies in the action of the unitary on the $l-$bits τ^z: $\mathcal{U}^\dagger \tau^z \mathcal{U} = \pm 1$. It can be verified (e.g., for the special point of the perfect spin flip) that the case $+1$ corresponds to the 0SG, and -1 to the π SG. These two cases cannot be smoothly be deformed into one another. Now, when an Ising symmetry-breaking field is applied, the quasi-degeneracy of the doublet of the 0SG is lifted, with the cat states $|\psi_\pm\rangle$ being replaced by the conventional, unentangled states $|\uparrow\downarrow\downarrow\uparrow\downarrow\rangle$ and $|\downarrow\uparrow\uparrow\downarrow\uparrow\rangle$ as eigenstates. Crucially, these are no longer eigenstates of $\mathcal{P}$, which therefore can no longer be used as a generator of an Ising symmetry. By contrast, there is no degenerate perturbation theory to be done in the case of the π SG, and the cat states remain eigenstates.

The second crucial ingredient to establish the absolute stability of the πSG is supplied by many-body localization, which supplies the fact that the flip operator $\mathcal{P}$ can be perturbatively continued to $\tilde{\mathcal{P}}$ as *any* local symmetry breaking term is added to the Hamiltonian. This proceeds the same way that the l-bits evolve upon addition of those terms. Like for the l-bit, the outcome is nonuniversal in the sense that the resulting operator depends explicitly on the Hamiltonian, unlike the spin flip operator $\mathcal{P}_\pi$ which implements a global Ising symmetry. In this sense, $\tilde{\mathcal{P}}$ in general reflects an *emergent* Ising symmetry, that is, one not present in the starting Hamiltonian.

It is worth noting how far outside the familiar realm of statistical physics this result lies: Landau–Ginzburg theory states from the very outset that symmetries which are not there cannot be broken. Indeed, the very notion may seem as absurd as the persistence of the smile of the Cheshire cat, even once the cat has vanished. As Alice mused, "I've often seen a cat without a grin, but a grin without a cat! It's the most curious thing I ever saw in my life!" (Carroll, 1865).

This feature, termed absolute stability, is rather reminiscent of the stability of the topological order in the fractional quantum Hall effect (Section 5.1.4) and the RVB liquid or Kitaev's toric code, which are also stable to an arbitrary small perturbation. Not unsurprisingly, it is topology which underlies both phenomena.

In all cases, one finds an emergent discrete degeneracy, involving states threaded by fluxes in the above examples of topological order. For the DTC, the Ising symmetry was the crutch with which to discover this emergent Ising symmetry, which however can stand on its own. What basically happens is that the physics of MBL ensures that any variation in the properties of the system induced by the perturbation takes place smoothly. This is the essence of the l-bits, which get dressed continuously. As long as there is a discrete (i.e., topological) distinction between the objects under consideration, such as a global even or odd parity index, this distinction will persist even as everything else changes smoothly. In this way, the πSG is special. The PM is trivial to start with; the 0SG is not protected on account of its quasi-degeneracy, and the parity of the 0π paramagnet pertains to the isolated edges which are sensitive to a local symmetry-breaking perturbation, as may be checked by explicitly solving for the eigenstates of the two spins at the edges under application of a symmetry-breaking field. The formal demonstration of these properties proceeds by constructing the relevant perturbation theory and examining its impact on the eigenstate-ordered multiplets (Khemani et al., 2019).

We note, however, that even absolute stability has its limits. The one symmetry one cannot discard is the discrete time translational symmetry of the Floquet drive. Without it, there will be Fourier components to the drive which can generate matrix elements between states differing by quasienergy π. This leads to the same situation as the static field applied to the 0SG, which again proves fatal to the eigenstate order, thereby merging the πSG with the trivial PM. A Cheshire cat without a smile will not leave one behind.

We do remark on the similarity of this topology-induced stability to the case of the quantum Hall effect. Our present treatment started by providing a cartoon picture of the DTC, where the period doubling seemed to be put in entirely by hand, by splitting a full rotation into two pieces. This is just like the case of the quantum Hall effect, where it is known that a clean system always exhibits a Hall conductivity $\sigma_{xy} = \nu e^2/h$, where ν is an integer at integer filling. The amazing topological stability of the Hall effect appears when ν is tuned away from an integer or rational fraction, ν_0, but the Hall conductivity remains pinned at $\sigma_{xy} = \nu_0 e^2/h$. In the same way, in the DTC the period doubling remains robust even if the rotation angle is changed continuously away from half a rotation. While the quantum Hall effect enables one to determine the conductance quantum e^2/h to unprecedented accuracy, in the case of the discrete time crystal, the corresponding stability is that of (the admittedly prequantized) integers.

Spatiotemporal Order

We now return to the phenomenology of the DTC, in particular to the form of the correlations it encodes: there are long-range correlations both in space and in time – it represents an entirely new form of *spatiotemporal* order. This is probed most crisply by taking the limit of large distances and long times simultaneously, where the only signal in the spin correlations is the period doubled one. By contrast, as outlined above, the long-range spin glass order corresponds to the limit of large distances at equal times. Similarly, if one measures the spin autocorrelation function, that is, the limit of long times at short distances, one is sensitive to the oscillations of the spins on account of the difference between physical variables, σ^α, and the l-bits, so that one will observe Rabi-type oscillations set by the local environment (e.g., the effective field experienced by the l-bits). This behavior is called a time glass. This is illustrated in Figure 10.4.

Several experiments have been undertaken to look for discrete time crystals, on platforms as varied as nitrogen-vacancy centers in diamond, in various NMR platforms, and in a chain of trapped ions; recent reviews are Khemani et al. (2019) and Else et al. (2020). These also tend to use stepwise drives for clarity and ease of implementation. The central ingredients for the demonstration of a Floquet many-body DTC are the following. First, a spin rotation by an angle close to, but not at, half a rotation. Second, a dose of disorder to induce MBL; and third, interactions to lock the response robustly into period doubling.

The various experiments have seen such a frequency locking, which is a remarkable feat, and the fact that behavior of this type has been seen across platforms is a great experimental achievement for each of these. This is all the more remarkable as the experiments arrived only a short period after the theoretical work. Indeed, they have clarified considerably the understanding of discrete time crystals. In particular, a detailed analysis of these experiments shows that none of them has as yet realized the Floquet DTC *sensu stricto*. However, while not yet *bona fide* nonequilibrium phases of matter, they have unearthed a set of very interesting related phenomena, which now go under headings like symmetry-protected, algebraic or prethermal time crystals. The latest proposal in this realm is to use noisy intermediate-scale quantum (NISQ) technology to realize the Floquet DTC, as such a platform is ideally suited for the emulation of a stepwise drive, with locally addressable quantum gates naturally implementing the disordered pairwise interactions and fields appearing in the model Hamiltonian, and local initialization and read-out enabling a detailed analysis of the DTC signatures in the correlations for arbitrary initial states.

The theorist's dream experiment is shown in Figure 10.5. The data there shows the polarization of four different spins in a disordered interacting spin chain of length $L = 16$ as a function of stroboscopic time. This means that the value of the spin is obtained at a sequence of times offset by an integer number of drive periods $T = 2\pi/\Omega$. It is obtained from numerical simulations as follows.

The leftmost panel shows the effect of a uniform field effecting not quite half a rotation. This leads to peaks in the Fourier transform of the correlation function away from π. The second part supplies a disordered field which puts the individual spins out of step and leads to differences between the peaks, which still do not exhibit precise period doubling. The third part of the drive adds a disordered Ising exchange. It is only when this is added that the response locks in at period doubling, and the peak robustly shifts to frequency π. Note that the starting state was a random product state: the DTC signal is present for all initial states.

11

Symmetry, Topology, and Information

The capacity of topologically robust quantities to distinguish different phases even in absence of local order parameters is the central subject of this book. The topological degeneracy in the fractional quantum Hall effect thus reflects the topological order present there, which does not require the breaking of any local symmetry. This has supplemented the idea of the symmetry distinction of different phases, the bedrock of the Landau–Ginzburg–Wilson paradigm of phases and transitions between them, as outlined in the opening chapters.

What then about the interplay between those ideas? On one hand, there is the question of their compatibility, for example, can topologically ordered phases also exhibit conventional symmetry breaking? This item we have already touched upon in the context of quantum Hall ferromagnets in Section 3.7. On the other hand, one can ask whether there exist any phenomena which constitutively rely on a combination of ingredients from symmetry and topology. The first part of this chapter is devoted to taking a closer look at that question.

The second part addresses how quantum information concepts are useful to understand quantum wavefunctions, particularly those arising in topological states. A major impetus for work on non-Abelian states is the goal of quantum memories and computers, as described in Chapter 9. There has also been a useful flow in the opposite direction, and we sketch one way to quantify information in a single quantum wavefunction: the entanglement entropy with respect to a bipartition of Hilbert space. This turns out to help place topological states, particularly those with fractional particles, in a broader context, and has also led to a number of revolutionary numerical techniques. This chapter is of necessity more of a survey of an actively evolving area than the preceding ones, and we encourage readers who wish to delve more deeply to consult the reviews cited. We close with a few general comments on the continuing search for topological phases, both in real materials and in the mind's eye.

11.1 Symmetry-Protected Topological Phases

One of the central phenomena in topological condensed matter physics – the topological insulator – requires a helping of symmetry to exhibit its topology: as explained in Chapter 3, it is the presence of time reversal symmetry that eliminates the scattering between the counterpropagating edge states, and hence leads to the quantized transport coefficient.

In one dimension, we have so far encountered a variety of different topological systems hosting a number of interesting phenomena. These include the chains involving the names Peierls, Su–Schrieffer–Heeger, Majumdar–Ghosh, Haldane/AKLT, and Kitaev. Going beyond their individual properties, this discussion is devoted to identifying the more formal structure underpinning their existence, with a focus on genesis, distinctiveness and stability of their topological properties. Indeed, it turns out that nontrivial topological properties in $d = 1$ can essentially only occur in the presence of symmetries; in their absence, all states are topologically identical. In the following, we outline the theory underpinning this symmetry protected topology.

The notion of symmetry-protected topological phases in one dimension was advanced in particular in the context of studying the $S = 1$ Heisenberg chain with nearest-neighbor interactions, also known as the Haldane chain (Pollmann et al., 2012). A soluble relative of this model, the $S = 1$ AKLT chain, was introduced in the context of Klein models in Section 5.2.4. There, its basic phenomenology was discussed – a nondegenerate, short-range entangled state exhibiting fractional spin $S = 1/2$ edge states exponentially localized on a lengthscale set by the bulk gap. It had been noted early on that there existed deformations of this model which connected it to a trivial band insulator without encountering a gap closing *en route* (Anfuso and Rosch, 2007). An account placing this material in a broad information-theoretic context is available in Zeng et al. (2015).

11.1.1 Symmetry Fractionalization

The central observation for the notion of symmetry protected topology in $d = 1$ is that a restriction on the paths through the space of Hamiltonians to ones respecting certain symmetries does provide a notion of topological stability to the Haldane phase. The mathematical framework for capturing the underlying idea, which now goes by the name of symmetry fractionalization, involves representation theory, in particular the projective representations of the appropriate symmetry groups.

The central ingredient is simply stated: the action of a symmetry can act *independently* on the two edges of the chain, provided the bulk is gapped. This allows for the representations of the symmetry at the edge to acquire a relative phase, which under certain conditions there may take on only a discrete set of possible values. These discrete possibilities can then not be *continuously* deformed into one another, and they are hence topologically stable. When the underlying symmetry is removed, however, this structure disappears entirely.

The remainder of this section fleshes out these statements following the account of Verresen et al. (2017); it applies these insights to a family of Kitaev chains, christened α-chains, to provide a concrete unifying framework for a number of previously encountered models, and to generate insights into the overall richness of the resulting classification. We start by explaining the simplest setting before adding various generalizations until we are in a position to discuss the α-chains in general.

We consider a chain of length L with open boundary conditions and a local Hilbert space, $\mathcal{H}_i$, of dimension d, such that the total Hilbert space, $\mathcal{H} = \otimes_i \mathcal{H}_i$, has dimension d^L.

Let the system Hamiltonian H be symmetric with respect to a global symmetry group G. The action of this group on states in Hilbert space is encoded by a set of unitary matrices U; we identify the set of representations provided by these unitary matrices with the group itself, allowing us to write $U \in G$ as a shorthand.

Next, assume that we are dealing with a so-called on-site symmetry, that is, one whose members can be written as a tensor product over unitaries acting on individual sites i:

$$U = \otimes_i U_i . \tag{11.1}$$

We restrict our attention to the case of the symmetry G not being broken, so that the ground state in the presence of periodic boundary conditions is unique; the action of G must therefore be trivial in the bulk as it cannot convert different ground states into one another. This still leaves the possibility of its action being nontrivial at the edges, provided that there is (as is the case in the AKLT chain) an edge state degeneracy.

We thus define two operators, $U_{L,R}$ to act on the left and right edges of the system, respectively, such that $U = U_L U_R$. Again, as noted for the edge states of the AKLT chain, the support of $U_{L,R}$ will extend into the bulk by a distance set by the (inverse) bulk gap, so that in the limit of a long chain, $L \to \infty$, their support will be disjoint.

The essence of symmetry fractionalization is that $U_{L,R}$ are, *individually*, symmetries of H, that is, that $[U_L, H] = [U_R, H] = 0$. This can be seen by decomposing $H = H_L + H_R$ with the support of H_L chosen such that it is disjoint with that of U_R, and similarly for H_R and U_L. Then, $0 = [U, H_L] = [U_L U_R, H_L] = [U_L, H_L]U_R$. Since U_R is invertible, it follows that

$$[U_L, H_L] = [U_L, H_L + H_R] = [U_L, H] = 0, \tag{11.2}$$

as desired.

To extract the projective nature of the resulting edge representations, we continue to restrict ourselves to the simplest setting, namely, a "bosonic" system, that is, one in which operators acting with disjoint support commute; and we consider a commutative pair of symmetry operations $U, V \in G$, that is, $[U, V] = 0$, such that $UVU^{-1}V^{-1} = \mathbb{I}$. Then

$$\mathbb{I} = (U_L U_R)(V_L V_R)(U_L^{-1} U_R^{-1})(V_L^{-1} V_R^{-1}) = (U_L V_L U_L^{-1} V_L^{-1})(U_R V_R U_R^{-1} V_R^{-1}). \tag{11.3}$$

Since the sole action of one of these factors must be proportional to the identity in its region of support, it follows that

$$(U_L V_L U_L^{-1} V_L^{-1}) = \exp(i\alpha); \quad (U_R V_R U_R^{-1} V_R^{-1}) = \exp(-i\alpha). \tag{11.4}$$

A nontrivial value of $\exp(i\alpha) \neq 1$ implies that the symmetry operations are represented *projectively* at the edges.

The dimension of such a projective representation has an immediate physical interpretation. A d-dimensional projective representation is associated with a d-dimensional edge mode. The AKLT chain should thus go along with a $d = 2$-dimensional projective representation of the appropriate protecting symmetry.

This also means that a $d = 1$-dimensional representation is trivial in that it does not host a protected edge state. This is reflected in the twin facts that, first, such a situation does not permit nontrivial values of α and, second, that the accompanying phase factor in the projective representation case can be gauged away, leaving behind a trivial nonprojective symmetry representation via $\tilde{U}_L = \exp(i\alpha)U_L$.

Thus, the values of α for products in Eq. 11.4 do not fix the phases of the representation of the $U_{L,R}$ entirely. Like the magnetic field corresponding to different gauge choices of vector potential, there is a gauge-invariant content to these phases, and it is this which is used to group SPT phases into classes.

To use this as a basis for a topological classification scheme, one needs to determine which values of α can be deformed into one another continuously, and which cannot – the latter can then be said to have topological stability. Clearly, this is the case if the values are *discrete*, as this forbids a continuous deformation between them.

Perhaps the simplest instance is provided for a group consisting of a pair U, V of $\mathbb{Z}_2$ symmetries, that is, $G = \mathbb{Z}_2 \times \mathbb{Z}_2$. As U^2 is just the identity, $U_{L,R}^2$ can only be simple phase factors, which therefore commute with $V_{L,R}$. As $V_L^2 U_L^2 = V_L U_L^2 V_L = \exp(i\alpha) V_L U_L V_L U_L = \exp(2i\alpha) V_L^2 U_L^2$, it follows that $\exp(2i\alpha) = 1$, so that α can only take on two values, $\alpha = 0$ or π.

The object encoding a general classification scheme of SPT phases is then supplied by algebraic topology. The quantity in question is the second group cohomology group with coefficients in $U(1)$, denoted by $H^2(G, U(1))$. (Group cohomology is an abstract mathematical structure analogous to the cohomology of differential forms in Chapter 2.) For the example above, $H^2(\mathbb{Z}_2 \times \mathbb{Z}_2, U(1)) = \mathbb{Z}_2$. The symmetry group $SO(3)$ turns out to have the same property, $H^2(SO(3), U(1)) = \mathbb{Z}_2$, identifying half-integer and integer spins as topologically distinct. In particular, this implies that the subgroup of π-rotations also protects the Haldane phase. We note that the case of the Floquet 0π paramagnet, discussed in Section 10.5.2, only has a single $\mathbb{Z}_2$ symmetry; there, it is the temporal aspect of the drive – which also underpins the possibility of period doubling – which supplies the remaining ingredient.

At the same time, in the absence of a nontrivial symmetry group G, it is clear that this classification scheme yields only one, the topologically trivial, outcome. This observation underpins the statement that topological stability in one spatial dimension is predicated on symmetry protection.

An instance which does not yield a discrete set of outcomes is provided by $H^2(\mathbb{Z} \times \mathbb{Z}, U(1)) = U(1)$. This amounts to the possibility of a *continuous* set of phases: any given value would therefore not correspond to a topologically stable class. However, if one is considering a periodic system with a unit cell containing degrees of freedom with a finite-dimensional Hilbert space, the corresponding symmetry group will be finite dimensional, or a compact Lie group, both of which yield discrete outcomes and thence permit a topologically stable outcome.

The discussion of the previous paragraphs applies to unitary on-site symmetries (in particular excluding spatial symmetries such as translations) for "bosonic" systems. The word *bosonic* refers to systems where operators acting on different edges of the system commute. Both of these conditions are restrictive in the sense that there are generic physical situations which violate them. The first is provided by the case of antiunitary symmetries; and the second for fermionic systems, discussed in the following section, where operators with support on spatially disjoint regions need not commute on account of the *anticommutation* properties of fermions: the phases arising due to quantum statistics *can* be probed nonlocally.

The case of antiunitary symmetries, T, such as the time-reversal symmetry discussed in Section 3.3, is quite analogous to the above discussion. The symmetry again fractionalizes over the left and the right edge, where it thus acts independently.

Considering the case where $T = UK$, where U is an on-site symmetry and K is complex conjugation, and restricting ourselves to the case $T^2 = 1$, the ensuing treatment then makes use of the operator $\overline{U} = TUT$, to obtain $\mathbb{I} = (\overline{U}_L U_L)(\overline{U}_R U_R)$. Thence, $\overline{U}_L U_L = \exp(i\kappa)$, which in particular implies that $\overline{U}_L$ is proportional to the inverse of U_L, and hence commutes with it. It follows from complex conjugation that $\exp(i\kappa) = \exp(-i\kappa)$, so that $\kappa = 0$ or π. It follows that there are only two – hence topologically distinct – possibilities, $\overline{U}_L U_L = \pm 1$.

11.1.2 Fermionic Symmetry Fractionalization

The above exposition has explicitly relied on the possibility of defining sets of operators on two ends of a chain which commute with each other, the gap of the bulk acting as an effective barrier, keeping the gapless modes localized at the edge. The issue is that quantum statistics is not strictly local in this sense, and single fermion operators have an intrinsic nonlocal character to them. This idea is not at all an unfamiliar one – exchanging *well-separated* particles can lead to nontrivial phase accumulation, precisely the -1 factor distinguishing fermions from bosons. A more elaborate version of this has appeared in this book already in the context of topological quantum computing, where braiding well-separated quasiparticles is used to manipulate wavefunctions of non-Abelian anyons.

The former item is taken care of by allowing for the possibilities of operators on the two edges to either commute or anticommute: $U_L U_R = \pm U_R U_L$. One thus obtains what is known as a *graded* projective representation.

A further issue which has appeared in the context of topologically protected quantum computing, is the possibility of a *delocalized* complex Fermion mode fractionalizing into a pair of Majorana edge modes, the nonlocal character of which also provides topological protection against local sources of noise.

This item is more involved for the case of antiunitary symmetries, in that it precludes the possibility of even writing the operators in a way which factorizes over the two edges in the first place. This happens because factorizing complex conjugation requires a local basis for the fractionalized Hilbert space, which is unavailable for a single Majorana mode. If one wants to retain this separation as much as possible, one can allow for the possibility of such a Majorana mode by splitting the low-energy space into a part localized on the right, one on the left, and an optional extra nonlocalized complex fermion mode encoding presence or absence of the Majorana pair. For antiunitary symmetries, the value of $\overline{U}_L U_L$ will in general not be invariant under unitary (basis) transformations involving the nonlocal part. However, this freedom does not lead to a collapse of the topological distinctions.

At this point, we do not penetrate more deeply into the mathematical underpinnings and instead turn to a classification of a concrete set of one-dimensional models, the generalized Kitaev α-chains. These will provide illustrations of the above points, as well as a classification scheme for interacting fermionic symmetry protected phases. In addition, via a set of mappings, they will also turn out to represent, and thence provide a more unified framework for, a number of models which have already appeared elsewhere in this book.

11.1.3 General Kitaev Chains

We consider a fermionic chain. The complex fermion at each site, created by $c_j^\dagger$, is represented by the pair of Majorana operators $\gamma_j = c_j + c_j^\dagger$ and $i\tilde{\gamma}_j = c_j - c_j^\dagger$. The Hamiltonian now simply couples each $\tilde{\gamma}_j$ with the $\gamma_{j+\alpha}$ operator belonging to the site displaced by α:

$$H_\alpha = i \sum_{j=1}^{N-\alpha} \tilde{\gamma}_j \gamma_{j+\alpha} \,. \tag{11.5}$$

This amounts to a family of noninteracting models which, for open boundary conditions, manifestly have α Majorana fermion pairs corresponding to localized zero-energy modes at the endpoints of an open chain, as these lack an interaction partner.

The $\alpha = 0$ case is trivial – it is a simple (spinless) band insulator, with the Hamiltonian being of purely on-site form, enforcing a local energy cost for the removal of a fermion: $H_0 = \sum_j \left(1 - 2c_j^\dagger c_j\right)$. The first nontrivial case is $\alpha = 1$, which corresponds to the Kitaev chain discussed in Section 9.4. Next, $\alpha = 2$, with two Majoranas on each edge, corresponds to two-dimensional degree of freedom at each edge, and so on for larger values of α. These noninteracting chains can thus be classified according to $\mathbb{Z}$, a single integer. They can simply be interpreted as straightforward stacks of chains.

However, the resulting degeneracy is not a priori stable to the addition of further terms to the Hamiltonian: clearly, allowing for general *local* terms coupling the α Majoranas at a given edge can lift these degeneracies at least partially.

Nonetheless, there are restrictions imposed by the underlying symmetries of the problem. The simplest to enforce is fermionic parity symmetry. Any term in the Hamiltonian needs to include a product of an even number of (complex or Majorana) fermion operators; otherwise, locality is lost, as two odd-fermion terms in the Hamiltonian with spatially well-separated support will not commute. This symmetry operation is related to the operator

$$P = \prod_j \left(1 - 2c_j^\dagger c_j\right) = \prod_j \left(i\tilde{\gamma}_j \gamma_j\right) \,. \tag{11.6}$$

This immediately makes the chains with even and odd values of α distinct – those which do, or do not, host a Majorana edge mode, as mentioned above. This follows from writing

$$P = i^\alpha \left[\prod_{j=1}^{\alpha} \gamma_j \right] \left(\prod_{j=1}^{N-\alpha} i\tilde{\gamma}_j \gamma_{j+\alpha} \right) \left[\prod_{j=N-\alpha+1}^{N} \tilde{\gamma}_j \right] . \tag{11.7}$$

The term in the middle equals $(-1)^{N-\alpha}$ in the ground state. Defining the left and right operators in square brackets as $P_{L,R}$, respectively, then yields $P = (-1)^{N-\alpha} P_L P_R$ with the two topologically distinct possibilities α even or odd in

$$P_L P_R = (-1)^\alpha P_R P_L . \tag{11.8}$$

For the case of $\alpha = 2$, there even exist bilinear terms acting on the two Majorana pairs, $i\gamma_1\gamma_2$ and $i\tilde{\gamma}_{N-1}\tilde{\gamma}_N$ which would move the respective edge modes to a finite energy. However, these terms are in turn forbidden by a combination of time-reversal symmetry, T, and parity symmetry, P. For the case of spinless Fermion operators, T simply amounts to complex conjugation. Thus, demanding for the local complex Fermion operators that $Tc_jT = c_j$ leads to $T\gamma_jT = \gamma_j$ and $T\tilde{\gamma}_jT = -\tilde{\gamma}_j$, so that the term $Ti\tilde{\gamma}_{N-1}\tilde{\gamma}_NT = -i\tilde{\gamma}_{N-1}\tilde{\gamma}_N$ is forbidden by this symmetry. Similarly, the term $i\gamma_1\gamma_2$ is forbidden. These symmetries hence protect the respective edge degeneracies. The resulting ground-state degeneracy is then fourfold, encoded by one complex Fermion at each edge of the chain.

For $\alpha = 4$, another four-Fermion term becomes available for each edge, $i\gamma_1\gamma_2\gamma_3\gamma_4$. Adding such a term only lifts the degeneracy partially, as can be verified by a direct computation, leaving behind a twofold degeneracy of each edge, which is in turn protected by time-reversal symmetry. Since $i\gamma_1\gamma_2\gamma_3\gamma_4$ can be interpreted as an edge parity operator, the remaining edge mode is "bosonic."

Finally, for $\alpha = 8$, one reaches the end of the road: generic many-fermion interactions can now lift the edge degeneracy completely. This identifies the interacting 8-chain with the trivial 0-chain band insulator in the topological classification. These results underpin the eightfold classification of interacting fermionic models in one dimension: each such model is labeled by a number in $\mathbb{Z}_8$, as per the original classification of Fidkowski and Kitaev (2011).

We continue this discussion with a number of observations the aim of which is to embed these results in the broader fabric of this book (and this field).

This concerns the connection between the general α-chains, and other more familiar models. One can already try to guess what these may be from considering the edge degeneracies. We have already come across the SSH-chain and the Haldane/AKLT chain, both of which have an edge degeneracy of 4, due to a twofold degeneracy at each edge individually.

These could thus correspond to the $(\pm)2$-chain or the 4-chain–indeed, it turns out that the SSH model corresponds to the former, while the AKLT chain corresponds to the latter. This can be established by an explicit mapping between the models, which we provide explicitly for the case of the SSH chain. Given the basic degrees of freedom are distinct, we first need to identify a connection between the Majoranas of the Kitaev 2-chain, and the complex fermions of featuring in the Hamiltonian of the SSH model:

$$H_{\text{SSH}}/2 = (1 - \lambda) \sum_n (c_{A,n}^\dagger c_{B,n} + h.c.) + \lambda \sum_n (c_{A,n+1}^\dagger c_{B,n} + h.c.) , \qquad (11.9)$$

where $J \in \{A, B\}$ labels the sublattice.

Collecting the Majoranas into quartets, and endowing them with an additional sublattice label:

$$\gamma_{A,n} = \gamma_{2n}; \ \tilde{\gamma}_{A,n} = \gamma_{2n-1}; \ \gamma_{B,n} = \tilde{\gamma}_{2n-1}; \ \tilde{\gamma}_{B,n} = -\tilde{\gamma}_{2n} , \qquad (11.10)$$

and defining complex fermions in the usual way, $2c_{J,n} = \gamma_{J,n} + i\tilde{\gamma}_{J,n}$, yields the Hamiltonian as a sum of 0- and 2-chain Hamiltonians:

$$H_{\text{SSH}} = (1 - \lambda) \sum_n i\tilde{\gamma}_n \gamma_n + \lambda \sum_n i\tilde{\gamma}_n \gamma_{n+2} . \qquad (11.11)$$

Note that these transformations are all strictly local, involving only the fermions in one unit cell of the SSH chain. This thus presents an explicit identification of the Kitaev 2-chain, that is, $\lambda = 1$, with the topological SSH chain, while the 0-chain, $\lambda = 0$, corresponds to the trivial version. Following through the mapping of the symmetries in turn identifies the T and PT-symmetries with the particle-hole and sublattice symmetries.

There is no end to further facets, and considerable subtleties, that SPT physics present. We close this short account by mentioning a few of them.

First, the constructions establishing the equivalence of the 4-chain with the Haldane/AKLT chain and of the 8-chain with the trivial band insulator proceed along similar lines of arguments as the one given for the SSH chain above. Indeed, one can write a stack of two SSH chains as a single SSH chain for spinful fermions. In an appropriate Mott limit, this then becomes the Haldane chain. This chimes with the idea that two 2-chains give a 4-chain. Then, a stack of two Haldane chains is topologically trivial, as it corresponds to two stacked 4-chains, and hence an 8-chain, and hence a 0-chain. This illustrates how the group structure of the classification corresponds to stacking, thereby providing a physically rather intuitive picture.

Some of the subtleties involved are mathematical in nature, while others are physical. For instance, for any given model, it is a question of choice which of its symmetries one considers as indispensable, and hence candidates for providing symmetry protection; and which as incidental, and hence to be broken by generic interaction terms. A case in point is that a generic disorder term to which the Kitaev chain is stable will in general not preserve the sublattice symmetry of the "equivalent" SSH chain; to do so, would require fine-tuning in the Kitaev picture. There is a connection in this model to the matrix product states discussed in the next section, where symmetry fractionalization occurs naturally. Finally, the present treatment has been restricted to one-dimensional systems, in particular (stacked) chains. Much remains to be done in higher dimensions, where in particular the question of what is a complete classification jointly involving symmetry and topology remains an open question.

11.2 Entanglement Entropy in Topological States

There are many kinds of quantum information, and many ways to quantify them, but probably the most influential to date for basic questions in condensed matter physics has been the entanglement entropy of a pure state of a bipartite system.[1] A bipartite system means that the Hilbert space AB is a product of two smaller Hilbert spaces A and B, which frequently represent different spatial regions. Having described in Chapter 9 how non-Abelian topological platforms can be used to construct complex quantum states and protect them from decoherence, we now turn to the more basic question of what makes a quantum state complex in classical terms, and how that reveals properties of some topological states.

We can see from considering the wavefunction of a spin singlet that, even though product states (i.e., states that can be represented as a product of a pure state of subsystem A with a pure state of subsystem B) can be used to construct a basis for AB, not every state in AB is a product state. There is no way to write the singlet $(|\uparrow_A\rangle \otimes |\downarrow_B\rangle - |\downarrow_A\rangle \otimes |\uparrow_B\rangle)/\sqrt{2}$ as a product of a pure state of spin A and one of spin B.

We can quantify that the singlet is not a product state by introducing the entanglement entropy of a pure quantum state $|\Psi\rangle$, defined as the von Neumann entropy of the reduced density matrix created by a partition of the system into parts A and B:

$$S = -\text{Tr}\,\rho_A \log_2 \rho_A = -\text{Tr}\,\rho_B \log_2 \rho_B. \tag{11.12}$$

Here the base-2 logarithm means that the entropy is measured in bits. The reduced density matrix ρ_A, which is sufficient to compute expectation values of operators localized on A, is obtained by a partial trace over a basis in B. Its matrix elements between two states $\phi_{1,2}$ in the Hilbert space A are

$$\langle \phi_1 | \rho_A | \phi_2 \rangle = \sum_j (\langle \phi_1 | \times \langle \psi_j |) | \Psi \rangle \langle \Psi | (| \phi_2 \rangle \times | \psi_j \rangle). \tag{11.13}$$

---

[1] The usage of the word bipartite here differs from that of bipartite lattice, where each site on one of two sublattices only has neighbors on the opposite sublattice.

So S quantifies to what degree ρ_A describes a mixed state even though $|\Psi\rangle$ was a pure state.

Additional information on entanglement measures for a newcomer to quantum information can be found in the textbook by Nielsen and Chuang (2010). If the original state cannot be written as a product state, then the entanglement entropy is nonzero. Since it is determined by the reduced density matrix for a subsystem, which characterizes all physical measurements on that subsystem, there is no way to distinguish through measurements only on A whether an entropy arose from partition of an entangled pure state of AB or from a mixed state of AB.

In classical mechanics, a complete description of a system of N particles implies a complete description of any subset of the particles. Entanglement entropy captures the remarkable property of quantum mechanics that knowing the state of the full system maximally, in the sense that we do not think there is a better description than a pure state, does not imply a maximal description of the parts of the system. Entanglement entropy turns out to be essential in understanding how pure quantum systems may or may not thermalize, which is important for the dynamical questions in Chapter 10: if a pure state shows thermal behavior in local experiments, it must mean that the apparent thermal entropy is generated by the process of looking at part of the system, since the whole state has zero entropy.

Here we focus on another use of entanglement entropy, which is in the analysis of ground states of extended systems. To compute Eq. 11.12, we need to specify both a pure state and a partition of the Hilbert space. We will focus on real-space partitions; for example, in a spin model with two states per site, we might take A to be the Hilbert space built from the basis states of a contiguous block of spins, and B to be the Hilbert space of all the other spins. Most low-energy states arising in condensed matter are quite atypical or nongeneric from the point of view of the full Hilbert space AB. We would first like to explain is that fractionalization and topological order have a subtle signature in the entanglement entropy with respect to certain spatial partitions. This result goes under the heading of long-range entanglement, and fractionalized phases are at times called long-range entangled, in distinction to short-range entangled phases such as the AKLT chain, or symmetry-protected phases more generally.

The lowest-energy states of a local Hamiltonian (one with short-ranged interactions, say) inherit a degree of locality: one typically finds an area law for entanglement.[2] This means that for a simple real-space partition defining A, say a hypercube of side L, the entanglement entropy does not scale with the volume L^d but rather with the boundary L^{d-1}. For a state of dimers representing spin singlets, each spin singlet crossing the boundary generates one bit of entanglement

---

[2] Not to be confused with the area law of the Wilson loop (Section 6.2.1).

entropy, so short-ranged dimers yield an area law, and this result survives generically in ground states of gapped local theories even while there can be some degree of entanglement between far-away sites. There are many exceptions known to the area law even in translation-invariant local systems when the gap vanishes, for example in quantum critical systems in one spatial dimension described by conformal field theories, treated in detail by Calabrese and Cardy (2004); in systems with a Fermi surface in any dimension (Gioev and Klich, 2006; Wolf, 2006) via special properties of entanglement for free fermions (Peschel, 2003); and in the Rokhsar–Kivelson quantum dimer model (previously seen in Section 5.3.1) on the square lattice (Fradkin and Moore, 2006). Note also that finite-energy-density states of thermalizing systems should be expected to have volume-law rather than area-law entanglement entropy, in order for the partial trace to generate extensive thermal entropies as mentioned above.

The surprise in fractional topological states with an energy gap, such as the $\mathbb{Z}_2$ spin liquid or the Laughlin state at $\nu = 1/m$, is that there is a characteristic subleading correction to the area law. By careful consideration of topologically distinct partitions of the plane, it was shown independently in Kitaev and Preskill (2006) and Levin and Wen (2006) that the entanglement entropy of the ground state of a topological state, generated by partitioning a *simply connected* region out of the plane, behaves as

$$S_L = \alpha L - \gamma + \mathcal{O}(L^{-\nu}), \quad \text{for some } \nu > 0. \quad (11.14)$$

Here L is the linear size of the region. The first subleading correction γ is given by the logarithm of the quantum dimension of the topological state. The quantum dimension is the number that determines topological degeneracy on various manifolds, and is a sum over the dimension d_i of different quasiparticle types (Nayak et al., 2008):

$$\gamma = \log \mathcal{D} = \log \sqrt{\sum_{i=1}^{n} d_i^2}. \quad (11.15)$$

For the toric code, $\mathcal{D} = 2$, and for the Laughlin state, $\mathcal{D} = \sqrt{m}$. The integer Hall effect has no fractional particles or ground-state degeneracy and consequently $\mathcal{D} = 1$, $\gamma = 0$. The logarithm is taken to the same base as in the definition of S. As γ is a property of the topological quasiparticles, it is thus more universal (insensitive to short-distance physics) than the area law coefficient α. The entanglement spectrum, which is the set of eigenvalues of the reduced density matrix that make up the von Neumann entropy, is another tool to diagnose topological order by revealing structures similar to topological edge states (Li and Haldane, 2008).

The main use of relations such as Eq. 11.14 to date has not been in experiments; entanglement entropy is difficult to measure even in few-body systems and γ would be difficult to discern in the presence of a dominant area law contribution in any event. The topological entanglement entropy has been very important in numerics, however, as a way to distinguish different topological states. Indeed, a fitting note on which to close this section is to explain the significant recent advances in one category of numerical methods related to these quantum information concepts.

Historically, exact diagonalization was the most influential source of numerical information about strongly correlated topological states of both electrons and spins, followed closely by quantum Monte Carlo methods, which in cases without a sign problem (such as the quantum dimer model) can treat very large systems. A third class of methods originate in the density-matrix renormalization group (DMRG) invented by White (1992). These work by representing the many-particle quantum state as a particular kind of wavefunction, variously known as a tensor network or matrix product state (Schollwöck, 2011). We previously saw the compactness of this representation for states such as the AKLT state in Section 5.2.4 around Eq. 5.39.

The virtue of this representation is that it preserves entanglement at short distances while rejecting the weak long-distance entanglement that may not be physically very significant. In recent years, DMRG-type methods have come into their own as a source of information on where fractional topological phases appear in the phase diagram of physically motivated Hamiltonians. Applying perturbations such as fluxes to the two-dimensional space where the calculations are performed allows the quasiparticles and other properties to be determined (Zaletel et al., 2013). The matrix product state representation also provides insight into the structure of more complicated wavefunctions than the AKLT ground state, such as the Moore–Read state (Zaletel and Mong, 2012). DMRG and other numerical methods are providing an increasingly substantial and valuable bridge between the ideal models that are analytically tractable and the complexities of an actual experiment.

11.3 The Universe of Topological Materials; Closing Remarks

The weakly interacting topological materials, topological insulators and semimetals, are also seeing the impact of numerical methods in a new way. We have not discussed until now in any serious way the impact of crystalline symmetries on topological band structure. One simple reason is that there are 230 space groups of crystalline symmetries when time-reversal symmetry is unbroken, and 1651 magnetic (Shubnikov) space groups once time-reversal-breaking is included as a possibility, and the interaction between these symmetries and possible topological invariants is difficult to sketch concisely.

That relationship has been worked out, however, to the extent that it is now possible with density functional theory methods to search for candidate topological materials in databases containing tens of thousands of known compounds. An introduction to the theoretical underpinnings used in this category of topological materials search is Bradlyn et al. (2017). A number of candidate topological crystalline insulators (i.e., topological insulators protected by crystal symmetries) have been found using this approach, complementing those found by traditional means, along with various types of semimetals.

A simple example not requiring such sophisticated analysis is the antiferromagnetic topological insulator phase: even if time-reversal Θ is broken, a material could be invariant under a mathematically similar antiunitary symmetry $\Theta T_{1/2}$, where $T_{1/2}$ is a translational symmetry (Mong et al., 2010). In the simplest case of a bipartite uniaxial antiferromagnet, the translation $T_{1/2}$ goes between the two sublattices, and moving from one sublattice to the other while reversing time to flip spin is an antiunitary symmetry that squares to -1. This phase was recently discovered in $MnBi_2Te_4$ (Otrokov et al., 2019), which can be viewed roughly as layers of the canonical topological insulator Bi_2Te_3 separated by single layers of the magnet MnTe, with magnetic order alternating between the layers. Many candidates for various kinds of independent-electron topological materials, including semimetals, have now been proposed.

Computational materials searches can be useful for more strongly correlated materials such as spin liquids as well. Here the idea is to use electron counting and a sophisticated analysis of symmetry to rule out gapped single-particle insulators, whether ordinary or topological, so that if a gapped insulator is observed in a material in this symmetry class, it must be a consequence of strong correlations, as in a spin liquid (Watanabe et al., 2015). The approach is based on generalizing the familiar notion that there are no independent-electron band insulators with an odd number of electrons per unit cell. Finally, it is possible for topological single-particle bands to arise as an effective description via strong correlations; an influential example is the topological Kondo insulator phase (a topological insulator generated by Kondo-type correlations), possibly relevant to SmB_6 (Dzero et al., 2010).

Through discoveries like these, topological materials have led to a rapprochement between what had started to become two distinct communities in the theory of solids, one based on the sort of analysis and simplified models in this book, and one based on computational approaches to real materials. It is clear that the Berry-curvature gauge fields in the Brillouin zone, and related ideas like the quantum metric and its associated Christoffel symbols, are now here to stay in electronic structure calculations (Vanderbilt, 2018). Another link is between topology in condensed matter physics and problems in high energy physics such as anomalies and dualities in field theory. Work on dualities, which are deep connections between seemingly different field theories, was stimulated by a new proposal about how to look at the metallic state at $\nu = 1/2$ by Son (2015).

Many challenges remain for both theory and experiment, but correspondingly major efforts are starting to address some of these, such as searches for clearly established Majorana fermions in a topological superconductor. The number of discoveries in this field even during the writing of this book serves as a reminder that historically, whenever people start to feel that an area of physics is nearing completion, surprises and new applications have emerged. We feel confident that, however impressive we find the discoveries recorded in this book, equally profound discoveries await where the complexity of real materials meets the austere beauty of topology.

Appendix: Useful Sources, Quantities, and Equations

We have tried not to assume that readers have an unnecessarily high level of preparation, and attempted to make our presentation self-contained. Nonetheless, as it is our hope that this book will be read by a diverse audience, it seems worthwhile to provide sources for background material such as standard solid-state physics that may not be familiar to all readers. An excellent recent introduction to the solid state, approachable for advanced undergraduates, is Simon (2013). The classic comprehensive volume used in many graduate courses is Ashcroft and Mermin (1976), published just before the first topological phases were discovered; recently a long-awaited alternative has appeared that covers the conventional material with an eye to topological applications (Girvin and Yang, 2019).

A pedagogical little book that is useful both for background on the underlying physics of spin models and for some of the key advanced results is Auerbach (1994). People wishing to understand more of the mathematical structures from Chapter 2 without getting bogged down in proofs or unnecessary mathematical rigor will find Nakahara (1998) valuable; a proper mathematical introduction to algebraic topology is Fulton (1995). A treatment of topological insulators and superconductors, including many concrete lattice models, is (Bernevig and Hughes, 2013), and a review article on this subject is Qi and Zhang (2011), which like the other reviews mentioned here has a much more comprehensive bibliography than we can provide.

Readers with an interest in electronic structure concepts and methods for topological materials are highly encouraged to look at the recent book by Vanderbilt (2018). A recent graduate solid-state textbook that incorporates uses of the Berry phase is Cohen and Louie (2016). Many important properties of metals coming from the Berry curvature and other geometric objects of the Bloch states are best understood through the semiclassical approach described in the review by Xiao et al. (2010). A recent review article on Dirac and Weyl semimetals with experimental context is Armitage et al. (2018), and optical properties related to wavefunction geometries are reviewed in Orenstein et al. (2020).

We are aware that for many topics, particularly in the later chapters, our treatment will come across to an expert working in the field as decidedly incomplete. There are several books focused on the quantum Hall effects that go into considerably more detail than we can here (Sarma and Pinczuk, 1996; Jain, 2007). An overview of many-body physics by one of the first to identify the nature of topological order is (Wen, 2007). The notion of fractional charge is explored in a review by Rajaraman (2001).

The field of spin liquids has grown and morphed over the years, and there exist various reviews. An accessible short and longer introduction are (Balents, 2010) and (Knolle and Moessner, 2019); Kitaev spin liquids are covered from a theoretical and materials perspective in Hermanns et al. (2018) and Takagi et al. (2019), respectively.

An excellent review article on non-Abelian states for quantum computation is Nayak et al. (2008), and there is an online set of lecture notes by one of its co-authors (Simon, 2020). There are also books focused specifically on this area for both physicists (Pachos, 2012) and mathematicians (Wang, 2010). A very clear set of lecture notes on the various Kitaev models is by Kitaev and Laumann (2009). Finally, the article of Read and Rezayi (1999) on topological aspects of the $p + ip$ superconductor and Moore–Read state is neither a textbook nor a review but is highly recommended.

There are by now many applications of field theory in condensed matter, and topological phases are just one piece of that large field. Two classics in this area are by Fradkin (2013) and Tsvelik (2003). There are excellent recent works as well (Altland and Simons, 2010; Mudry, 2014), and the first of these treats many topics related to disordered systems. A new addition to the literature is Shankar (2017), which has particularly clear explanations of many of the key field-theory methods.

One minor challenge for the reader is that different sources listed here choose different systems for the quantities and equations of electromagnetism. The differences between the four most common systems are essentially fixed by two binary choices: whether one writes the dielectric constant in Coulomb's law with a non-trivial ϵ_0 or absorbs ϵ_0 into the definition of charge, and whether the electric and magnetic fields in the Lorentz force law have the same units or differ by a factor c. Two of the four possible systems are most commonly used in condensed matter physics and are shown in Table A.1, in the hope of minimizing confusion. We have chosen to use SI units in most places in this book as the key experimental results are typically reported in that system;[1] indeed, for two standard derived quantities used primarily in transport experiments the table lists their customary units.

---

[1] For example, at Bell Labs people spoke for many years with respect verging on awe about the voltmeters in Horst Störmer's lab, not about his statvoltmeters.

Table A.1 *Some useful quantities and equations*

Quantity or equation	SI (this book)	CGS-Gaussian
Speed of light c	2.99792468×10^8 m/s	$2.99792468 \times 10^{10}$ m/s
Reduced Planck constant $\hbar = h/(2\pi)$	$1.0545718 \times 10^{-34}$ J s	$1.0545718 \times 10^{-27}$ erg s
Elementary charge e	$1.602176634 \times 10^{-19}$ C	$4.80320425 \times 10^{-10}$ esu
Fine structure constant α	$\dfrac{e^2}{(4\pi\epsilon_0)\hbar c} = \dfrac{1}{137.05991}$	$\dfrac{e^2}{\hbar c} = \dfrac{1}{137.05991}$
Flux quantum[a] Φ_0	$\dfrac{h}{e} = 4.1356676 \times 10^{-15}$ Wb	$\dfrac{hc}{e} = 4.1\ldots \times 10^{-7}$ G cm^2
Josephson constant K_J	$\dfrac{2e}{h} = 483597.8484$ GHz/V	
von Klitzing constant R_K	$\dfrac{h}{e^2} = 25812.80745\ \Omega$	
Lorentz force law	$\mathbf{F} = q(\mathbf{E} + \mathbf{v} \times \mathbf{B})$	$\mathbf{F} = q(\mathbf{E} + (\mathbf{v}/c) \times \mathbf{B})$
Vacuum permittivity ϵ_0	$8.85418781 \times 10^{-12}$ F/m	$\dfrac{1}{4\pi}$
Vacuum permeability μ_0	$\dfrac{1}{c^2\epsilon_0}$	$\dfrac{4\pi}{c^2}$
Maxwell's equations	$\nabla \cdot \mathbf{E} = \dfrac{\rho}{\epsilon_0}$	$\nabla \cdot \mathbf{E} = 4\pi\rho$
	$\nabla \cdot \mathbf{B} = 0$	$\nabla \cdot \mathbf{B} = 0$
	$\nabla \times \mathbf{E} = -\dfrac{\partial \mathbf{B}}{\partial t}$	$\nabla \times \mathbf{E} = -\dfrac{1}{c}\dfrac{\partial \mathbf{B}}{\partial t}$
	$\nabla \times \mathbf{B} = \mu_0 \mathbf{j} + \mu_0\epsilon_0 \dfrac{\partial \mathbf{E}}{\partial t}$	$\nabla \times \mathbf{B} = \dfrac{4\pi}{c}\mathbf{j} + \dfrac{1}{c}\dfrac{\partial \mathbf{E}}{\partial t}$
Fields and potentials	$\mathbf{E} = -\nabla\phi - \dfrac{\partial \mathbf{A}}{\partial t}$	$\mathbf{E} = -\nabla\phi - \dfrac{1}{c}\dfrac{\partial \mathbf{A}}{\partial t}$
	$\mathbf{B} = \nabla \times \mathbf{A}$	$\mathbf{B} = \nabla \times \mathbf{A}$
Gauge transformation	$\phi \to \phi - \dfrac{\partial \chi}{\partial t}$	$\phi \to \phi - \dfrac{1}{c}\dfrac{\partial \chi}{\partial t}$
	$\mathbf{A} \to \mathbf{A} + \nabla\chi$	$\mathbf{A} \to \mathbf{A} + \nabla\chi$
	$\psi \to \psi e^{iq\chi/\hbar}$	$\psi \to \psi e^{iq\chi/\hbar c}$
One-particle Hamiltonian	$\dfrac{1}{2m}\left(\dfrac{\hbar}{i}\nabla - q\mathbf{A}\right)^2 + q\phi$	$\dfrac{1}{2m}\left(\dfrac{\hbar}{i}\nabla - \dfrac{q\mathbf{A}}{c}\right)^2 + q\phi$

[a]In the context of superconductivity, the term "flux quantum" usually denotes $\Phi_0/2$.

References

Abanin, Dmitry A., De Roeck, Wojciech, and Huveneers, François. 2015. Exponentially slow heating in periodically driven many-body systems. *Phys. Rev. Lett.*, **115**(25), 256803.

Abrahams, E., Anderson, P. W., Licciardello, D. C., and Ramakrishnan, T. V. 1979. Scaling theory of localization: Absence of quantum diffusion in two dimensions. *Phys. Rev. Lett.*, **42**(10), 673–676.

Agrawal, Manindra, Kayal, Neeraj, and Saxena, Nitin. 2004. PRIMES is in P. *Ann. Math.*, **160**, 781–793.

Alet, F., and Sorensen, E. S. 2000. Magnetization profiles and NMR spectra of doped Haldane chains at finite temperatures. *Phys. Rev. B*, **62**(21), 14116–14121.

Alicea, Jason, Oreg, Yuval, Refael, Gil, von Oppen, Felix, and Fisher, Matthew P. A. 2011. Non-Abelian statistics and topological quantum information processing in 1D wire networks. *Nat. Phys.*, **7**(5), 412–417.

Altland, Alexander, and Simons, Ben D. 2010. *Condensed Matter Field Theory*. 2nd ed. Cambridge University Press.

Altland, Alexander, and Zirnbauer, Martin R. 1997. Nonstandard symmetry classes in mesoscopic normal-superconducting hybrid structures. *Phys. Rev. B*, **55**(Jan), 1142–1161.

Anderson, P. W. 1952. An approximate quantum theory of the antiferromagnetic ground state. *Phys. Rev.*, **86**(5), 694–701.

Anderson, P. W. 1956. Ordering and antiferromagnetism in ferrites. *Phys. Rev.*, **102**(4), 1008–1013.

Anderson, P. W. 1972. More is different. *Science*, **177**(4047), 393–396.

Anderson, P. W. 1977. Local moments and localised states. Nobel lecture.

Anderson, P. W. 1987. The resonating valence bond state in La_2CuO_4 and superconductivity. *Science*, **235**(Mar), 1196–1198.

Anderson, P. W., and Rowell, J. M. 1963. Probable observation of the Josephson superconducting tunneling effect. *Phys. Rev. Lett.*, **10**(6), 230–232.

Anfuso, F., and Rosch, A. 2007. String order and adiabatic continuity of Haldane chains and band insulators. *Phys. Rev. B*, **75**(14), 144420.

Armitage, N. P., Mele, E. J., and Vishwanath, Ashvin. 2018. Weyl and Dirac semimetals in three-dimensional solids. *Rev. Mod. Phys.*, **90**(Jan), 015001.

Arovas, Daniel, Schrieffer, J. R., and Wilczek, Frank. 1984. Fractional statistics and the quantum Hall effect. *Phys. Rev. Lett.*, **53**(Aug), 722–723.

Ashcroft, N. W., and Mermin, N. D. 1976. *Solid State Physics*. Philadelphia: Saunders College.

Auerbach, A. 1994. *Interacting Electrons and Quantum Magnetism*. New York: Springer.

Avron, J. E., Seiler, R., and Simon, B. 1983. Homotopy and quantization in condensed matter physics. *Phys. Rev. Lett.,* **51**(1), 51–53.

Balents, L., Fisher, M. P., and Girvin, S. M. 2002. Fractionalization in an easy-axis Kagome antiferromagnet. *Phys. Rev. B,* **65**(22), 224412.

Balents, L. 2010. Spin liquids in frustrated magnets. *Nature,* **464**(7286), 199–208.

Bardarson, Jens H., and Moore, Joel E. 2013. Quantum interference and Aharonov-Bohm oscillations in topological insulators. *Rep. Prog. Phys.,* **76**(5), 056501.

Barrett, S. E., Dabbagh, G., Pfeiffer, L. N., West, K. W., and Tycko, R. 1995. Optically pumped NMR evidence for finite-size skyrmions in GaAs quantum wells near Landau level filling $\nu = 1$. *Phys. Rev. Lett.,* **74**(25), 5112–5115.

Beenakker, C. W. J. 2013. Search for Majorana fermions in superconductors. *Annu. Rev. Condens. Matter Phys.,* **4**(1), 113–136.

Bernevig, B. Andrei, and Hughes, Taylor L. 2013. *Topological Insulators and Topological Superconductors*. Princeton, NJ: Princeton University Press.

Bernevig, B. Andrei, Hughes, Taylor L., and Zhang, Shou-Cheng. 2006. Quantum spin Hall effect and topological phase transition in HgTe quantum wells. *Science,* **314**(5806), 1757–1761.

Berry, M. V. 1984. Quantal phase-factors accompanying adiabatic changes. *Proc. R. Soc. A,* **392**, 45–57.

Bishop, D. J., and Reppy, J. D. 1978. Study of the superfluid transition in two-dimensional ^{4}He films. *Phys. Rev. Lett.,* **40**, 1727–1730.

Blaizot, J. P., and Ripka, G. 1986. *Quantum Theory of Finite Systems*. Cambridge, MA: MIT Press.

Bonderson, Parsa, Kitaev, Alexei, and Shtengel, Kirill. 2006. Detecting non-abelian statistics in the $\nu = 5/2$ fractional quantum Hall state. *Phys. Rev. Lett.,* **96**(1), 016803.

Bradlyn, Barry, Elcoro, L., Cano, Jennifer, Vergniory, M. G., Wang, Zhijun, Felser, C., Aroyo, M. I., and Bernevig, B. Andrei. 2017. Topological quantum chemistry. *Nature,* **547**(7663), 298–305.

Bramwell, Steven T., and Gingras, Michel J. P. 2001. Spin Ice State in Frustrated Magnetic Pyrochlore Materials. *Science,* **294**(5546), 1495–1501.

Bravyi, Sergey. 2006. Universal quantum computation with the $\nu = 5/2$-fractional quantum Hall state. *Phys. Rev. A,* **73**(Apr), 042313.

Bychkov, Yu. A., and Rashba, E. I. 1984. Oscillatory effects and the magnetic susceptibility of carriers in inversion layers. *J. Phys. C: Solid State Phys.,* **17**(33), 6039–6045.

Calabrese, P., and Cardy, J. 2004. Entanglement entropy and quantum field theory. *J. Stat. Mech.,* **06**, P06002.

Cardy, J. L. 1996. *Scaling and Renormalization in Statistical Physics*. Cambridge: Cambridge University Press.

Carroll, Lewis. 1865. *Alice's Adventures in Wonderland*. New York: Macmillan.

Castelnovo, C., Moessner, R., and Sondhi, S. L. 2008. Magnetic monopoles in spin ice. *Nature,* **451**(7174), 42–45.

Castelnovo, C., Moessner, R., and Sondhi, S. L. 2012. Spin ice, fractionalization, and topological order. *Annu. Rev. Condens. Matter Phys.,* **3**(1), 35–55.

Castelnovo, Claudio, and Chamon, Claudio. 2007. Entanglement and topological entropy of the toric code at finite temperature. *Phys. Rev. B,* **76**(18), 184442.

Chaikin, P. M., and Lubensky, T. C. 1995. *Principles of Condensed Matter Physics*. Cambridge: Cambridge University Press.

Chalker, J. T., and Coddington, P. D. 1988. Percolation, quantum tunnelling and the integer Hall effect. *J. Phys. C: Solid State Phys.,* **21**(14), 2665–2679.

Chalker, J. T., and Dohmen, A. 1995. Three-dimensional disordered conductors in a strong magnetic field: Surface states and quantum Hall plateaus. *Phys. Rev. Lett.*, **75**(24), 4496–4499.

Chamon, C., Goerbig, M. O., Moessner, R., and Cugliandolo, L. F. 2017. *Topological Aspects of Condensed Matter Physics*. Lecture Notes of the Les Houches Summer School 103. Oxford: Oxford University Press.

Chamon, Claudio. 2005. Quantum glassiness in strongly correlated clean systems: An example of topological overprotection. *Phys. Rev. Lett.*, **94**(4), 040402.

Cho, Gil Young, and Moore, Joel E. 2011. Topological BF field theory description of topological insulators. *Ann. Phys.*, **326**(6), 1515–1535.

Cohen, Marvin L., and Louie, Steven G. 2016. *Fundamentals of Condensed Matter Physics*. Cambridge: Cambridge University Press.

Comtet, A., Jolicoeur, T., Ouvry, S., and David, F. (Eds.). 1999. *Aspects topologiques de la physique en basse dimension*. [Topological aspects of low dimensional systems]. Berlin: Springer.

de Juan, Fernando, Grushin, Adolfo G., Morimoto, Takahiro, and Moore, Joel E. 2017. Quantized circular photogalvanic effect in Weyl semimetals. *Nat. Commun.*, **8**(1), 15995.

DiFrancesco, P., Mathieu, P., and Senechal, D. 1997. *Conformal Field Theory*. Berlin: Springer.

Dunlap, D. H., and Kenkre, V. M. 1986. Dynamic localization of a charged particle moving under the influence of an electric field. *Phys. Rev. B*, **34**(Sep), 3625–3633.

Dzero, Maxim, Sun, Kai, Galitski, Victor, and Coleman, Piers. 2010. Topological Kondo insulators. *Phys. Rev. Lett.*, **104**, 106408.

Eckardt, André. 2017. Colloquium: Atomic quantum gases in periodically driven optical lattices. *Rev. Mod. Phys.*, **89**(1), 011004.

Elitzur, Shmuel, Moore, Gregory, Schwimmer, Adam, and Seiberg, Nathan. 1989. Remarks on the canonical quantization of the Chern–Simons–Witten theory. *Nucl. Phys. B*, **326**(1), 108–134.

Else, Dominic V., Bauer, Bela, and Nayak, Chetan. 2016. Floquet time crystals. *Phys. Rev. Lett.*, **117**(9), 090402.

Else, Dominic V., Monroe, Christopher, Nayak, Chetan, and Yao, Norman Y. 2020. Discrete time crystals. *Ann. Rev. Condens. Matter Phys.*, **11**(1), 467–499.

Essin, A. M., and Moore, J. E. 2007. Topological insulators beyond the Brillouin zone via Chern parity. *Phys. Rev. B*, **76**, 165307.

Essin, Andrew M., Moore, Joel E., and Vanderbilt, David. 2009. Magnetoelectric polarizability and axion electrodynamics in crystalline insulators. *Phys. Rev. Lett.*, **102**(14), 146805.

Essin, Andrew M., Turner, Ari M., Moore, Joel E., and Vanderbilt, David. 2010. Orbital magnetoelectric coupling in band insulators. *Phys. Rev. B*, **81**, 205104.

Fazekas, P., and Anderson, P. W. 1974. On the ground state properties of the anisotropic triangular antiferromagnet. *Philos. Mag.*, **30**(2), 423–440.

Feng, Xiao-Yong, Zhang, Guang-Ming, and Xiang, Tao. 2007. Topological characterization of quantum phase transitions in a S=1/2 spin model. *Phys. Rev. Lett.*, **98**, 087204.

Fennell, T., Deen, P. P., Wildes, A. R., Schmalzl, K., Prabhakaran, D., Boothroyd, A. T., Aldus, R. J., McMorrow, D. F., and Bramwell, S. T. 2009. Magnetic Coulomb phase in the spin ice $Ho_2Ti_2O_7$. *Science*, **326**(5951), 415.

Fertig, H. A., and Halperin, B. I. 1987. Transmission coefficient of an electron through a saddle-point potential in a magnetic field. *Phys. Rev. B*, **36**(Nov), 7969–7976.

Fidkowski, Lukasz, and Kitaev, Alexei. 2011. Topological phases of fermions in one dimension. *Phys. Rev. B*, **83**(Feb), 075103.

Fradkin, E., and Moore, J. E. 2006. Entanglement entropy of 2D conformal quantum critical points: Hearing the shape of a quantum drum. *Phys. Rev. Lett.*, **97**, 050404.

Fradkin, Eduardo. 2013. *Field Theories of Condensed Matter Physics*. 2nd ed. Cambridge: Cambridge University Press.

Fradkin, Eduardo, and Shenker, Stephen H. 1979. Phase diagrams of lattice gauge theories with Higgs fields. *Phys. Rev. D*, **19**(Jun), 3682–3697.

Fredenhagen, Klaus, and Marcu, Mihail. 1986. Confinement criterion for QCD with dynamical quarks. *Phys. Rev. Lett.*, **56**(3), 223–224.

Freed, Daniel S., and Moore, Gregory W. 2013. Twisted equivariant matter. *Ann. Henri Poincaré*, **14**(8), 1927–2023.

Friess, B. 2016. *Spin and Charge Ordering in the Quantum Hall Regime*. Berlin: Springer International.

Fu, L., Kane, C. L., and Mele, E. J. 2007. Topological insulators in three dimensions. *Phys. Rev. Lett.*, **98**, 106803.

Fu, Liang, and Kane, C. L. 2006. Time reversal polarization and a Z_2 adiabatic spin pump. *Phys. Rev. B*, **74**(Nov), 195312.

Fu, Liang, and Kane, C. L. 2007. Topological insulators with inversion symmetry. *Phys. Rev. B*, **76**, 045302.

Fulton, W. 1995. *Algebraic Topology: A First Course*. Berlin: Springer.

Gaitan, F. 2013. *Quantum Error Correction and Fault Tolerant Quantum Computing*. Boca Raton, FL: CRC Press.

Gioev, D., and Klich, I. 2006. Entanglement entropy of fermions in any dimension and the Widom conjecture. *Phys. Rev. Lett.*, **96**, 100503.

Girvin, S. M., MacDonald, A. H., and Platzman, P. M. 1986. Magneto-roton theory of collective excitations in the fractional quantum Hall effect. *Phys. Rev. B*, **33**(Feb), 2481–2494.

Girvin, S. M., and Yang, K. 2019. *Modern Condensed Matter Physics*. Cambridge: Cambridge University Press.

Goerbig, M. O. 2011. Electronic properties of graphene in a strong magnetic field. *Rev. Mod. Phys.*, **83**(4), 1193–1243.

Guggenheim, E. A. 1945. The principle of corresponding states. *J. Chem. Phys.*, **13**(7), 253–261.

Haldane, F. D. M. 1981. "Luttinger liquid theory" of one-dimensional quantum fluids. I. Properties of the Luttinger model and their extension to the general 1D interacting spinless Fermi gas. *J. Phys. C: Solid State Phys.*, **14**(19), 2585–2609.

Haldane, F. D. M. 1983a. Fractional quantization of the Hall effect–a hierarchy of incompressible quantum fluid states. *Phys. Rev. Lett.*, **51**, 605–608.

Haldane, F. D. M. 1983b. Nonlinear field theory of large-spin Heisenberg antiferromagnets: Semiclassically quantized solitons of the one-dimensional easy-axis Néel state. *Phys. Rev. Lett.*, **50**(Apr), 1153–1156.

Haldane, F. D. M. 1988. Model for a quantum Hall effect without Landau levels: Condensed-matter realization of the "parity anomaly." *Phys. Rev. Lett.*, **61**(Oct.), 2015–2018.

Halperin, B. I. 1982. Quantized Hall conductance, current-carrying edge states, and the existence of extended states in a two-dimensional disordered potential. *Phys. Rev. B*, **25**, 2185.

Halperin, B. I. 1984. Statistics of quasiparticles and the hierarchy of fractional quantized Hall states. *Phys. Rev. Lett.*, **52**, 1586.

Halperin, B. I., Lee, Patrick A., and Read, Nicholas. 1993. Theory of the half-filled Landau level. *Phys. Rev. B*, **47**(Mar), 7312–7343.

Hansson, T. H., Oganesyan, Vadim, and Sondhi, S. L. 2004. Superconductors are topologically ordered. *Ann. Phys.*, **313**(2), 497–538.

Harper, Fenner, Roy, Rahul, Rudner, Mark S., and Sondhi, S. L. 2020. Topology and broken symmetry in Floquet systems. *Annu. Rev. Condens. Matter Phys.*, **11**(1), 345–368.

Harris, M. J., Bramwell, S. T., McMorrow, D. F., Zeiske, T., and Godfrey, K. W. 1997. Geometrical frustration in the ferromagnetic pyrochlore $Ho_2Ti_2O_7$. *Phys. Rev. Lett.*, **79**(13), 2554–2557.

Heller, P., and Benedek, G. B. 1962. Nuclear magnetic resonance in MnF_2 near the critical point. *Phys. Rev. Lett.*, **8**(11), 428–432.

Henley, Christopher L. 2010. The "Coulomb phase" in frustrated systems. *Annu. Rev. Condens. Matter Phys.*, **1**(Apr), 179–210.

Hermanns, M., Kimchi, I., and Knolle, J. 2018. Physics of the Kitaev model: Fractionalization, dynamic correlations, and material connections. *Annu. Rev. Condens. Matter Phys.*, **9**(Mar), 17–33.

Hermele, Michael, Fisher, Matthew P., and Balents, Leon. 2004. Pyrochlore photons: The U(1) spin liquid in a S=1/2 three-dimensional frustrated magnet. *Phys. Rev. B*, **69**(6), 064404.

Hofstadter, Douglas R. 1976. Energy levels and wave functions of Bloch electrons in rational and irrational magnetic fields. *Phys. Rev. B*, **14**(Sep), 2239–2249.

Hoyos, Carlos, and Son, Dam Thanh. 2012. Hall viscosity and electromagnetic response. *Phys. Rev. Lett.*, **108**(Feb), 066805.

Hsieh, D., Qian, D., Wray, L., Xia, Y., Hor, Y. S., Cava, R. J., and Hasan, M. Z. 2008. A topological Dirac insulator in a quantum spin Hall phase. *Nature*, **452**, 970.

Huckestein, Bodo. 1995. Scaling theory of the integer quantum Hall effect. *Rev. Mod. Phys.*, **67**(Apr), 357–396.

Isakov, S. V., Gregor, K., Moessner, R., and Sondhi, S. L. 2004. Dipolar spin correlations in classical pyrochlore magnets. *Phys. Rev. Lett.*, **93**(16), 167204.

Ivanov, D. A. 2001. Non-abelian statistics of half-quantum vortices in *p*-wave superconductors. *Phys. Rev. Lett.*, **86**(Jan), 268–271.

Jackeli, G., and Khaliullin, G. 2009. Mott insulators in the strong spin-orbit coupling limit: From Heisenberg to a quantum compass and Kitaev models. *Phys. Rev. Lett.*, **102**(1), 017205.

Jackiw, R., and Rebbi, C. 1976. Solitons with fermion number 1/2. *Phys. Rev. D*, **13**(Jun), 3398–3409.

Jain, Jainendra K. 2007. *Composite Fermions*. Cambridge: Cambridge University Press.

José, Jorge V., Kadanoff, Leo P., Kirkpatrick, Scott, and Nelson, David R. 1977. Renormalization, vortices, and symmetry-breaking perturbations in the two-dimensional planar model. *Phys. Rev. B*, **16**(3), 1217–1241.

Josephson, B. D. 1962. Possible new effects in superconductive tunnelling. *Phys. Lett.*, **1**(7), 251–253.

Kane, C. L., and Mele, E. J. 2005a. Quantum spin Hall effect in graphene. *Phys. Rev. Lett.*, **95**, 146802.

Kane, C. L., and Mele, E. J. 2005b. Z_2 topological order and the quantum spin Hall effect. *Phys. Rev. Lett.*, **95**, 146802.

Kasahara, Y., Ohnishi, T., Mizukami, Y., et al. 2018. Majorana quantization and half-integer thermal quantum Hall effect in a Kitaev spin liquid. *Nature*, **559**(7713), 227–231.

Kasteleyn, P. W. 1961. The statistics of dimers on a lattice : I. The number of dimer arrangements on a quadratic lattice. *Physica*, **27**(12), 1209–1225.

Kaufmann, Ralph M., Li, Dan, and Wehefritz-Kaufmann, Birgit. 2016. Notes on topological insulators. *Rev. Math. Phys.*, **28**(10), 1630003.

Kharzeev, Dmitri E. 2014. The chiral magnetic effect and anomaly-induced transport. *Prog. Part. Nucl. Phys.*, **75**, 133–151.

Khemani, Vedika, Lazarides, Achilleas, Moessner, Roderich, and Sondhi, S. L. 2016. Phase structure of driven quantum systems. *Phys. Rev. Lett.*, **116**(25), 250401.

Khemani, Vedika, Moessner, Roderich, and Sondhi, S. L. 2019. A brief history of time crystals. arXiv:1910.10745.

Khmelnitskii, D. E. 1984. Quantum hall effect and additional oscillations of conductivity in weak magnetic fields. *Phys. Lett. A*, **106**, 182.

King-Smith, R. D., and Vanderbilt, D. 1993. Theory of polarization of crystalline solids. *Phys. Rev. B*, **47**, 1651.

Kitaev, A., and Preskill, J. 2006. Topological entanglement entropy. *Phys. Rev. Lett.*, **96**, 110404.

Kitaev, A. Yu. 1997. Quantum computations: Algorithms and error correction. *Russ. Math. Surv.*, **52**(6), 1191–1249.

Kitaev, A. Yu. 2001. Unpaired majorana fermions in quantum wires. *Phys.-Usp.*, **44**, 131–136.

Kitaev, A. Yu. 2003. Fault-tolerant quantum computation by anyons. *Ann. Phys.*, **303**(1), 2–30.

Kitaev, A. Yu. 2009. Periodic table for topological insulators and superconductors. *AIP Conf. Proc.*, **1134**, 22.

Kitaev, Alexei. 2006. Anyons in an exactly solved model and beyond. *Ann. Phys.*, **321**(1), 2–111.

Kitaev, Alexei, and Laumann, Chris. 2009. Topological phases and quantum computation. arXiv:0904.2771.

Kitagawa, K., Takayama, T., Matsumoto, Y., et al. 2018. A spin-orbital-entangled quantum liquid on a honeycomb lattice. *Nature*, **554**(7692), 341–345.

Klitzing, K. v., Dorda, G., and Pepper, M. 1980. New method for high-accuracy determination of the fine-structure constant based on quantized Hall resistance. *Phys. Rev. Lett.*, **45**(Aug), 494–497.

Knolle, J., and Moessner, R. 2019. A field guide to spin liquids. *Annu. Rev. Condens. Matter Phys.*, **10**(Mar), 451–472.

Knolle, J., Kovrizhin, D. L., Chalker, J. T., and Moessner, R. 2015. Dynamics of fractionalization in quantum spin liquids. *Phys. Rev. B*, **92**(11), 115127.

Koenig, M., Wiedmann, S., Bruene, C., Roth, A., Buhmann, H., Molenkamp, L. W., Qi, X.-L., and Zhang, S.-C. 2007. Quantum spin Hall insulator state in HgTe quantum wells. *Science*, **318**(Oct), 766.

Kogut, John B. 1979. An introduction to lattice gauge theory and spin systems. *Rev. Mod. Phys.*, **51**(4), 659–714.

Kosterlitz, J. M., and Thouless, D. J. 1973. Ordering, metastability, and phase transitions in two-dimensional systems. *J. Phys. C*, **6**, 1181–1203.

Kuwahara, Tomotaka, Mori, Takashi, and Saito, Keiji. 2016. Floquet–Magnus theory and generic transient dynamics in periodically driven many-body quantum systems. *Ann. Phys.*, **367**(Apr), 96–124.

Laughlin, R. B. 1983. Anomalous quantum Hall effect: An incompressible quantum fluid with fractionally charged excitations. *Phys. Rev. Lett.*, **50**(May), 1395–1398.

Lee, P. A., and Ramakrishnan, T. V. 1985. Disordered electronic systems. *Rev. Mod. Phys.*, **57**(2), 287.

Leggett, A. J. 1984. Nucleation of ^{3}He-B from the A phase: A cosmic-ray effect? *Phys. Rev. Lett.*, **53**(11), 1096–1099.

Leinaas, J. M., and Myrheim, J. 1977. On the theory of identical particles. *Nuovo Cimento B Serie*, **37**(1), 1–23.

Levin, M., and Wen, X. G. 2006. Detecting topological order in a ground state wave function. *Phys. Rev. Lett.*, **96**, 110405.

Li, H., and Haldane, F. D. M. 2008. Entanglement Spectrum as a generalization of entanglement entropy: Identification of topological order in non-abelian fractional quantum Hall effect states. *Phys. Rev. Lett.*, **101**, 010504.

Liang, S., Doucot, B., and Anderson, P. W. 1988. Some new variational resonating-valence-bond-type wave functions for the spin-1/2 antiferromagnetic Heisenberg model on a square lattice. *Phys. Rev. Lett.*, **61**(Jul), 365–368.

Lieb, Elliott H. 1994. Flux phase of the half-filled band. *Phys. Rev. Lett.*, **73**(16), 2158–2161.

Limot, L., Mendels, P., Collin, G., Mondelli, C., Ouladdiaf, B., Mutka, H., Blanchard, N., and Mekata, M. 2002. Susceptibility and dilution effects of the Kagomé bilayer geometrically frustrated network: A Ga NMR study of $SrCr_{9p}Ga_{12-9p}O_{19}$. *Phys. Rev. B*, **65**(14), 144447.

Lindner, Netanel H., Refael, Gil, and Galitski, Victor. 2011. Floquet topological insulator in semiconductor quantum wells. *Nat. Phys.*, **7**(6), 490–495.

Longuet-Higgins, Hugh Christopher, Öpik, U., Pryce, Maurice Henry Lecorney, and Sack, R. A. 1958. Studies of the Jahn–Teller effect .II. The dynamical problem. *Proc. R. Soc. London, Ser. A*, 244116.

Ludwig, Andreas W. W., Fisher, Matthew P. A., Shankar, R., and Grinstein, G. 1994. Integer quantum Hall transition: An alternative approach and exact results. *Phys. Rev. B*, **50**(Sep), 7526–7552.

Ma, Jing, and Pesin, D. A. 2015. Chiral magnetic effect and natural optical activity in metals with or without Weyl points. *Phys. Rev. B*, **92**(Dec), 235205.

Majumdar, Chanchal K., and Ghosh, Dipan K. 1969. On next-nearest-neighbor interaction in linear chain. I. *J. Math. Phys.*, **10**(8), 1388–1398.

Malashevich, A., Souza, I., Coh, S., and Vanderbilt, D. 2010. Theory of orbital magneto-electric response. *New J. Phys.*, **12**, 053032.

Maxwell, J. Clerk. 1864. L. On the calculation of the equilibrium and stiffness of frames. *London Edinburgh Dublin Philos. Mag. J. Sci.*, **27**(182), 294–299.

Mermin, N. D. 1979. The topological theory of defects in ordered media. *Rev. Mod. Phys.*, **51**(Jul), 591–648.

Moessner, R. 1997. *Two Systems with Macroscopically Degenerate Ground States*. PhD thesis, Oxford University.

Moessner, R., and Raman, K. S. 2008. Quantum dimer models. arXiv:0809.3051.

Moessner, R., and Sondhi, S. L. 2001a. Ising models of quantum frustration. *Phys. Rev. B*, **63**(22), 224401.

Moessner, R., and Sondhi, S. L. 2001b. Resonating valence bond phase in the triangular lattice quantum dimer model. *Phys. Rev. Lett.*, **86**(9), 1881–1884.

Moessner, R., and Sondhi, S. L. 2003a. Theory of the [111] magnetization plateau in spin ice. *Phys. Rev. B*, **68**(6), 064411.

Moessner, R., and Sondhi, S. L. 2003b. Three-dimensional resonating-valence-bond liquids and their excitations. *Phys. Rev. B*, **68**(18), 1881.

Moessner, R., Sondhi, S. L., and Fradkin, Eduardo. 2002. Short-ranged resonating valence bond physics, quantum dimer models, and Ising gauge theories. *Phys. Rev. B*, **65**(2), 024504.

Moessner, R., Tchernyshyov, Oleg, and Sondhi, S. L. 2004. Planar pyrochlore, quantum ice and sliding ice. *J. Stat. Phys.*, **116**(1-4), 755–772.

Moll, Philip J. W., Nair, Nityan L., Helm, Toni, Potter, Andrew C., Kimchi, Itamar, Vishwanath, Ashvin, and Analytis, James G. 2016. Transport evidence for Fermi-arc-mediated chirality transfer in the Dirac semimetal Cd_3As_2. *Nature*, **535**(7611), 266–270.

Mong, R., Essin, A. M., and Moore, J. E. 2010. Antiferromagnetic topological insulators. *Phys. Rev. B*, **81**, 245209.

Moore, G., and Read, N. 1991. Non-abelions in the fractional quantum Hall effect. *Nucl. Phys. B*, **360**, 362.

Moore, Gregory W., and Seiberg, Nathan. 1989. Taming the conformal zoo. *Phys. Lett. B*, **220**, 422–430.

Moore, J. E. 2010. The birth of topological insulators. *Nature*, **464**, 194.

Moore, J. E., and Balents, L. 2007. Topological invariants of time-reversal-invariant band structures. *Phys. Rev. B*, **75**, 121306(R).

Moore, J. E., Ran, Y., and Wen, X.-G. 2008. Topological surface states in three-dimensional magnetic insulators. *Phys. Rev. Lett.*, **101**(18), 186805.

Moore, J. E., and Orenstein, J. 2010. Confinement-induced Berry phase and helicity-dependent photocurrents. *Phys. Rev. Lett.*, **105**(2), 026805.

Morimoto, Takahiro, and Furusaki, Akira. 2013. Topological classification with additional symmetries from Clifford algebras. *Phys. Rev. B*, **88**(Sep), 125129.

Morris, D. J. P., Tennant, D. A., Grigera, S. A., et al.. 2009. Dirac strings and magnetic monopoles in the spin ice $Dy_2Ti_2O_7$. *Science*, **326**(5951), 411.

Motrunich, Olexei, Damle, Kedar, and Huse, David A. 2002. Particle-hole symmetric localization in two dimensions. *Phys. Rev. B*, **65**(6), 064206.

Mudry, Christopher. 2014. *Lecture Notes on Field Theory in Condensed Matter Physics*. Singapore: World Scientific.

Murakami, S., Nagaosa, N., and Zhang, S.-C. 2004. Spin-Hall insulator. *Phys. Rev. Lett.*, **93**, 156804.

Murakami, Shuichi. 2007. Phase transition between the quantum spin Hall and insulator phases in 3D: Emergence of a topological gapless phase. *New J. Phys.*, **9**(9), 356.

Nagaosa, Naoto, Sinova, Jairo, Onoda, Shigeki, MacDonald, A. H., and Ong, N. P. 2010. Anomalous Hall effect. *Rev. Mod. Phys.*, **82**(May), 1539–1592.

Nakahara, M. 1998. *Geometry, Topology and Physics*. Bristol: Institute of Physics.

Nandkishore, Rahul, and Huse, David A. 2015. Many-body localization and thermalization in quantum statistical mechanics. *Annu. Rev. Condens. Matter Phys.*, **6**(1), 15–38.

Nasu, J., Knolle, J., Kovrizhin, D. L., Motome, Y., and Moessner, R. 2016. Fermionic response from fractionalization in an insulating two-dimensional magnet. *Nat. Phys.*, **12**(10), 912–915.

Nayak, Chetan, and Wilczek, Frank. 1996. 2n-quasihole states realize 2^{n-1}-dimensional spinor braiding statistics in paired quantum Hall states. *Nucl. Phys. B*, **479**(3), 529–553.

Nayak, Chetan, Simon, Steven H., Stern, Ady, Freedman, Michael, and Das Sarma, Sankar. 2008. Non-Abelian anyons and topological quantum computation. *Rev. Mod. Phys.*, **80**(3), 1083–1159.

Nielsen, Michael A., and Chuang, Isaac L. 2010. *Quantum Computation and Quantum Information*. 10th Anniversary ed. Cambridge: Cambridge University Press.

Niu, Q., Thouless, D. J., and Wu, Y.-S. 1985. Quantized Hall conductance as a topological invariant. *Phys. Rev. B*, **31**, 3372.

Nussinov, Zohar, and van den Brink, Jeroen. 2015. Compass models: Theory and physical motivations. *Rev. Mod. Phys.*, **87**(1), 1–59.

Obradors, X., Labarta, A., Isalgué, A., Tejada, J., Rodriguez, J., and Pernet, M. 1988. Magnetic frustration and lattice dimensionality in $SrCr_8Ga_4O_{19}$. *Solid State Commun.*, **65**(3), 189–192.

Oka, Takashi, and Aoki, Hideo. 2009. Photovoltaic Hall effect in graphene. *Phys. Rev. B,* **79**(8), 081406.

Oka, Takashi, and Kitamura, Sota. 2019. Floquet engineering of quantum materials. *Annu. Rev. Condens. Matter Phys.*, **10**(1), 387–408.

Oreg, Yuval, Refael, Gil, and von Oppen, Felix. 2010. Helical liquids and Majorana bound states in quantum wires. *Phys. Rev. Lett.*, **105**(Oct), 177002.

Orenstein, J. W., et al. Forthcoming. *Ann. Condens. Matter Phys.*

Ortiz, Gerardo, and Martin, Richard M. 1994. Macroscopic polarization as a geometric quantum phase: Many-body formulation. *Phys. Rev. B*, **49**(May), 14202–14210.

Otrokov, M. M., Klimovskikh, I. I., Bentmann, H., et al. 2019. Prediction and observation of an antiferromagnetic topological insulator. *Nature*, **576**(7787), 416–422.

Pachos, Jiannis K. 2012. *Introduction to Topological Quantum Computation*. Cambridge: Cambridge University Press.

Panati, G., Spohn, H., and Teufel, S. 2003. Effective dynamics for Bloch electrons: Peierls substitution and beyond. *Commun. Math. Phys.*, **242**, 547.

Pancharatnam, S. 1956. Generalized theory of interference, and its applications. *Proc. Ind. Acad. Sci. Sect. A*, **44**(5), 247–262.

Parameswaran, Siddharth A., Sondhi, S. L., and Arovas, Daniel P. 2009. Order and disorder in AKLT antiferromagnets in three dimensions. *Phys. Rev. B*, **79**(2), 024408.

Pedrocchi, Fabio L., Chesi, Stefano, and Loss, Daniel. 2011. Physical solutions of the Kitaev honeycomb model. *Phys. Rev. B*, **84**(16), 165414.

Peierls, Rudolf. 1955. *Quantum Theory of Solids*. Oxford: Oxford University Press.

Peschel, Ingo. 2003. Calculation of reduced density matrices from correlation functions. *J. Phys. A*, **36**(14), L205–L208.

Pollmann, Frank, Berg, Erez, Turner, Ari M., and Oshikawa, Masaki. 2012. Symmetry protection of topological phases in one-dimensional quantum spin systems. *Phys. Rev. B*, **85**(7), 075125.

Polyakov, A. M. 1987. *Gauge Fields and Strings*. London: Harwood Academic.

Pretko, Michael, and Radzihovsky, Leo. 2018. Fracton-elasticity duality. *Phys. Rev. Lett.*, **120**(19), 195301.

Pretko, Michael, Chen, Xie, and You, Yizhi. 2020. Fracton phases of matter. *Int. J. Mod. Phys. A*, **35**(6), 2030003.

Qi, Xiao-Liang, and Zhang, Shou-Cheng. 2011. Topological insulators and superconductors. *Rev. Mod. Phys.*, **83**(Oct), 1057–1110.

Qi, Xiao-Liang, Hughes, Taylor L., and Zhang, Shou-Cheng. 2008. Topological field theory of time-reversal invariant insulators. *Phys. Rev. B*, **78**(19), 195424.

Quay, C. H. L., Hughes, T. L., Sulpizio, J. A., Pfeiffer, L. N., Baldwin, K. W., West, K. W., Goldhaber-Gordon, D., and de Picciotto, R. 2010. Observation of a one-dimensional spin–orbit gap in a quantum wire. *Nat. Phys.*, **6**(5), 336–339.

Rachel, Stephan, Fritz, Lars, and Vojta, Matthias. 2016. Landau levels of Majorana fermions in a spin liquid. *Phys. Rev. Lett.*, **116**(16), 167201.

Rahmani, Armin, and Franz, Marcel. 2019. Interacting Majorana fermions. *Rep. Prog. Phys.*, **82**(8), 084501.

Rajaraman, R. 2001. Fractional charge arxiv: cond-mat/0103366.

Ralko, Arnaud, Ferrero, Michel, Becca, Federico, Ivanov, Dmitri, and Mila, Frédéric. 2006. Dynamics of the quantum dimer model on the triangular lattice: Soft modes and local resonating valence-bond correlations. *Phys. Rev. B,* **74**(13), 134301.

Ramirez, A. P., Hayashi, A., Cava, R. J., Siddharthan, R., and Shastry, B. S. 1999. Zero-point entropy in "spin ice." *Nature*, **399**(6734), 333–335.

Ran, Y., Zhang, Y., and Vishwanath, A. 2009. One-dimensional topologically protected modes in topological insulators with lattice dislocations. *Nat. Phys.*, **5**, 298.

Read, N. 1989. Order parameter and Ginzburg–Landau theory for the fractional quantum Hall effect. *Phys. Rev. Lett.*, **62**(Jan), 86–89.

Read, N., and Green, Dmitry. 2000. Paired states of fermions in two dimensions with breaking of parity and time-reversal symmetries, and the fractional quantum Hall effect. *Phys. Rev. B*, **61**, 10267.

Read, N., and Rezayi, E. 1999. Beyond paired quantum Hall states: Parafermions and incompressible states in the first excited Landau level. *Phys. Rev. B*, **59**, 8084.

Rees, Dylan, Manna, Kaustuv, Lu, Baozhu, Morimoto, Takahiro, Borrmann, Horst, Felser, Claudia, Moore, J. E., Torchinsky, Darius H., and Orenstein, J. 2020. Helicity-dependent photocurrents in the chiral Weyl semimetal RhSi. *Sci. Adv.*, **6**(29), eaba0509.

Resta, R. 1992. Theory of the electric polarization in crystals. *Ferroelectrics*, **136**, 51.

Rice, M. J., and Mele, E. J. 1982. Elementary excitations of a linearly conjugated diatomic polymer. *Phys. Rev. Lett.*, **49**(Nov), 1455–1459.

Rokhsar, Daniel S., and Kivelson, Steven A. 1988. Superconductivity and the quantum hard-core dimer gas. *Phys. Rev. Lett.*, **61**(20), 2376–2379.

Roman, E., Mokrousov, Y., and Souza, I. 2009. Orientation dependence of the intrinsic anomalous Hall effect in hcp cobalt. *Phys. Rev. Lett.*, **103**, 097203.

Roy, R. 2009. Topological phases and the quantum spin Hall effect in three dimensions. *Phys. Rev. B*, **79**, 195322.

Rudner, Mark S., and Lindner, Netanel H. 2020. The Floquet engineer's handbook. arXiv:2003.08252.

Rudner, Mark S., Lindner, Netanel H., Berg, Erez, and Levin, Michael. 2013. Anomalous edge states and the bulk-edge correspondence for periodically driven two-dimensional systems. *Phys. Rev. X*, **3**(Jul), 031005.

Ryu, Shinsei, Schnyder, Andreas P., Furusaki, Akira, and Ludwig, Andreas W. W. 2010. Topological insulators and superconductors: Tenfold way and dimensional hierarchy. *New J. Phys.*, **12**(6), 065010.

Ryu, Shinsei, Moore, Joel E., and Ludwig, Andreas W. W. 2012. Electromagnetic and gravitational responses and anomalies in topological insulators and superconductors. *Phys. Rev. B*, **85**(Jan), 045104.

Sachdev, S. 1999. *Quantum Phase Transitions*. London: Cambridge University Press.

Saminadayar, L., Glattli, D. C., Jin, Y., and Etienne, B. 1997. Observation of the e/3 fractionally charged Laughlin quasiparticle. *Phys. Rev. Lett.*, **79**(13), 2526–2529.

Sarma, Sankar Das, and Pinczuk, Aron. 1996. *Perspectives in Quantum Hall Effects*. Hoboken, NJ: John Wiley.

Sau, J. D., Lutchyn, R. M., Tewari, S., and Sarma, S. Das. 2010. Generic new platform for topological quantum computation using semiconductor heterostructures. *Phys. Rev. Lett.*, **104**, 040502.

Savary, Lucile, and Balents, Leon. 2017. Disorder-induced quantum spin liquid in spin ice pyrochlores. *Phys. Rev. Lett.*, **118**(8), 087203.

Savit, Robert. 1980. Duality in field theory and statistical systems. *Rev. Mod. Phys.*, **52**(Apr), 453–487.

Schiffer, P., and Daruka, I. 1997. Two-population model for anomalous low-temperature magnetism in geometrically frustrated magnets. *Phys. Rev. B*, **56**(Dec), 13712–13715.

Schlom, Darrell G., and Pfeiffer, Loren N. 2010. Oxide electronics: Upward mobility rocks! *Nat. Mater.*, **9**(11), 881–883.

Schollwöck, Ulrich. 2011. The density-matrix renormalization group in the age of matrix product states. *Ann. Phys.*, **326**(1), 96–192.

Schottky, W. 1918. Über spontane Stromschwankungen in verschiedenen Elektrizitätsleitern. *Ann. Phys.*, **362**(23), 541–567.

Senthil, T., Vishwanath, Ashvin, Balents, Leon, Sachdev, Subir, and Fisher, Matthew P. A. 2004. Deconfined quantum critical points. *Science*, **303**(5663), 1490–1494.

Sethna, James P. 2006. *Statistical Mechanics: Entropy, Order Parameters, and Complexity.* Oxford: Oxford University Press.

Shankar, Ramamurti. 2017. *Quantum Field Theory and Condensed Matter: An Introduction.* Cambridge: Cambridge University Press.

Shapiro, Sidney. 1963. Josephson currents in superconducting tunneling: The effect of microwaves and other observations. *Phys. Rev. Lett.*, **11**(2), 80–82.

Shastry, B. Sriram, and Sutherland, Bill. 1981. Excitation spectrum of a dimerized next-neighbor antiferromagnetic chain. *Phys. Rev. Lett.*, **47**(13), 964–967.

Shor, Peter W. 1994. Algorithms for quantum computation: Discrete logarithms and factoring. Pages 124–134 of *Proceedings 35th Annual Symposium on Foundations of Computer Science.* Santa Fe, NM: IEEE.

Shor, Peter W. 1995. Scheme for reducing decoherence in quantum computer memory. *Phys. Rev. A*, **52**(Oct), R2493–R2496.

Simon, Steven H. 2013. *The Oxford Solid State Basics.* Oxford: Oxford University Press.

Simon, Steven H. 2020. *Topological Quantum: Lecture Notes and Proto-Book.* http://www-thphys.physics.ox.ac.uk/people/SteveSimon/.

Sodemann, Inti, and Fu, Liang. 2015. Quantum nonlinear Hall effect induced by Berry curvature dipole in time-reversal invariant materials. *Phys. Rev. Lett.*, **115**(Nov), 216806.

Solzhenitsyn, Aleksandr Isaevich. 1968. *The First Circle.* New York: Harper and Row.

Son, Dam Thanh. 2015. Is the composite fermion a Dirac particle? *Phys. Rev. X*, **5**(Sep), 031027.

Sondhi, S. L., Karlhede, A., Kivelson, S. A., and Rezayi, E. H. 1993. Skyrmions and the crossover from the integer to fractional quantum Hall effect at small Zeeman energies. *Phys. Rev. B*, **47**(24), 16419–16426.

Stern, Ady. 2008. Anyons and the quantum Hall effect: A pedagogical review. *Ann. Phys.*, **323**(1), 204–249.

Stern, Ady, von Oppen, Felix, and Mariani, Eros. 2004. Geometric phases and quantum entanglement as building blocks for non-abelian quasiparticle statistics. *Phys. Rev. B*, **70**, 205338.

Stone, M. 1994. *Bosonization.* Singapore: World Scientific.

Stone, Michael, and Goldbart, Paul. 2009. *Mathematics for Physics: A Guided Tour for Graduate Students.* Cambridge: Cambridge University Press.

Su, W. P., Schrieffer, J. R., and Heeger, A. J. 1979. Solitons in polyacetylene. *Phys. Rev. Lett.*, **42**(Jun), 1698–1701.

Svore, Krysta M., Divincenzo, David P., and Terhal, Barbara M. 2007. Noise threshold for a fault-tolerant two-dimensional lattice architecture. *Quantum Info. Comput.*, **7**(4), 297–318.

Takagi, Hidenori, Takayama, Tomohiro, Jackeli, George, Khaliullin, Giniyat, and Nagler, Stephen E. 2019. Concept and realization of Kitaev quantum spin liquids. *Nat. Rev. Phys.*, **1**(4), 264–280.

Tedoldi, F., Santachiara, R., and Horvatić, M. 1999. ^{89}Y NMR imaging of the staggered magnetization in the doped haldane chain $Y_2BaNi_{1-x}Mg_xO_5$. *Phys. Rev. Lett.*, **83**(2), 412–415.

Teo, Jeffrey C. Y., and Kane, C. L. 2014. From Luttinger liquid to non-abelian quantum Hall states. *Phys. Rev. B*, **89**(Feb), 085101.

Thonhauser, T., Ceresoli, Davide, Vanderbilt, David, and Resta, R. 2005. Orbital magnetization in periodic insulators. *Phys. Rev. Lett.*, **95**(Sep), 137205.

Thouless, D. J., Kohmoto, M., Nightingale, M. P., and den Nijs, M. 1982. Quantized Hall conductance in a two-dimensional periodic potential. *Phys. Rev. Lett.*, **49**(6), 405–408.

Titum, Paraj, Berg, Erez, Rudner, Mark S., Refael, Gil, and Lindner, Netanel H. 2016. Anomalous Floquet–Anderson insulator as a nonadiabatic quantized charge pump. *Phys. Rev. X*, **6**(May), 021013.

Tserkovnyak, Yaroslav, and Simon, Steven H. 2003. Monte Carlo evaluation of non-abelian statistics. *Phys. Rev. Lett.*, **90**(Jan), 016802.

Tsui, D. C., Stormer, H. L., and Gossard, A. C. 1982. Two-dimensional magnetotransport in the extreme quantum limit. *Phys. Rev. Lett.* , **48**(22), 1559–1562.

Tsvelik, Alexei M. 2003. *Quantum Field Theory in Condensed Matter Physics*. 2nd ed. Cambridge: Cambridge University Press.

Vanderbilt, David. 2018. *Berry Phases in Electronic Structure Theory: Electric Polarization, Orbital Magnetization and Topological Insulators*. Cambridge: Cambridge University Press.

Verresen, Ruben, Moessner, Roderich, and Pollmann, Frank. 2017. One-dimensional symmetry protected topological phases and their transitions. *Phys. Rev. B,* **96**(16), 165124.

Vijay, Sagar, Haah, Jeongwan, and Fu, Liang. 2016. Fracton topological order, generalized lattice gauge theory, and duality. *Phys. Rev. B,* **94**(23), 235157.

Volovik, G. E. 2012. *The Universe in a Helium Droplet*. Oxford: Oxford University Press.

von Keyserlingk, C. W., Khemani, Vedika, and Sondhi, S. L. 2016. Absolute stability and spatiotemporal long-range order in Floquet systems. *Phys. Rev. B* , **94**(8), 085112.

von Klitzing, Klaus. 2017. Metrology in 2019. *Nat. Phys.*, **13**(2), 198.

Vozmediano, M. A. H., Katsnelson, M. I., and Guinea, F. 2010. Gauge fields in graphene. *Phys. Rep.*, **496**(4–5), 109–148.

Wan, Xiangang, Turner, Ari M., Vishwanath, Ashvin, and Savrasov, Sergey Y. 2011. Topological semimetal and Fermi-arc surface states in the electronic structure of pyrochlore iridates. *Phys. Rev. B*, **83**(May), 205101.

Wang, Y. H., Steinberg, H., Jarillo-Herrero, P., and Gedik, N. 2013. Observation of Floquet–Bloch states on the surface of a topological insulator. *Science*, **342**(6157), 453–457.

Wang, Zhenghan. 2010. *Topological Quantum Computation*. CBMS Regional Conference Series in Mathematics 112. Providence, RI: American Mathematical Society.

Watanabe, Haruki, Po, Hoi Chun, Vishwanath, Ashvin, and Zaletel, Michael. 2015. Filling constraints for spin-orbit coupled insulators in symmorphic and nonsymmorphic crystals. *Proc. Nat. Acad. Sci.*, **112**(47), 14551–14556.

Wegner, Franz J. 1971. Duality in generalized Ising models and phase transitions without local order parameters. *J. Math. Phys.*, **12**(10), 2259–2272.

Wen, X.-G. 1992. Theory of the edge states in fractional quantum Hall effects. *Int. J. Mod. Phys. B*, **6**, 1711.

Wen, X. G., and Niu, Q. 1990. Ground-state degeneracy of the fractional quantum Hall states in the presence of a random potential and on high-genus Riemann surfaces. *Phys. Rev. B*, **41**(May), 9377–9396.

Wen, Xiao-Gang. 2007. *Quantum Field Theory of Many-Body Systems: From the Origin of Sound to an Origin of Light and Electrons*. Oxford: Oxford University Press.

Weyl, Hermann. 1939. Invariants. *Duke Math. J.*, **5**(3), 489–502.

White, Steven R. 1992. Density matrix formulation for quantum renormalization groups. *Phys. Rev. Lett.*, **69**(Nov), 2863–2866.

Wilczek, F. 1987. Two applications of axion electrodynamics. *Phys. Rev. Lett.*, **58**, 1799.

Willans, A. J., Chalker, J. T., and Moessner, R. 2010. Disorder in a quantum spin liquid: Flux binding and local moment formation. *Phys. Rev. Lett.*, **104**(23), 237203.

Willett, R., Eisenstein, J. P., Störmer, H. L., Tsui, D. C., Gossard, A. C., and English, J. H. 1987. Observation of an even-denominator quantum number in the fractional quantum Hall effect. *Phys. Rev. Lett.*, **59**(Oct), 1776–1779.

Winter, Stephen M., Tsirlin, Alexander A., Daghofer, Maria, van den Brink, Jeroen, Singh, Yogesh, Gegenwart, Philipp, and Valentí, Roser. 2017. Models and materials for generalized Kitaev magnetism. *J. Phys. Condens. Matter*, **29**(49), 493002.

Witten, Edward. 2016. Fermion path integrals and topological phases. *Rev. Mod. Phys.*, **88**(Jul), 035001.

Wolf, M. M. 2006. Violation of the entropic area law for fermions. *Phys. Rev. Lett.*, **96**, 010404.

Wong, T. 2018. *Quantum Computing, Universal Gate Sets.* https://www.scottaaronson.com/qclec/16.pdf.

Wu, Liang, Salehi, M., Koirala, N., Moon, J., Oh, S., and Armitage, N. P. 2016. Quantized Faraday and Kerr rotation and axion electrodynamics of a 3D topological insulator. *Science*, **354**(6316), 1124–1127.

Wu, Liang, Patankar, Shreyas, Morimoto, Takahiro, Nair, et al.. 2017. Giant anisotropic nonlinear optical response in transition metal monopnictide Weyl semimetals. *Nat. Phys.*, **13**(4), 350.

Xia, Y., Qian, D., Hsieh, D., et al. 2009. Observation of a large-gap topological-insulator class with a single Dirac cone on the surface. *Nat. Phys.*, **5**(6), 398–402.

Xiao, Di, Chang, Ming-Che, and Niu, Qian. 2010. Berry phase effects on electronic properties. *Rev. Mod. Phys.*, **82**(Jul), 1959–2007.

Zak, J. 1964. Magnetic translation group. *Phys. Rev.*, **134**(Jun), A1602–A1606.

Zaletel, Michael P., and Mong, Roger S. K. 2012. Exact matrix product states for quantum Hall wave functions. *Phys. Rev. B*, **86**(Dec), 245305.

Zaletel, Michael P., Mong, Roger S. K., and Pollmann, Frank. 2013. Topological characterization of fractional quantum Hall ground states from microscopic Hamiltonians. *Phys. Rev. Lett.*, **110**(Jun), 236801.

Zeng, Bei, Chen, Xie, Zhou, Duan-Lu, and Wen, Xiao-Gang. 2015. *Quantum Information Meets Quantum Matter: From Quantum Entanglement to Topological Phase in Many-Body Systems*. Berlin: Springer.

Zhang, S. C., Hansson, T. H., and Kivelson, S. 1989. Effective-field-theory model for the fractional quantum Hall effect. *Phys. Rev. Lett.*, **62**(Jan), 82–85.

Zhong, Shudan, Moore, Joel E., and Souza, Ivo. 2016. Gyrotropic magnetic effect and the magnetic moment on the Fermi surface. *Phys. Rev. Lett.*, **116**(Feb), 077201.

Zhou, Jian-Hui, Jiang, Hua, Niu, Qian, and Shi, Jun-Ren. 2013. Topological invariants of metals and the related physical effects. *Chin. Phys. Lett.*, **30**(2), 027101.

Zirnbauer, Martin R. 2020. Particle-hole symmetries in condensed matter. arXiv:2004.07107.

Index

BF theory, 215
K matrix, 214
α-chain, 342
β-function, 244
$\hat{E} \times \hat{B}$ drift, 248
  edge state, 249
θ-angle, 120
g-factor, 89
  exchange-enhanced, 90
p-wave superconductor, 296
$p + ip$ superconductor, 271
  weak and strong coupling, 303

absolute stability, 335
AC/DC, 266
adiabatic approximation, 13
adiabatic transport, 63, 66
Aharonov-Bohm effect, 13, 70, 165, 166, 176, 194,
    247, 266, 303, 312
  and flux attachment, 204
AKLT chain, 150, 341, 350
Alice in Wonderland, 337
Altland-Zirnbauer classes (tenfold way), 254
analog computer, 290
Anderson localization, 241
angle-resolved photoemission spectroscopy (ARPES),
    3, 86
anomalies, 354
anomalous Floquet–Anderson insulator, 326
anomalous Hall effect, 218
  in T-breaking Weyl semimetal, 230
anomalous velocity, 220
  as motion within the unit cell, 221
antiferromagnet, 36
antiferromagnetic topological insulator, 353
antiunitary operator, 76, 79
anyon, 6, *see also* fractional statistics
  non-Abelian, 142
arrow of time, 77

axion electrodynamics, 119, *see* quantized
    magnetoelectric polarizability

band insulator, 95, 346
band structure, 29
BCS theory, 254
BEC-BCS crossover, 303
Berezinskii–Kosterlitz–Thouless transition, 12, 53,
    213, 250
Berry curvature, 15
  as magnetic field in momentum space, 221
  monopole of, 225
  resulting phase space volume correction, 223
Berry flux, 15, 91
Berry phase, 7, 11, 13, 88
  Abelian, 17
  discrete, 17, 105
  non-Abelian, 17, 96, 110, 224
Betti number, 48
bipartite lattice, 349
bismuth, 225
Bloch oscillation, 324
Bloch's theorem, 7, 12, 24, 27
Bogoliubov–de Gennes formalism, 254, 308
Boltzmann equation, 232
boron nitride, 73, 226
bosonization, 213, 299
Bott periodicity, 259
Bragg peak, 172
braid group, 141
  representation in Pfaffian state, 305
braiding, 140
  in Majorana wire T-junctions, 311
Brillouin zone, 7, 15, 24
  effective, 113, 115
  time-reversal invariant planes in, 118
bulk photovoltaic effect (BPVE), 97
bulk-edge correspondence, 3
  in Chern-Simons theory, 211
  in Floquet systems, 329

371

Cartan, 258
cat
  Cheshire, 337
  Schrödinger, 323
Chalker-Coddington model, 251
charge density
  coarse grained, 147
Chern band, 74
Chern form in even dimensions, 102
Chern insulator, 29, 74, 81
Chern number, 71, 72, 118
  additivity, 73
  as obstruction to globally defined wavefunctions, 106
  connection to polarization pumping, 100
  from gauge-invariant project operators, 104
  from homotopy, 105, 107
  in three dimensions, 73
  sum rule, 72, 105
  two-band example, 105
Chern-Simons form
  in odd dimensions, 102
  of band structure, 120, 122
  of electromagnetic field, 203
Chern-Simons theory, 8, 139, 203
  and BF theory, 8
  and gauge invariance, 205
  and Hall effect, 209
  as internal or statistical gauge theory, 203
  gauge fixing, 210
  multicomponent Abelian, 214
  non-Abelian, 215
  spectrum on torus, 209
  topological degeneracy of, 209
  vs. Maxwell theory, 211
chiral anomaly, 232
chiral Luttinger liquid, 213
chiral metal, *see* surface state
chiral photocurrent, 233
Church-Turing thesis, 288
circular photogalvanic effect (CPGE), 233
classification
  of interacting fermions in $d = 1$, 347
Clifford gate, 293
clock model, 57
coarse graining, 32, 75, 147, 171, 195, 196, 201, 276
coherence length
  superconductor, 303
coherent state, 17
cohomology, 7, 12, 40, 106
  de Rham, 47
commensurate flux, 71
compass model, 234
complexity class, 288
composite bosons, 204
composite fermion, 204, 268, 280
  p=wave superconductor of, 302
  metal, 270, 271

Compton wavelength, 177
conductance, 243
conductivity, 243
conformal field theory, 301
  and non-Abelian edge states, 306
    central charge, 112
    conformal blocks, 301
constant of motion, 319
constraint counting, 275
Cooper pair, 303
Coulomb interaction
  magnetic, 176
  pseudopotentials, 131
Coulomb phase, 189
coupled wire construction, 213
critical exponent, 5, 38
critical field, 261
critical point, 4
critical temperature, 33
crystal field, 168
crystal momentum, 12, 24
crystalline symmetries, 98
curvature, 41
cutoff, 37
cyclotron frequency, 21

decoherence, 9, 289
deconfined quantum criticality, 162
deconfinement, 164, 176, 187, 190
  diagnostic, 190
defect
  crystalline, 266
  flux, 266
degeneracy, 126, 170, 175
  as diagnostic, 322
  continuous, 275
  extensive, 275
density matrix renormalization group (DMRG), 155, 352
differential form, 44
  1-form, 44
  2-form, 46
  covector and vector, 44
dimerization, 145
dipolar interaction, 168, 172
dipole moment, 149
  conservation, 199
Dirac cone, 3, 86
  in Kitaev honeycomb model, 236
  odd number on 3D TI surface, 87
Dirac monopole, 197
Dirac point, 73
  as four-band crossing in three dimensions, 228
  gap, 283
Dirac sea, 29
Dirac spectrum, 306
Dirac string, 20, 106, 175, 197
disclination, 200

dislocation, 200, 266
disorder, 9
  dilution, 272, 277
  distortion, 272, 277
  graphene, 272
  Kitaev honeycomb model, 277
  spin liquid, 271
domain wall, 185
driving, 316
  periodic, 316
Drude theory, 248

edge mode
  internal, 268
edge state, 3, 8, 59, 66, 82, 144
  degeneracy, 342
  from $\hat{E} \times \hat{B}$ drift, 249
  helical, 239, 268
  hydrodynamical picture of, 213
  of Kitaev chain, 298
  tunneling into, 214
eigenstate order, 316, 322
elasticity, 275
electrical polarization, 7, 96
  ambiguity by polarization quantum, 99
  and symmetry, 98
  change under adiabatic process, 101
  connection to Chern number, 100
  in one-dimensional model, 99
  physical interpretation, 99
  practical calculation, 105
Elitzur's theorem, 181
emergence, 1, 37
emergent gauge field, 276
energy conservation, 319
entanglement
  long-range versus short-range, 350
entanglement entropy, 349
  and numerical methods, 352
  and thermalization, 350
  area law, 351
  topological, 351
  volume law, 351
entanglement spectrum, 351
entropy
  maximization, 319
equilibrium, 316
Euler angles, 17
Euler characteristic, 48
exact sequence of a fibration, 107

factoring and primality, 288
Fermi arc, 228
  and quantum oscillations, 229
Fermi liquid
  non-, 125
Fermi surface

number of sheets as topological, 85
Fermi velocity, 31
fermion
  Bogoliubov, 256, 263
  bond, 307
  matter, 307
fermion parity, 346
ferromagnet, 275
  XY, 36, 50, 53
  flat-band, 89
  Ising, 5, 186, 321
  isotropic, 35
  quantum Hall, 90
  uniaxial, 33, 35
fidelity, 291
filling factor, 2, 93, 126, 247
fine structure constant, 177, 198, 233
  of quantum spin ice, 198
Floquet
  Brioullin zone, 318
  engineering, 316, 324
  ensembles, 319
  ETH, 319
  periodic Gibbs ensemble, 319
  system, 318
  topological insulator, 324
flux binding, 166, 272, 282
flux insertion, 136, 138, 263, 266, 269
flux line
  tension, 190
flux quantum, 61, 263
  superconducting, 264
flux trick, 107
fractional charge, 136
fractional quantum number, 146
fractional statistics, 6, 140
  1D vs. 2D, 300
  from flux attachment, 205
  in Chern-Simons theory, 205
  in three dimensions, 216
  non-Abelian, 9, 142, 286
fractionalization, 6, 123, 126, 174, 285
fracton, 198, 200
Fubini-Study metric, 104
fusion rule, 193

gauge field, 6, 7, 11
  artificial, 278
  emergent, 174
gauge invariance, 15, 263
gauge symmetry, 181
gauge transformation, 15, 264
  large or non-null-homotopic, 99
Gauss's law, 187
Gauss–Bonnet theorem, 40
generalized Gibbs ensemble, 319, 321
geodesic, 16
geometric phase, 11, 13

Ginzburg–Landau equation, 262
glass, 316
glassiness, 200
glide reflection, 98
Goldstone mode, 90, 196
Gottesman-Knill theorem, 293
gradient expansion, 36, 262, 277
graphene, 23, 29, 73, 118, 225
    artificial gauge field, 278
    intervalley scattering in, 226
    Kane-Mele model of, 82, 83
ground-state degeneracy, 8
group cohomology, 344
Grover's search algorithm, 287
Gugelhupf, 43
guiding center, 22
gyrotropic magnetic effect, 231

Haldane gap, 151
Haldane model, 73, 74, 83
Haldane pseudopotential, 130, 270, 302, 305
Hamiltonian
    effective, 317
    Floquet, 318
    nonlocal, 318
    periodic, 317
    time dependent, 317
heat capacity, 278
Heisenberg chain, 341
Hessian, 41
hierarchy of quantum Hall states (Haldane-Halperin),
    269
Higgs phase, 188
high-temperature superconductor, 125, 156
Hofstadter model, 69, 70
holon, 158, 164
homotopy, 7, 9, 12, 40, 50
    definition, 49
    homotopy groups π_n, 49
    of nondegenerate Hamiltonians, 105
    of time-reversal-invariant Hamiltonians, 112
homotopy theory, 200
honeycomb lattice, 83
Hopf insulator, 106
Hopf invariant, 106
hopping integral, 26
Hund's rule, 89
Hurewicz theorem, 50

ice rule, 169
iceberg, 186, 289
imaginary time, 39
incompressibility, 134
interference, 241
inversion layer, 67
inversion symmetry, 96, 97
irrational charge, 148

Ising flux, 234
Ising gauge theory (IGT)
    odd, 188
    pure, 181
Ising model, 33, 181, 185, 188
    transverse-field, 56, 299

Jain sequence, 269
Jordan–Wigner transformation, 299
Josephson effect, 258, 263
    4π, 265
Josephson frequency, 265

K-theory, 260
kaons, 75
Karplus-Luttinger velocity, *see* anomalous velocity
Kasteleyn matrix, 161
Killing form, 110
Kitaev chain, 294, 297, 342
    edge Majorana mode of, 298
    robustness to local perturbations, 298
    solution of, 298
Kitaev honeycomb model, 29, 310
    anisotropic gapped phase, 306
    connection to $RuCl_3$, 237
    exact solution of, 306
    flux excitations, 236
    in magnetic field, 236
    non-Abelian gapped phase, 307
Klein model, 151
Klitzing constant, 58
Kramers pair, 76, 109
    no mixing from time-reversal, 79
    proof from antiunitarity of time-reversal, 78
Kramers point, 81

ladder operator, 31
Landau level, 7, 12, 20, 69, 89, 90
    central, 32
    filled, 128
    form factor, 133
    Landau gauge, 23, 24
    projection, 133
    relativistic, 31
    symmetric gauge, 23
Landau–Ginzburg theory, 5, 32, 34, 37
lattice
    honeycomb, 29
    kagome, 275
    pyrochlore, 168, 275
Laughlin state, 268
Laughlin wavefunction, 129
    for bosons, 132
level repulsion, 322
level statistics
    Poissonian, 322
Lifshits tail, 250

link variable, 181
Liouville's theorem, 223
liquid–gas transition, 4
localization, 80
  dynamic, 324
  in one dimension, 109
  of even number of edge states, 79, 109
  Wannier-Stark, 201
localization length, 242
logic gate, 292, 293
  fidelity of, 291
Lorentz transformation, 247
Luttinger liquid, 213

magnetic monopole
  Dirac, 176
magnetic field
  pseudoscalar, 92
magnetic length, 31
magnetic monopole, 8, 20, 126, 175, 197
magnetic translation group, 23, 25, 133
  many-body, 139
magnetism, 4
magnetoelectric response, 96
  as higher-dimensional version of polarization, 122
Magnus expansion, 324
main sequence of quantum Hall states, 269
Majorana fermion, *see also* Majorana mode
  π-, 329
  and neutrinoless double β-decay, 295
  in high energy physics, 294
Majorana mode, 87, 258, 295
  braiding of, 304
  edge, 345
  experimental status, 311
  Hilbert space, 296
  in proximitized wire, 311
Majorana operator, 295
  anticommutation relation of, 295
Majumdar-Ghosh chain, 149
manifold, definition of, 40
many-body localization (MBL), 316, 321
matrix product state, 352
Maxwell electromagnetism, 195
mean level spacing, 243
mean-field theory, 33
measurement-based quantum computing, 290
Meissner effect, 261
metal, 241
metallic transport, 218
mobility, 68
mobility edge, 242
mobility gap, 218
modulation doping, 67
Moore–Read state, *see* Pfaffian state
Mott insulator, 95
multicritical point, 35
multifractality, 278

Néel state, 156, 167
natural optical activity, 231
neutrino as candidate Majorana fermion, 296
neutron scattering, 4
Nielsen-Ninomiya theorem, 228
no-cloning theorem, 290
Noether's theorem, 318
noisy intermediate-scale quantum (NISQ) computing, 291, 339
non-Abelian statistics, 311
nonlinear σ-model, 38, 90, 110
nonlinear Hall effect, 233
nonlinear optics in Weyl semimetals, 230
  quantized, 233
NP-completeness, 289
nuclear magnetic resonance (NMR), 154
  SCGO, 276
  skyrmions, 93

off-diagonal long range order, 255
one-parameter scaling, 243
  in Anderson localization, 243
optical rotation, 231
orbital magnetic moment, 223, 232
order
  spatiotemporal, 338
order parameter, 12
  manifold, 35, 38, 50, 90, 110
  soft and hard fluctuations, 50
organizing principle, 1
oscillator strength, 134

parallel transport, 16
paramagnet, 186
  0π, 332, 344
partial trace, 349
path independence, 47
path integral, 37, 39
  spin, 39
Peierls instability, 145
Peierls substitution, 222
penetration depth, 261
percolation, 250
perpetuum mobile, 334
Pfaffian state, 271, 302
phase of matter, 1
phase structure, 316
  nonequilibrium, 320
phase transition, 4, 33
photon
  emergent, 195
photovoltaic effect, 97
pinch-point, 4, 173, 200
point group vs. space group, 98
polar crystal, 97
polarization quantum, 99
  connection to Chern number, 102
precision, 2

prethermalization, 326
projection operator, 17
projective symmetry group, 23, 236
projector Hamiltonian, 151, 160
proximity effect, 258
pseudopotential, *see* Haldane pseudopotential
pseudospin, 226
  Ising, 273
pumping, 63, 66, 102, 107, 116, 267
  and integer quantum Hall effect, 109
  by flux *within* system, 123

quantization, 2, 58, 88, 263
  Dirac magnetic monopole, 176
quantized magnetoelectric polarizability, 4, 7, 96
  and half-quantized Hall effect, 88
  in interacting system, 123
quantum computer, 287, 289
  gates, 292
quantum computing, 9
quantum critical point, 299, 351
quantum dimer model, 8, 57, 159
  entanglement in, 351
quantum error correction, 290
  threshold and overhead, 291
  why difficult, 290
quantum geometric tensor, 104
quantum Hall effect, 2, 58
  fractional, 6, 126
  integer, 109
  of surface state, 119
quantum Hall ferromagnet, 92
quantum information, 290, 349
quantum metric, 104, 354
quantum order, 196
quantum point contact, 138
quantum quench, 316
quantum speedup, 293
quantum spin Hall effect, 77
quasi-energy, 318
  and crystal momentum, 318
quasihole, 268
quasiparticle, 6, 177
qubit, 287
  flux, transmon, 265
  physical vs. logical, 291
  topological, 292, 293

Raman scattering, 238
random bipartite hopping, 258, 272
random matrix theory, 253
  Wigner-Dyson ensembles, 253
random vector potential, 272, 278
Read–Rezayi state, 310
real fermion, *see* Majorana operator
reference wavefunction, 14
resonating valence bond (RVB) liquid, 156

triangular lattice, 160
reversible computing, 290
RhSi, 233
Rice-Mele model, 102
Rokhsar-Kivelson point, 160, 162
$RuCl_3$, 237, 315
  Raman scattering and fermionic excitations, 238

scanning tunneling spectroscopy, 239
self-averaging, 240
semiclassical equations of motion, 220
  dependence on wavefunctions, 95
semiclassical equilibrium, 224
semimetal, 219
  Dirac vs. Weyl, 225, 228
  nodal line, 225
  surface Fermi arc in Weyl case, 228
  topological, 9
Shor's factoring algorithm, 288
simplex, 276
single mode approximation, 133, 162
skyrmion, 12, 51, 89
  in quantum Hall ferromagnet, 91
  via stereographic projection, 93
Solovay-Kitaev theorem, 293
space group
  magnetic, 353
specific heat, 272
spectral flow, 64
spin
  Heisenberg, 275
  Ising, 275
  orphan, 272, 276
spin glass
  π-, 331
  0-, 331
spin ice, 2, 4, 167, 194
  materials, 168
  quantum, 195, 274
spin liquid, 8, 157
  Coulomb, 168, 271
  gapless, 219, 234
  proximate, 237
spin stiffness, 90
spin–charge separation, 126, 150, 164
spin-orbit coupling, 3, 34, 59, 311
  and S_z conservation, 83, 84, 87
  as relativistic effect, 76
  in Kane-Mele model, 83
  is time-reversal even, 77
  Rashba, 83, 87
  vanishing g-factor, 89
spinon, 6, 150, 158, 165, 197
$SrCr_8Ga_4O_{19}$ (SCGO), 274
state of matter, 1
statistical transmutation, 166
statistics
  anyon, 141

non-Abelian, 310
relative, 193
steady state, 316
stiffness, 250
universal jump, 250
Stokes's theorem, 49
string breaking, 191
stroboscopy, 334, 339
structure factor, 134
Su-Schrieffer-Heeger (SSH) chain (model), 102, 103,
    146, 347
spinful, 348
supercell, 70, 71
superconducting quantum interference device
    (SQUID), 265
superconductor
inhomogeneous, 257
parity sector, 299
topological, 286
triplet, 258
type I, 261
superfluid
stiffness, 56
two-dimensional, 50
superfluidity, 4
superselection sector, 167, 194
surface state, 3, 59
chiral metal, 85, 116
of 3D topological insulator, 87, 88, 118
symmetry
antiunitary, 78, 345
broken, 4, 12, 32, 34, 37
group, 34
inversion, 96
nonsymmorphic, 97
on-site, 342
particle-hole, 258
projective, 343
time-reversal, 78, 347
symmetry fractionalization, 341
fermionic, 345
symmetry-protected topology (SPT), 260, 341
synchronization, 333

tangent bundle, 16
tantalum arsenide (TaAs), 228
Tao-Thouless state, 139
temperature, 316, 319
infinite, 319
tenfold way (Altland-Zirnbauer classes), 253
tensor gauge theory, 198
thermal conductivity of Majorana edge mode, 315
thermalization, 316
tight binding, 26
time crystal, 334
many-body Floquet, 334
time evolution
operator, 317

unitary, 317
time translation symmetry
discrete, 318
time-reversal invariant momenta, 81, 82, 112
time-reversal symmetry, 81, 87
as antiunitary operator, 76
classical, 75
integer versus half-integer spin, 78
macroscopic, 75
TKNN integer, *see also* Chern number, 72
topological band structure, 71, 72
topological crystalline insulator, 353
topological defect, 9, 12, 40, 50, 200
topological defect-free configurations, 52
topological degeneracy, 6, 139
topological entanglement entropy, 351
topological insulator, 2, 59, 96
antiferromagnetic, 353
Floquet, 324
Kondo, 353
materials, 84
pumping interpretation of, 116
strong and weak, 118
three-dimensional, 85, 88, 118
two-dimensional, 78, 80, 84
weak, 117, 268
with inversion symmetry, 110
topological invariant, 8, 40, 59, 88, 96
$\mathbb{Z}_2$, 59, 78, 109
abstract methods for, 97
continuous vs. discrete computation of, 48
of many-particle wavefunction, 107
topological magnon, 234
topological order, 6, 138
Thouless-type, 59
Wen-type, 59
topological quantum computer, 293
topological quantum computing, 285
fault tolerance, 291
universality of, 310
topological quantum field theory (TQFT), 8, *see also*
    Chern-Simons theory, 179
topological quantum memory, 291
and nonlocal information storage, 292
topological superconductor, 286
topology, 7, 12, 40
as integral of geometry, 40
toric code, 192, 216
transverse-field Ising model, 299
Trotter-Suzuki decomposition, 182
Turing machine
classical, 288
quantum, 288
two fluid model, 260
two-dimensional electron gas (2DEG), 137
two-parameter scaling, 243

underconstraint, 170, 275

universality, 1, 3, 34, 38

vison, 165
von Neumann entropy, 349
vortex, 9, 12, 39
    half-quantum, 301
    in p-wave superconductor, 301
    in superconductor, 260
    quantum Hall, 268
    superfluid versus superconducting, 52
    trapped particle, 52
    vortex loop, 51
    vortex-antivortex pair, 53, 55

Wannier orbital, 26
wavefunction geometry, 218
    and perturbation theory, 97
    definition of, 95
weak localization, 246
    negative magnetoresistance, 247
Wegner-Wilson loop, *see* Wilson loop
Wess–Zumino term, 110

Wess–Zumino–Witten model, 110
Weyl
    fermion, 295
    fermion of definite helicity, 226
    Hermann, 11
Weyl point
    and jump of Chern number, 230
    as monopole of Berry flux, 227
    model Hamiltonian of, 227
    multifold, 233
    optical transitions across, 233
    stability of, 226
    topological charge, 227
Wigner crystal, 130
Wilson loop, 15, 70, 189
    area law, 190
    perimeter law, 190
winding number, 46, 50, 51, 140, 165

XX spin chain, 300
XY model (*see also* ferromagnet, xy)
    with clock term, 37, 57, 187